A Historical Geography Research of Canals in China

Nianhai Shi

A Historical Geography Research of Canals in China

山东人民出版社
国家一级出版社 全国百佳图书出版单位

Springer

Nianhai Shi
Jinan, China

Translated by Dianying Wang

ISBN 978-981-95-3516-3 ISBN 978-981-95-3517-0 (eBook)
https://doi.org/10.1007/978-981-95-3517-0

Jointly published with Shandong People's Publishing House

Translation from the Chinese language edition: "《中国的运河》" by Nianhai Shi, © Shandong People's Publishing House 2022. Published by Shandong People's Publishing House. All Rights Reserved.

Previously Published by Shandong People's Publishing House

This Springer imprint is published by the registered company Springer Nature Singapore Pte Ltd.
The registered company address is: 152 Beach Road, #21-01/04 Gateway East, Singapore 189721, Singapore

Preface

This book, *Canals in China*, originated in my early career. Though colleagues in the nascent People's Republic of China suggested a reprint, other commitments prevented me from revising it at the time. However, in recent years, the arguments and information in this text have been referenced within the field, indicating its continued relevance and prompting this revised edition.

As an early work, it required substantial updating. My decades of studying historical geography, which often dealt with canal-related issues, provided a solid foundation for corrections and expansions. Furthermore, the field of canal research has advanced significantly in the intervening years, with new, insightful studies that I've incorporated to address the original work's shortcomings.

Many have asked about my original motivations for writing this book. The origins of this work are closely tied to the relationship between historical geography and diachronic geography.[1] While historical geography has long-standing scholarly roots, it was traditionally known as "diachronic geography," with "historical geography" only gaining widespread recognition in more recent times. However, this shift in terminology reflects more than a simple change of name, instead signifying a fundamental difference in academic approach. Diachronic geography primarily examines the administrative changes of geographical entities. But this raises two critical questions: First, if the focus is mainly on the evolution of administrative structures, how can this approach account for more expansive geographical transformations over historical periods? Second, what is its practical relevance? These questions weighed heavily on my mind during the War of Resistance Against Japan.

When placed under scrutiny, diachronic geography struggles to fully account for many historical changes in geographical phenomena, and cannot adequately explain numerous historical events related to geography. As for practical relevance, while

[1] Translator's Note: The term "diachronic geography" is a translation of the Chinese (沿革地理学), a subfield that has no direct equivalent in the geography discipline within the Anglophone world. The author explains more about their historic relationship in the subsequent sentences. Broadly, diachronic geography focuses on the process through which political and administrative space is transformed over time, such as the gradual evolution of administrative divisions, place names, and trade networks.

not entirely devoid of applications, the usefulness of diachronic geography is rather limited. Since the founding of the People's Republic of China, it has become clear through collective scholarly effort that diachronic geography is merely one component of the broader field of historical geography, dedicated to studying geographical phenomena in historical periods. Geographical landscapes are constantly changing, and these transformations inevitably impact human activity, particularly in production and development. At the same time, human interventions alter and reshape the environment, influencing geographical changes in turn. This process is continuous, reciprocal, and governed by identifiable patterns. The objective of historical geography is to study these transformations, their effects, and the related patterns, so that humanity can better utilize and modify the natural environment. This broader mission cannot be fully achieved within the confines of traditional diachronic geography.

During the War of Resistance Against Japan, these ideas had not yet fully taken shape in my mind and I was only able to explore the following two points:

1. While diachronic geography certainly holds a place within historical geography, research in historical geography should not be confined to this subfield alone. But how could this limitation be overcome? At the time, the discipline's nature and scope had not yet been clearly defined, making it difficult to provide a precise answer. Nevertheless, I had already begun to recognize the importance of studying geographical change and the need to explain the process of this change as well as its causes and consequences. This realization allowed me to move ever-so-slightly beyond the traditional framework of diachronic geography.
2. I gradually came to understand that a discipline like historical geography should not only be broadly useful for society but should also carry more practical applications across multiple fields. If a field of study lacks practical relevance, its long-term survival becomes uncertain. History provides numerous examples of disciplines that have faded into obscurity. While the reasons for their decline vary, failure to remain useful to society is often a significant factor.

With these ideas in mind, I sought to put them into practice. At the time, I was living in Beibei, Chongqing, near Professor Gu Jiegang. I shared my thoughts with him, and he expressed both approval and encouragement. This led me to embark on a study of China's canal systems. The title *China's Canals* was personally suggested by Professor Gu, and he also reviewed the completed manuscript and arranged for its publication. Now, as I revise the book for republication, I deeply regret that he is no longer with us. Remembering the days when I had the privilege of following in his footsteps along the banks of the Jialing River, I cannot help but feel overwhelmed with emotion.

Given my objectives at the time, my research on canals could not be limited to tracing the historical evolution of individual waterways. Nevertheless, my approach remained grounded in historical scholarship. While I attempted to analyze the causes, processes, and effects of change, my focus was predominantly on social and human factors, with less attention given to natural influences. Through subsequent years of fieldwork, I have come to realize the need to address this gap. However, this would require more than minor revisions. More comprehensive reworking is therefore necessary, and some aspects will need to be more fully explored in future studies.

Although historical geography should serve practical purposes, at the time of my initial research, it was difficult to expect immediate real-world applications from a study of canals. The War of Resistance Against Japan had reached a critical stage, and most of the regions historically capable of canal construction had already fallen into enemy hands. Moreover, given the prevailing social conditions, undertaking canal development was virtually impossible. Any practical application of this research would have to wait until the founding of the People's Republic of China. Indeed, it was only after the establishment of New China that the restoration of old canals and the construction of new ones could become a national priority.

Since then, I have had the opportunity to visit historical canal sites and witness the improvements made to the Grand Canal, which has been a source of great inspiration. I have conducted fieldwork in various locations, tracing the ancient course of the He River in Dingtao District and Juye County in Shandong, examining the site of Fangtou in Qi and Xun Counties in Henan where Cao Cao diverted the Qi River into the Bai Canal, exploring the remains of Han and Tang dynasty transport canals in the Guanzhong Plain, and investigating the relics of Sui and Tang canals in counties across Henan and Anhui. In Luoyang, I studied the impressive scale of the Tang Dynasty's Hanjia Granary, and in Kaifeng, I searched for the remnants of the Bian River bridges depicted in Zhang Zeduan's *Along the River During the Qingming Festival*. Outside Tong County, I examined the confluence of the Tonghui and Lu Rivers. Each site provided a vivid testament to the remarkable achievements of the working people of past generations. I have also traveled extensively around Tai Lake, investigated the historic canal attributed to Wu Zixu in Dongba Town, Gaochun, Jiangsu, and traced the course of the Jiangnan Canal to its intersection with the Qiantang River south of Hangzhou. Following the canals near Jiaxing, I reached Suzhou's Fengmen Gate, where I saw an unimpeded flow of boats, their white sails crowding the horizon. Travelling between the Yangzi and Huai rivers, I observed the newly widened canals south of Gaoyou running up and down the river, jade waters stretched out to the horizon. Endless fleets of wooden ships, towed by steamers, moved in seamless succession, resembling dragons crossing the sea before vanishing into the distance. These experiences have strengthened my conviction that only under socialism can canals truly fulfill their potential, reinforcing my determination to revise this book.

Today, as the entire nation works in unison to revitalize China, remarkable progress is being made in modernization, visible in an endless stream of new achievements. The expansion and development of canal systems are an important part of this progress. Just a few days ago, I learned that the widening and dredging of the Gaoyou-to-Lincheng section of the Grand Canal had been completed. For those of us engaged in canal research, such news is truly exhilarating! If this revised edition of my book can make even a modest contribution to the country's modernization efforts, it would be a source of great personal fulfillment.

Chinese New Year's Eve, 1985
Jinan, China

Nianhai Shi

Contents

About the Author

Nianhai Shi (1912–2001) was a pioneering founder of modern historical geography in China. He held key academic positions at Shaanxi Normal University, Northwest University, and Lanzhou University, where he served as Vice President and Director of the Institutes of Historical Geography and Tang Dynasty Studies. Shi contributed significantly to the study of China's territorial evolution and canals, co-authoring *A History of the Evolution of China's Territory* with Gu Jiegang. His major works include *China's Canals* and *Outline of Chinese Historical Geography*. He was also President of the Shaanxi Historical Society and the Chinese Society of Ancient Capitals.

List of Figures

Chapter 1
The Utilization of Natural Waterways in Ancient Times

The Advantages of Waterway Transportation

Waterborne transport has long been more economical and less labor-intensive than land transport, a principle that holds as true today as it did in antiquity. Even the simplest watercraft, so long as it can navigate the current, allows for efficient travel along waterways and can reach various destinations along their banks. While land transport is not constrained by the course of rivers and offers greater flexibility in route selection, it also presents significant challenges. Outside of the flat plains, mountainous and hilly terrain poses natural obstacles that hinder movement. Even on level ground, the force required to push or pull a cart carrying a given load is substantially greater than that needed to propel a boat of the same capacity. As a result, wherever both land and water routes were available, people often favored water transport. Even if a water route was more circuitous, it remained the more efficient and practical option.

This fundamental advantage of waterways was well understood in ancient times. Early human activity was closely tied to rivers, and settlements were commonly established along their banks. In contrast, mountainous regions often acted as barriers to expansion. The preference for riverside settlements stemmed from multiple factors: alluvial soils in river valleys were fertile and well-suited for agriculture, rivers provided abundant fish and aquatic resources for early fishing and hunting societies, and, most importantly, waterways facilitated transportation and communication. Archaeological discoveries in China over the past few decades further confirm this pattern. The majority of Neolithic cultural sites have been found near rivers. Notable examples include sites along the Wei River, Luo River, and the lower reaches of the Fen and Su Rivers, all tributaries of the Yellow River. The region east of the present-day Longshan area in Gansu, spanning both banks of the lower Wei River, has long functioned as a vital east–west transportation corridor, a route later followed by the Longhai (Lanzhou-Lianyungang) Railway. East of Xi'an, this corridor runs along the Wei River's south bank, while west of Xi'an, it shifts to the north bank.

N. Shi, *A Historical Geography Research of Canals in China*,
https://doi.org/10.1007/978-981-95-3517-0_1

Interestingly, the distribution of newly discovered Neolithic cultural sites closely corresponds to this shift in the corridor's alignment along the Wei River.

Transportation in the Three Dynasties

Though scattered across the entire country, known Neolithic sites are notably concentrated in the Yellow River basin—particularly along the middle reaches of the Yellow River. Even during the Xia, Shang, and Zhou dynasties, the range of early Han cultural activity remained largely confined to the Yellow River basin. More precisely, the "Yellow River Basin" here refers to the specific section of the river below the Longmen Gate, as well as the Wei and Ji tributaries, with no connection to the upper reaches of the Yellow River above Longmen. This distinction arose because the lower Yellow River (below Longmen) was navigable by boats, while the upper reaches were not yet utilized for transportation during that era.

The capitals of the Xia, Shang, and Zhou dynasties were frequently relocated. The exact reasons for each relocation varied, ranging from political motives, to economic considerations, or sudden disasters. Without delving into the exact causes, the chosen locations of their capitals alone substantiate the claims made above. The Xia people originally inhabited the Central Plains region. Why did their migrations consistently remain along both banks of the Yellow River? The Shang people, said to have originated in the east, similarly moved within proximity of the Yellow River. According to traditional accounts, Shang relocations were driven by Yellow River flooding. Yet if such disasters compelled the Shang to abandon their homeland, why did they persistently cling to the Yellow River basin rather than seek new territories? The Zhou people unquestionably emerged from the west. Their migratory trajectory further demonstrates the role of waterways in transportation. Initially settling between the Jing and Wei rivers, the Zhou migrated eastward along the Wei River, eventually clashing with and conquering the Shang. Why did they move eastward instead of westward? The reason is straightforward: the navigability of the Jing and Wei rivers increased toward the east. While some historical narratives attribute Zhou migration to pressure from western Rong tribes, this explanation alone is insufficient. Although the Rong harassed the Zhou, their threat was arguably less formidable than that of the Shang. If the Zhou could subdue the Shang, why would they be powerless against the Rong? Clearly, traditional interpretations warrant further scrutiny. Of course, by the late Western Zhou period, as Zhou power waned, they did in fact under Rong domination, but this is a separate matter altogether.

The migration patterns of the Xia, Shang, and Zhou dynasties, along with those of their neighboring tribes (whether hostile or allied) were largely confined to the Yellow River basin. The frequent conflicts and military campaigns that took place among these groups remained limited to this compact region. While more distant states existed as part of distinct regions with distinct transportation systems, interactions with these distant polities were sparse due to geographical barriers that naturally resulted in less frequent exchanges.

Enfeoffed States of the Zhou Dynasty

Among the Three Dynasties, the Zhou's territorial expansion surpassed that of the Xia and Shang. The paths taken by the Zhou expansion also followed the waterways. Originating from the west, where the terrain was higher, the Zhou expanded eastward, southeastward, and northeastward along rivers. Advancing eastward via the Yellow River, they conquered the Shang. After the fall of the Shang, the Zhou promptly established a number of feudal states governed by their own royal clans (a process referred to as "enfeoffment"). At that time, the Yellow River flowed into the Bohai Sea through what is now Hebei Province. Following its course, the Zhou founded the State of Yan. To the east of the Central Plains, the Ji River provided access to the sea, enabling the Zhou to expand eastward along its banks, where the States of Qi, Lu, and several smaller polities were established. Southeastward lay the Ying and Ru rivers, near which the Zhou established prominent states such as Chen, Cai, and Xu.

Luoyi (modern-day Luoyang) served as a convergence point for these three major water systems. The Duke of Zhou's transformation of Luoyi into the eastern capital of the Zhou dynasty was therefore a strategic move. The Han River stretching south became a thoroughfare for the Zhou expansion, leading to the establishment of down-river states further east along the Han. Additionally, in the eastern stretches of the Yellow River, the Fen and Su rivers facilitated Zhou expansion upstream, resulting in the founding of the States of Huo and Jin. Altogether, these waterways formed the transportation network of the era. By controlling this system, the Zhou consolidated their power and maintained stable rule. Later, during the Rebellion of the Three Guards and revolts of the Huaiyi tribes, the Zhou were able to swiftly suppress these uprisings, offering a testament to the strategic advantages afforded by their mastery of waterway logistics.

Utilization of Waterways in the Spring and Autumn Period

Military conflicts among states became increasingly frequent during the Spring and Autumn period, yet the large-scale utilization of waterways remained limited. When the Zhou first enfeoffed states, they were initially concerned that newly-conquered territories in the former Shang territories to the east would be unstable and they therefore established a series of important feudal domains in the Central Plains, all while overlooking the southern state of Chu, then dismissed as "uncivilized barbarians." Unexpectedly, Chu later emerged as a major threat to the Zhou court. Had the Zhou capital not been moved eastward, military campaigns against a strengthened Chu could have nonetheless been launched via the Han River. However, after the relocation, this direct water route to Chu lost its strategic value. The states of Qi and Jin, as hegemons of the Central Plains, championed the slogan "Honor the King and Expel the Barbarians." Beyond confronting northern Rong tribes, their primary target was Chu. The prolonged rivalry between Chu and the Central Plains states unfolded along a north–south axis, intersecting perpendicular to the predominantly eastward-flowing river systems of the time.

Throughout the Spring and Autumn period, Qi and Jin engaged in numerous battles with Chu for supremacy, all of which were fought with chariots rather than naval forces. In the north, where Qi and Jin alternated as hegemons, smaller states aligned with the dominant power without major issues. Conflicts between Qi and Jin themselves also occurred. Jin, situated to the west, and Qi to the east, were divided by the Yellow River, which flowed diagonally from southwest to northeast. Neither state actively exploited this waterway. In contrast, the state of Qin developed robust waterborne communication with Jin. Jin was based in Jiang (south of modern Yicheng, Shanxi) near the Fen River, and Qin in Yong (south of modern Fengxiang, Shaanxi) close to the Wei River, allowing the two to be linked via each waterway's respective connection to the Yellow River. The historic "Qin-Jin Naval Engagement" of 647 BCE took place along this waterway. Prior to the conflict, Qin transported grain to famine-stricken Jin via the same route, exemplifying this connectivity.[1] However, due to the route's circuitous nature, subsequent military conflicts between Qin and Jin primarily followed the Su River south of the Fen River.[2] Nonetheless, the main military conflicts continued to rely on chariot warfare rather than naval forces.

Despite prolonged conflicts between the northern states of Qi and Jin and southern Chu, the latter never suffered complete defeat. In the late Spring and Autumn period, however, the rise of Wu to Chu's east was accompanied by relentless attacks that exhausted Chu's defenses, culminating in Wu's capture of the Chu capital Ying.[3] Both Chu and Wu, situated in marsh-rich regions along the Yangtze and Huai River basins, demonstrated strategic use of natural waterways. However, the Yangtze's navigational value remained limited, possibly due to the vast and treacherous Jiujiang section. Consequently, military engagements between Chu and Wu primarily followed the Huai River.[4] During this period, chariot warfare dominated

---

[1] *Zuo commentary*, 13th year of Duke Xi.

[2] *Zuo commentary*, 24th year of Duke Xi: "(The Earl of Qin gave asylum to Chong'er), crossed the Yellow River, besieged Linghu, and took Jiuchu… and entered Quwo." According to Du Yu's commentary: "The Sang Spring is located west of Xie County, Hedong Commandery; southeast of Xie County lies Jiucheng." Linghu is located west of present-day Linyi County. All these places are near the Su River. Therefore, when the Earl of Qin took in Chong'er, he entered through the Su valley. Again, in the 12th year of Duke Wen, "The Earl of Qin attacked Jin and took Jima; the people of Jin resisted." This campaign corresponds to the one mentioned in the 13th year of Duke Cheng, during which Prime Minister Lü severed ties with Qin, stating: "(They) entered our Hequ, attacked our Su, captured our royal officials, and destroyed our horses at Jima; hence the Battle of Hequ." This battle likely also took place near the Su River. The 24th year of Duke Xi corresponds to 636 BCE, and the 12th year of Duke Wen corresponds to 615 BCE.

[3] The event of Wu's invasion of Ying is recorded in *Zuo Commentary (Zuo Zhuan)*, 4th year of Dinggong (506 BCE).

[4] The numerous military campaigns between the states of Wu and Chu are recorded in *Records of the Grand Historian* (*Shiji*), vol. 31, "Hereditary House of Wu Taibo," and vol. 40, "Hereditary House of Chu." The theaters of war between the two states were all situated in the Huai River basin.

as the benchmark of military strength. However, Wu initially lacked chariot expertise. Jin therefore sent instructors to Wu, highlighting their proficiency in waterborne logistics.[5]

The Transportation Network in the *Yu Gong*

By the early Warring States period, all navigable waterways within China had already been extensively utilized. Consequently, efforts were made to link these major waterways together to form a comprehensive transportation network. The scale and structure of this ancient network can still be partially reconstructed today. The extant text of the *Tribute of Yu* (*Yu Gong*), a chapter of the *Book of Documents* (*Shang Shu*), preserves an outline of this system and records various related matters. Although the *Yu Gong* is attributed to the legendary sage-king Yu the Great, it was in fact written by an anonymous scholar in the early Warring States period.[6] The pseudonymous attribution to Yu was a deliberate rhetorical strategy. The Warring States period was a time of intense political fragmentation, with numerous contending states locked in endless warfare, leaving the people in a state of perpetual unrest. The author, yearning for national unification, envisioned the use of the country's extensive water transportation system as a means to unify the realm. He even went so far as to plan a national transportation infrastructure to serve a unified empire. Since such ambitions were impractical under the conditions of his time, he chose to express his aspirations under the name of Yu the Great.

In his vision of a unified China, the country would be divided into nine administrative regions (or "provinces"), with the capital located in the southwestern part of Jizhou, not far from the Yellow River. From this central hub, the tribute from the other eight provinces could be transported via the waterways to the capital. The planned territory of Jizhou roughly corresponds to modern-day Shanxi, Hebei, and parts of Henan. It was a vast region, with the proposed capital situated in its southwestern area. At that time, the Yellow River flowed northeastward from the southwest of Jizhou, skirting its eastern boundary. Thus, the northeastern areas of Jizhou could send tribute down the river to the capital. The envisioned province of Yanzhou was located between the Yellow River and the Ji River, encompassing parts of today's Shandong and Hebei. In addition to bordering both rivers, the region also had the Luo River flowing through its interior. Tribute from Yanzhou could be transported via the Luo and Ji Rivers to the Yellow River and then upstream to the capital.

---

[5] *Zuo commentary*, 7th year of Duke Cheng: "Wuchen requested to serve as envoy to Wu, and the Marquis of Jin granted permission. King Shoumeng of Wu was pleased and thereby established diplomatic relations with Jin. One-third of a military unit was sent to Wu, and a portion was stationed there. They shared archery and chariotry, taught Wu how to use war chariots and how to conduct battle formations and rebellion against Chu." Wu's adoption of chariot warfare likely began from this point. The 7th year of Duke Cheng corresponds to 584 BCE.

[6] Shi Nianhai. *Collected essays on rivers and mountains, volume II: On the date of composition of the Yu Gong.*

The proposed province of Qingzhou lay east of Mount Tai and west of the East China Sea, corresponding to modern Shandong. The Wen River flowed from Qingzhou into the Ji River, making the tribute route: Wen → Ji → Yellow River. The proposed province of Xuzhou extended from the northern slope of Mount Tai to the Huai River in the south and bordered the sea to the east. It included the southern part of today's Shandong and the northern parts of Jiangsu and Anhui. Xuzhou contained the Si River and, to its south, the Huai River. The lower Si River flowed into the Huai, and the Si and Ji Rivers were connected by an artificial canal called the He River. Thus, tribute from Xuzhou could travel through the Huai and Si Rivers into the He, from there into the Ji, and then on to the Yellow River.

South of the Huai River lay the proposed Yangzhou, traversed by the Yangtze River. Tribute from Yangzhou would be shipped along the Yangtze, entering the Huai from the coast, and from there following the tribute route of Xuzhou northward. The modern provinces of Hubei and Hunan formed the region referred to as Jingzhou in the *Yu Gong*. Jingzhou was rich in rivers: the Yangtze, Tuo, Qian, and Han. Tribute from Jingzhou could be transported along these rivers to the borders of Yuzhou, then carried overland to the Luo River in Yuzhou, and subsequently into the Yellow River. Yuzhou, roughly corresponding to the southwestern half of modern Henan, was crisscrossed by the Luo, Yi, Chan, and Jian Rivers. The latter three all converged into the Luo River, which in turn emptied into the Yellow River—making Yuzhou exceptionally well-suited for tribute transport.

At that time, modern Shaanxi and Gansu were known as Yongzhou. The Yellow River flowed directly through this region, along with the Jing and Wei Rivers. Therefore, tribute from Yongzhou could be easily sent downstream to the capital. South of Yongzhou and west of Jingzhou lay the proposed region of Liangzhou, corresponding to present-day Sichuan. The author seemed to have only limited knowledge of Liangzhou's geography. He claimed that the upper reaches of the Han River and Jialing River were connected by a river called the Qian, which allowed tribute from Liangzhou to be transported via the Qian into the Han River, then moved overland to the Wei River, and finally into the Yellow River.

This transportation network may not seem particularly remarkable by today's standards. However, at a time when geographical knowledge was still quite limited, especially given that the realm was fragmented among rival states, conducting a detailed survey would have been extremely difficult. It is truly impressive that the author was able to devise such a comprehensive transportation system under such conditions. As for the proposed tribute route via the Qian River in Liangzhou, which does not correspond to geographical reality, he can hardly be blamed for this inaccuracy. Even in the centuries that followed—spanning two to three millennia up to the Qing dynasty—no definitive resolution to this problem was ever reached. With this in mind, there is no need to be overly critical.

This proposed transportation network demonstrates just how highly people of the time valued waterway transportation. Overland routes could be used on occasion, but they were considered only as a last resort. For example, between the proposed Liangzhou and Yongzhou regions lay the formidable barriers of the Bashan and Qinling mountain ranges. The author of the *Yu Gong*, in a bold flight of imagination, envisioned a nonexistent "Qianshui" (Hidden River) to overcome this obstacle. In reality, there was no feasible solution between the Han and Wei Rivers, so the only option was overland transport.

Another example is Yangzhou. If one were to travel overland from Yangzhou to Jizhou, the route would need to pass through only one intermediary region: Yuzhou. However, because navigable waterways were available, the tribute route instead took a long, circuitous path: from the Yangtze River to the sea, from the sea into the Huai River, then through the Si River, the He Canal, into the Ji River, and finally into the Yellow River before reaching the capital. This deliberate detour reflects the practical advantages of water transport, which was far more economical and labor-saving than overland alternatives. This transportation network also reveals that people of the time were already making full use of available waterways and were actively working to develop and expand them. There is evidence that conditions were beginning to permit the construction of artificial canals, and the idea of opening new waterways to facilitate transport was already taking root.

Chapter 2
Canal Construction in the Pre-Qin Period and Its Impact

The Emergence of Canals

While water transport was undoubtedly economical and efficient, it also had inherent limitations. Waterways typically follow fixed courses, and whether traveling upstream or downstream, the areas they can reach are ultimately limited. This posed little problem in ancient times when populations were sparse and political entities small. However, as human activity gradually expanded, the constraints of fixed water routes became increasingly inconvenient. At the time, the major rivers within China each formed separate systems. Traveling from one system to another often involved great difficulty—and in some cases was entirely impossible. To overcome this challenge, it became necessary to rely on overland routes to compensate for the gaps in waterway connectivity. Yet land transport was inherently uneconomical, and the repeated loading and unloading of goods only added further complications. As a result, people later sought ways to artificially excavate canals that would link previously unconnected waterways, allowing boats to travel directly between them. The construction of such canals supplemented the deficiencies of natural rivers and represented a significant advancement in transportation.

Sima Qian's *Records of the Grand Historian* offers a foundational overview of early canal construction in China: "At Xingyang, water was drawn from the Yellow River to create the Hong Canal, which flowed southeast to link the states of Song, Zheng, Chen, Cai, Cao, and Wei, joining with the Ji, Ru, Huai, and Si Rivers. In the state of Chu, channels were opened westward to connect with the Han River and the Yunmeng Marshes, and eastward to link the Yangtze and Huai Rivers. In Wu, canals were excavated to connect the Three Rivers and Five Lakes. In Qi, a canal was cut between the Zi and Ji Rivers. In Shu, the governor Li Bing cleaved Mount Lidui to mitigate the flooding of the Mo River, redirecting its waters into two channels that flowed through Chengdu. All of these canals were navigable by boat; surplus water

N. Shi, *A Historical Geography Research of Canals in China*,
https://doi.org/10.1007/978-981-95-3517-0_2

was used for irrigation, bringing substantial benefit to the people."[1] This detailed and expansive account by Sima Qian provides valuable insight into the systematic development of hydraulic engineering during early Chinese history.

The Earliest Canals

Sima Qian did not specify which canal was the earliest to be excavated. Traditional accounts often credit King Fuchai of Wu, who in 486 BCE—during the ninth year of Duke Ai of Lu—dug the Han Canal (*Hangou*, 邗沟) to connect the Yangtze and Huai Rivers. Yet the earliest canal was neither built by King Fuchai, nor was the Han Canal necessarily Wu's first. In fact, the earliest known canal construction took place in the state of Chu.

During the reign of King Zhuang of Chu (613–591 BCE), the minister Sunshu Ao is said to have diverted the Ju River to form a reservoir near Yunmeng Marsh. The Ju River rises in the southeast of present-day Yuan'an County, Hubei, flows through Dangyang, and enters the Yangtze River. Although this area was still some distance from Ying, Chu's capital (near modern Jiangling County), the river's lower reaches were already close to the Yunmeng Marsh.

The exact means by which Sunshu Ao redirected the Ju River is unclear, but the story appears in the *Imperial Survey* (*Huanglan*),[2] compiled by Miao Xi and others during the Three Kingdoms period. It is uncertain what sources they drew upon, and given the significant historical distance and the legendary quality of the material, some scholars remain skeptical. Nevertheless, Yunmeng was a land of marshes, and efforts to regulate the waterways along its shores for boat travel were entirely plausible. Sunshu Ao, known for his expertise in water management, had also improved the watercourses of Qisi (in present-day northwest Gushi County, Henan).[3] There is no reason to rule out his involvement in regulating the Ju River and creating a reservoir at Yunmeng. Still, the case should not be overstated. To

[1] Book of canals and waterways. In *Records of the grand historian,* vol. 29. *The Book* states: "In the east, canals were opened through the Hong Canal to connect the Yangtze and Huai Rivers." However, the character *Hong* (鸿) appears to be a scribal interpolation. In volume 29 of the *Book of Han* (*Han Shu*), in the *Treatise on Waterways* (*Gouxu Zhi*), this same passage is cited correctly as: "In the east, canals were cut to connect the Yangtze and Huai Rivers," omitting the erroneous character.

[2] Biographies of upright officials: Sunshu Ao, jijie commentary. In *Records of the grand historian,* vol. 119.

[3] *Universal geographic record of the Taiping era,* vol. 129: "The Quebei Reservoir lies 100 paces east of Anfeng County. *Huainanzi* states: 'The Chu Prime Minister constructed the Qisi Weir to irrigate the Yulou plains.' Additionally, Yudi Zhi cites Cui Shi's *Monthly Ordinances* (*Yueling*), attributing the Qisi Weir to Sunshu Ao." Qisi County was located northwest of modern Gushi County, Henan, implying the Qisi Weir was in its vicinity. Anfeng County corresponds to present-day Shouxian County, Anhui. Given the considerable distance between these locations, the Qisi Weir and Quebei Reservoir were distinct hydraulic projects.

exaggerate beyond the limits of the sources is something responsible historians must avoid.

Later, during the reign of King Ling of Chu (540–529 BCE), a canal was reportedly excavated near the capital, Ying, to facilitate grain transport to the newly constructed Zhanghua Terrace.[4] This terrace stood to the southeast of the capital. According to later records, the Yangs River originated northwest of Ying, flowed past the southern outskirts of the city, and continued eastward to join the Mian River (the modern Han River) near present-day Qianjiang in Hubei. On the southern bank of the Yang River was a sizable lake known as Li Lake, connected to the river and located adjacent to the Zhanghua Terrace.[5] Given the natural presence of both river and lake, construction in the area would have been relatively straightforward. The term "canal" as recorded in historical sources may refer to the dredging of existing waterways to improve navigability, or to the cutting of short channels—either between Ying and the Yang Rier, or between Li Lake and the terrace. Since the area already featured rivers and lakes, large-scale canal excavation was likely unnecessary. Compared with Sunshu Ao's project of diverting the Ju River to create a reservoir in the Yunmeng Marshlands, this undertaking may have been far more expansive in scale.

Sunshu Ao's redirection of the Ju River to form the Yunmeng reservoir can likely be identified with what Sima Qian described as the "canal linking the Han River to the Yunmeng marsh" in the *Records of the Grand Historian: Book of Canals and Waterways*. In that passage, Sima Qian writes: "In the state of Chu, channels were opened westward to connect with the Han River and the Yunmeng Marshes, and eastward to link the Yangtze and Huai Rivers through the Hong Canal." This passage was later cited by Ban Gu in the *Gouxu Zhi* in the *Book of Han*, which renders the sentence as: "In the east, canals were cut to connect the Yangtze and Huai Rivers," notably omitting the character "Hong" (鸿). If we use the *Gouxu Zhi* to cross-reference the *Treatise on Canals and Waterways*, it becomes apparent that the "Hong" in "Hong Canal" may be a later scribal interpolation.

The historically known Hong Canal lies between the Ji, Ru, Huai, and Si rivers, and cannot plausibly be relocated to the area between the Yangtze and Huai Rivers. The canal in question should instead refer to the Han Canal (邗沟), excavated by King Fuchai of Wu near Han City (邗城). However, an alternative interpretation has been proposed: that Sima Qian's reference to the Hong Canal between the Yangtze and Huai does not stem from a textual error but instead denotes the Fei River. How, then, could the Fei River be understood as the Hong Canal? The answer lies in its transformation through large-scale hydraulic engineering.

To the west of the Fei River once flowed the Bi River (沘水), now known as the Pi River (淠河). Between these two rivers lay the Qisi Reservoir, also known as the Quebei Pond (芍陂), an expansive man-made reservoir formed by damming the Bishui. The Quebei collected water from the west and discharged it eastward into the Fei River. Originally, the Fei River flowed northward into the Huai River. But with additional water from the Quebei, its volume increased, raising its water level

[4] Commentary on the Mian River. In *Water classic*.

[5] Ibid.

significantly. This elevation was said to have caused it to overflow its headwaters, crossing a local ridge and merging with the Shi River to the south. As a result of this confluence, the Shi River came to be regarded as a continuation of the Fei River. The Shi River itself emptied into Chao Lake, which connects to the Yangtze River, thereby forming a continuous waterway from the Huai to the Yangtze—resembling the configuration described by Sima Qian. Some accounts even attribute this "Hong Canal" to Sunshu Ao himself.

References to Sunshu Ao's hydraulic works first appear in the *Huainanzi*, specifically in the chapter *Renjianxun*, which states: "Sunshu Ao diverted the waters of Qisi to irrigate the Yulou plain." Qisi was located northwest of present-day Gushi County, Henan, while Yulou lay to its southeast—both distant from the Quebei Pond, situated between the Bi and Fei Rivers. The *Book of the Later Han Dynasty: Biography of Wang Jing* notes: "In the territory of Lujiang Commandery, there are rice fields reclaimed near the Quebei Pond by Sunshu Ao, the Chancellor of Chu." Likewise, the *Water Classic: Commentary of the Fei River* records: "The Fei River flows northeast, passing east of Baishao Pavilion, where it forms a lake known as Quebei… a project attributed to Sunshu Ao."

Scholars continue to debate whether the reservoir in Qisi is identical to the Quebei Pond. Focusing strictly on the Quebei, located in a relatively flat plain between the Bishui and Fei Rivers, the claim that it caused the Fei River to reverse course—overflowing its source, crossing elevated terrain, and flowing into the Shi River—is difficult to reconcile with hydrological and topographic evidence. My field investigation at the Fei River's headwaters confirmed that while the surrounding hills are not particularly high, the slopes are relatively steep. Regardless of how much the water level may have risen, it is unlikely that the Fei River could have flowed backward over such terrain. Thus, unless one accepts that Sima Qian's phrase "connecting the Yangtze and Huai Rivers by the Hong Canal" contains a scribal error, it is difficult to support the hypothesis that the so-called "Hong Canal" refers to the Fei River. Such a claim would likely yield conclusions inconsistent with historical and geographical realities.

Canals Attributed to Wu Zixu

The hydraulic interventions led by Sunshu Ao in regulating the Ju River, along with King Ling of Chu's canal construction, were concentrated around the Yunmeng Marsh. Historically, "Yunmeng" referred to the marshlands spanning both sides of the Yangtze, incorporating the Jiang-Han floodplain. This terrain featured intricate waterways and alternating elevations—expanding into contiguous lakes during high water and fragmenting into isolated shallows and mudbanks in droughts. The Chu State had long regarded this region as a vital natural bulwark, particularly given the presence of three strategic mountain passes—Dasui (Great Tunnel), Zhiyuan (Straight Shaft), and Ming'e (Dark Defile), along the eastern flank of the Dabie

Mountains (modern Huangxian, Wusheng, and Pingjing passes straddling the Henan-Hubei border). However, during the reign of King Zhao of Chu (515–489 BCE), Wu Zixu led the armies of Wu in their invasion of Chu, advancing precisely along this strategic route.[6] According to Wu Zixu's initial plan, he likely believed that once his forces breached these strategic passes, the Wu navy could swiftly advance into the Yunmeng Marsh. However, he had not anticipated that the Chu army would collapse so swiftly in this battle.

He arrived at the shores of the Yunmeng Marsh around the eleventh or twelfth month of the ancient lunar calendar, a time when water levels were at their lowest. He could not afford to keep his troops camped by the water's edge, waiting until the following spring floods. Thus, he had no choice but to dig a canal. The excavation advanced from the Han River toward the Yang River, which flowed through the heart of Yunmeng Marsh. Naturally, the canal was dug in the vicinity of the Yang River and its surrounding marshes. With the Wu fleet consisting of numerous ships, a larger canal was necessary to accommodate their passage, making the project extremely labor-intensive. Later, this canal was named after him and became known as the Zixu Ditch.[7]

The source of this waterway is said to have been the Yang River, and its location was likely to the southwest of the Ji'nan City (i.e., the Capital Ying). It flowed southeastward into the Ji'nan City, then exited the city as the Longbei Pond. The canal skirted the southern side of the Capital Ying before turning northeastward and becoming known as the Yang River.[8] This account, however, appears inconsistent with the actual wartime situation. Wu Zixu came from the east of Ying, so why was the canal dug west of the capital? To reach Ying from the Han River's banks required crossing part of Yunmeng Marsh—why wasn't the canal dug within the marsh itself, but rather west of Ying which was already outside the marsh boundaries? These questions still lack a satisfactory explanation.

The Canal Network Between the Three Rivers and Five Lakes

Tradition holds that Wu Zixu commissioned a canal within Wu's territory passing near present-day Dongba in Gaochun County, Jiangsu. Known historically as the Xu Stream (胥溪) or Xu River (胥河), its name likely honors its reputed builder. This waterway linked western lakes—Gucheng, Shijiu, and Danyang—with eastern water systems including the Santa Lake, Changdang Lake, Jingxi River, and Zhenze Lake. The critical engineering feat involved breaching a 3–5-*li* elevated terrain at Dongba, enabling Wu's navy to traverse from Tai Lake directly to the Yangtze near

[6] *Zuo commentary*: 4th year of Duke Ding.

[7] Commentary on the Mian River. *In Water Classic*.

[8] Ibid.

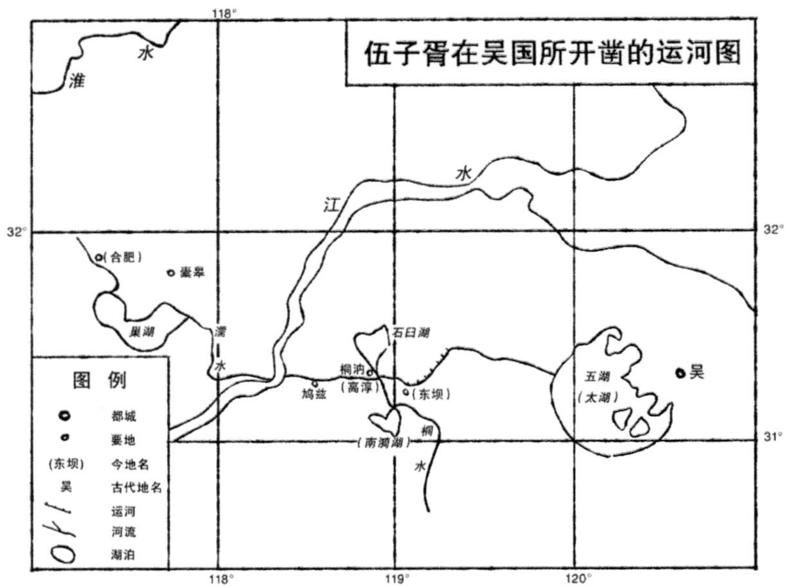

图例 Legend

都城 Capital Cities

要地 Strategic Locations

今地名 Present Names

古地名 Historical Names

运河 Canals

河流 Rivers

湖泊 Lakes

江水 Yangtze River

Fig. 2.1 Canals Dug under the supervision of Wu Zixu within the territory of the state of Wu

modern Wuhu, Anhui. Ming scholar Han Bangxian's *Study of the Guangtong Dam*[9] documents this route, though his sources remain unspecified. (See Fig. 2.1).

This waterway did indeed connect several lakes, but due to an elevated terrain segment requiring artificial excavation, it was definitely not a natural channel. During my field investigation at Dongba, I was grateful for local residents' explanations: several years ago, numerous ancient bowls and basins dating back to the pre-Qin period were discovered along the southern section of the waterway. The significant quantity of these artifacts found by the waterside strongly suggests they were left behind during the original canal construction. This evidence substantiates that Zixu's canal-digging during his expedition against Chu is not a unsupported hypothesis.

[9] Cited by Hu Wei in *The detailed commentary on the "Tribute of Yu"*.

Following Wu Zixu's campaign, this waterway continued to serve as a military route in subsequent Wu-Chu conflicts. In the 15th year of Duke Ai of Lu (480 BC), Chu generals Zixi and Ziqiu launched their attack against Wu via this very route. Their forces advanced as far as Tongrui. According to Du Yu's commentary, the Tong River—originating from the Baishi Mountain in the southwest of Guangde County—flowed northwest into the Danyang Lake.[10] The Guangde County of Wei-Jin period corresponds to present-day Guangde County in Anhui Province. Tongrui likely marked the confluence where the Tong River entered Danyang Lake, representing a strategic point along this military corridor. Before Wu Zixu's expedition against Chu, this route had likely never witnessed warfare.[11]

The canal, mentioned in the *Tribute of Yu* in connection with the "Three Rivers" of Yangzhou, was later interpreted by scholars as the central one of these rivers, thereby obscuring its original man-made nature.[12] As previously noted, the *Tribute of Yu* was actually composed during the Warring States period. If the text from that era considered this waterway one of the Three Rivers, it suggests that by then, the canal had already taken on the appearance of a natural watercourse, flowing smoothly without obstruction.

Here, it is worth addressing why the Wu state chose to construct this canal rather than simply using the Yangtze River for military transport. To understand this, we must first examine the strategic dynamics between Wu and Chu. As previously discussed, although the Yangtze River lies between the two states, it provided little advantage for transportation or warfare at the time. Consequently, most conflicts between Wu and Chu took place near the Huai River. Between the Huai and the Yangtze lay Chao Lake, whose waters flowed into the Yangtze via the Ru River. The point where the Ru entered the Yangtze was not far from the canal excavated by Wu Zixu. This meant that Wu's navy, coming from the south, could bypass the Yangtze and sail directly into Chao Lake. Without this canal, Wu's fleet would have had to depart from Gusu (modern-day Suzhou), enter the Yangtze from the south, and then laboriously sail upstream—a journey fraught with the perils of river currents and storms, as well as being considerably longer and more arduous.

---

[10] *Zuo commentary, 15th year of Duke Ai with Du's commentary.*

[11] *Zuo Commentary, 3rd Year of Duke Xiang*: "Chu's Zichong attacked Wu, leading a well-trained army, captured Jiuzi, and reached Hengshan." Du Yu's commentary states: "Jiuzi was a Wu settlement east of Wuhu County in Danyang (present-day Gaoyi). Hengshan was south of Wucheng County in Wuxing." If Du's annotation is accurate, Zichong's eastern campaign must have utilized this route. However, Gu Yanwu and Jiang Yong questioned Du's identification of Hengshan. Gu posited Hengshan should be in modern Danyang County, while Jiang argued for Dangtu County. According to Gu and Jiang's theories, Chu forces could not have used this waterway. From Zichong's attack on Wu to Wu's capture of Ying (506 BCE), over sixty years elapsed. If this waterway existed earlier, why wasn't it utilized in numerous Wu-Chu conflicts during those six decades? Thus, I contend Du Yu likely misidentified the Hengshan mentioned in Zichong's campaign; otherwise, there's no reason such a strategic route would remain unused for sixty years.

[12] *Book of Han, vol. 28: Treatise on Geography* states: "In Wuhu County of Danyang Commandery, the Middle River originates southwest and flows east to Yangxian before entering the sea." Wuhu corresponds to modern Wuhu County, Anhui; Yangxian is present-day Yixing County, Jiangsu. The course of the Middle River aligns precisely with this canal's waterway.

What was the route like from north of Chao Lake to the Huai River? North of Chao Lake lies Hefei. Near Hefei, the Fei River flows north into the Huai River,[13] while the Shi River flows south into Chao Lake. During the summer floods, the Shi River would connect with the Fei River, forming a natural waterway, which is the origin of the name "Hefei" (literally "joined Fei").[14] This implies that outside the summer flood season, the Fei and Shi Rivers remained separate.

This waterway between the Fei and Shi Rivers has been mentioned in numerous historical records, with some even suggesting, as previously noted, that it had been artificially modified to become a canal. I personally visited Hefei to investigate this waterway. With the support of the Anhui Water Resources Bureau and accompanied by two engineers, we conducted a joint survey. We reached the source of the South Fei River (known as the Shi River in the *Water Classic*), where we indeed found remnants of an attempted canal excavation, which was clearly abandoned.

The low-lying hills and scattered sources of the South Fei River result in weak flow in its main channel. Although the elevation is not extreme, the canal was never completed. Without a continuous channel and with insufficient water flow, navigation by boats or rafts would have been impossible. The earlier accounts of the Shi River receiving water from the Fei River, as recorded in historical texts, do not align with the actual conditions. The various claims about the excavation of a canal in this area appear to be based on hearsay and are difficult to substantiate.

As previously referenced, Sima Qian stated: "In Wu, canals were dug to connect the Three Rivers and Five Lakes". Subsequently, many other scholars have proposed various interpretations regarding the Three Rivers and Five Lakes. The term "Three Rivers" first appears in the *Tribute of Yu*. The text states: "When the Three Rivers had been led in, the Zhen Lake was brought to stability." The *Tribute of Yu* provides clear descriptions of the northern and southern rivers of the "Three Rivers," leaving no need for further interpretation. As for the middle river mentioned in the *Tribute of Yu*, it is described as: "Passing through Jiujiang to reach Dongling, then turning northeast to converge at Hui, and finally flowing east as the middle river into the sea." Sima Qian did not elaborate on the middle river, possibly because reliable evidence was already scarce in his time.

However, the *Tribute of Yu* first states "When the Three Rivers had been led in" before mentioning "Zhen Lake was brought to stability", suggesting that the Three Rivers were not downstream of Zhen Lake. Instead, the canals dug by Wu Zixu might correspond to a waterway in the Three Rivers region.

By the Han dynasty, the eastward branch of the Yangtze River diverging from Wuhu was identified as the middle river.[15] Since Wu Zixu's canal was located precisely where this "middle river" supposedly flowed eastward, the canal itself gradually fell into obscurity. Nevertheless, archaeological findings confirm that this canal was neither a natural watercourse nor an ancient channel of the mighty Yangtze River.

---

[13] Commentary on the Fei River. *In Water classic.*

[14] Commentary on the Shi River. *In Water classic.*

[15] Treatise on geography. In *Book of Han*, vol. 28.

The term "Five Lakes" has been subject to varied interpretations. Some scholars attempt to identify five distinct lakes by name, while others argue that the phrase refers collectively to a single body of water—Zhenze Lake, known today as Tai Lake. These differing perspectives reflect a broader lack of consensus. In fact, the Tai Lake basin and its surrounding region form an expansive lake district dotted with numerous interconnected bodies of water. Given this complexity, it is difficult—if not impossible—to trace the precise origins or referents of each historical name with certainty.

Modern researchers have reasonably identified the ancient watercourse of Wu State and the Baichi River, as recorded in *Chorography of the Wu State, A Concise History of the Yue and Wu States*, as canals excavated between the Five Lakes. According to the text, this ancient waterway "exited the Ping Gate northward into the region of Guo, joined the Baichi River, flowed out of Chao Lake, turned north through Li territory, passed Meiting, entered Yang Lake, then exited Yupu before emptying into the Yangtze River, ultimately connecting to Guangling".

Historical evidence suggests that the Ping Gate was the original northern gate of Suzhou, while Chao Lake corresponds to Cao Lake. Meiting was likely located in today's Wuxi, and Yang Lake (杨湖) may refer to Yanghu Lake (阳湖), situated between Changzhou and Wuxi along the Jiangnan Canal. Yupu is identified as Xili Port in Jiangyin, and Guangling was located on the Shugang northwest of Yangzhou. This canal likely originated northwest from Suzhou, traversed Caohu Lake, followed the Taibo River and the Jiangnan Canal, passed through Yanghu Lake into the ancient Furong Lake, and finally entered the Yangtze via Ligang to reach Yangzhou. As for the Baichi River, *A Concise History of the Yue and Wu States* notes its critical role in "connecting to the Yangtze River, enabling Wu State to transport grain."

It is said that the Baichi River was located near Hezhuang Mountain, approximately forty *li* southwest of Yanguan Town in present-day Haining County, Zhejiang.[16] Situated in the southern part of the Wu State, this river served as a vital waterway connecting to the Yue State. Many of the battlefields between Wu and Yue were concentrated in this region, suggesting that the Baichi River played a crucial role in transporting military supplies. The construction of this ancient watercourse in Wu laid the groundwork for King Fuchai of Wu's later excavation of the Han Canal. Without this earlier waterway, Wu's vessels would have faced significant challenges reaching Hanxia, diminishing the strategic importance of the Han Canal. Thus, the records in *A Concise History of the Yue and Wu States* regarding these two canals remain credible. Moreover, the existence of such canals substantiates Sima Qian's assertion that "in Wu, canals were dug to connect the Three Rivers and Five Lakes," was not without historical basis.

[16] Wei Songshan & Wang Wenchu. (1979). Formation and evolution of the Jiangnan canal. *Zhonghua Literature and History Collections*, 2.

King Fuchai's Construction of the Han Canal

The construction of canals by Wu Zixu took place during the reign of King Helü of Wu. Under Helü's rule, Wu achieved dominance by crushing the state of Chu. However, his death from wounds sustained in a campaign against Yue cast a shadow over his legacy. Yet this regret was soon remedied by his son, King Fuchai, who conquered Yue, leaving Wu without rivals in the southeastern region. Filled with ambition, Fuchai believed that Wu's military strength entitled it to become the leader of the states.

However, the powerful states of Qi and Jin in the Central Plains still stood in his way. To claim supremacy, Wu had to confront them. Since Wu's main force was its navy, advancing northward by water posed a logistical challenge. While it was possible to enter the Huai River via the Yangtze by detouring through the sea, even ignoring the dangers of maritime storms, the distance from the Huai to Jin remained vast. Fortunately, Fuchai was a ruler of bold vision, and canal construction had already proven effective for Wu. Without hesitation, in the tenth year of his reign, he began digging a canal directly linking the Yangtze to the Huai—the aforementioned Han Canal, excavated in 486 BCE. This Han Canal diverted water from Guangling (modern-day Yangzhou, Jiangsu) northeastward into Sheyang Lake (located between present-day Xinghua, Jianhu, Yancheng, and Baoying counties in Jiangsu). From there, it connected to Mokou, where it merged with the Huai River.[17]

Its name derived from Han City, built by Fuchai at the canal's inlet. Over time, the canal acquired multiple regional names: Qu River (Qushui),[18] Han River (Hanjiang),[19] Hanmin Canal (Hanminggou), and Zhongdu River (Zhongdushui),[20] each reflecting local linguistic and historical variations.

The Han City is located in present-day Yangzhou, Jiangsu Province. Today, the straight-line distance from Yangzhou to the Yangtze River to the south is approximately thirty *li*. Given that the Han Canal was constructed to connect the Yangtze River and the Huai River, why was the Han City situated thirty *li* away from the riverbank? This discrepancy arises from the shifting course of the Yangtze River over time, and we cannot judge ancient decisions based on the current riverbank configuration. If the Han City was not close to the river, where would the water for the Han Canal have been sourced? According to field investigations, the riverbank at that time roughly followed a line from Xupu in the northwest of modern-day Yizheng County, to Wantou in the northeast of Yangzhou, and then to Yiling in the

[17] The Han Canal excavated by the King of Wu is merely recorded in *The Book of Han*: The 9th Year of Duke Ai as "Wu built the Han city, connecting the Yangtze and Huai." Du Yu's commentary elaborates: "They constructed a city on the Han River and dug a canal northeast to Sheyang Lake, then northwest to MoKou entering the Huai River, establishing a grain transport route. This corresponds to present-day Hanjiang in Guangling." MoKou, located north of modern Huai'an County, Jiangsu, is now called Beishen Weir.

[18] Treatise on geography. In *Book of Han*, vol. 28.

[19] Citing Du Yu's commentary on *Zuo Commentary*.

[20] Commentary on the Huai River. In *Water classic*.

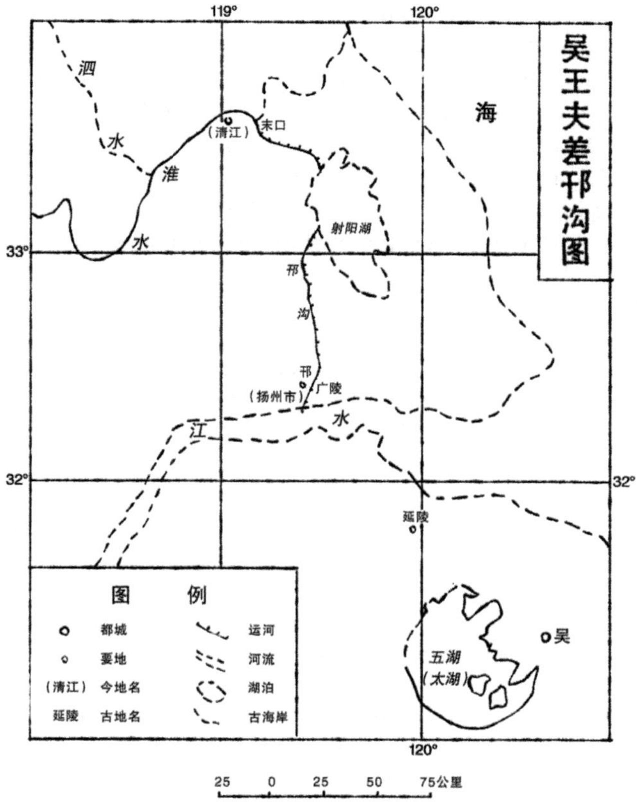

图例 Legend
都城 Capital Cities
要地 Strategic Locations
(清江) (Qingjiang)
延陵 Yanling
今地名 Present Names
古地名 Historical Names
运河 Canals
河流 Rivers
湖泊 Lakes
古海岸 Ancient Coastline

Fig. 2.2 King Fuchai of Wu's Han Canal

northeast of Jiangdu County.[21] The Han City was situated precisely along this line. After the excavation of the Han Canal, the northern confluence where it joined the Huai River underwent multiple changes, while the southern one with the Yangtze River remained relatively stable for a long period (See Fig. 2.2).

There are differing accounts regarding the starting and ending points of the Han Canal. In the Eastern Han dynasty, Ban Gu expressed his viewpoint in *The Book of Han*: "The canal water originates from the Yangtze River, flows north into the

[21] Yangzhou Regional Water Conservancy Bureau. *Yangzhou Water Conservancy.*

Sheyang Lake." This indicates that the Han Canal ran northward to Sheyang County and emptied into the Sheyang Lake. Based on this record, some scholars argue that the canal only connected Jiangdu to the Sheyang Lake, with no direct link to the Huai River.[22] The water entering the lake would then flow further into the sea. If this were true, King Fuchai of Wu's northward campaign would have required departing from the Han City, traveling via the canal and the Sheyang Lake, then entering the sea before detouring to the Huai River. Wouldn't this render the canal's construction entirely pointless? Moreover, the Sheyang Lake was not far from the Huai River. If King Fuchai could excavate a waterway from the Han City to the Sheyang Lake, why would he not extend it to the Huai River? The canal remained a vital transport route well into the Western Han and Three Kingdoms periods,[23] with no record of additional construction during that time. Had King Fuchai not connected it to the Huai River, how could it have remained navigable for centuries without obstruction?

When King Fuchai constructed the canal, he adopted the same method used by Wu Zixu, which involved utilizing the local lakes. Between the lower reaches of the Yangtze River and the Huai River lies a low-lying area dotted with lakes. Although the landscape has changed somewhat since ancient times, a glance at a modern map still reveals a patchwork of grayish-white markings indicating waterways. In antiquity, this region likely had even more and larger lakes. Take Sheyang Lake as an example. During the Song Dynasty, it spanned three hundred *li* in surface area,[24] yet today, no trace of it remains. Only the Sheyang River meanders through this area, winding south of Funing County before reaching Sheyang County and emptying into the sea. The Han Canal, starting from Yangzhou, did not run directly northward but instead veered northeast to make use of these lakes. Due to the topography, the canal's flow was not consistently southward or northward. Instead, it drew water from the Yangtze River in the south and the Huai River in the north, with the two flows converging in Sheyang Lake. This phenomenon of bidirectional flow within a single waterway was not unique to the Han Canal but was a common feature of several other canals as well.

Scholars hold varying interpretations regarding how both the Yangtze and Huai River waters could be diverted into Sheyang Lake. Mencius stated: "Dredge the Ru and Han Rivers, channel the Huai and Si Rivers, and direct their flow into the Yangtze."[25] By Mencius' time, the Han Canal had already been constructed. This clearly suggests that the canal was built to divert the Huai and Si Rivers southward into the Yangtze. However, Ban Gu recorded in the *Book of Han*: "(In Jiangdu County of Guangling State), the canal water originates from the [Yangtze] River, flows northward, and enters Sheyang Lake."[26] The author of the *Water Classic* further noted: "(The Huai River) flows east past the northern side of Huaiyin County, where the

[22] Chen Li. *Cartographic commentary on waterways in 'Treatise on geography of the Book of Han'.*

[23] Biography of Liu Bi, King Wu. In *Records of the grand historian*, vol. 106; Book of Wei: Annals of Emperor Wen. In *Records of the Three Kingdoms*, vol. 2.

[24] Chuzhou. In *Universal geography of the Taiping era*, vol. 124.

[25] *Mencius: Duke Wen of Teng.*

[26] Treatise on geography. In *Book of Han*, vol. 28.

Middle Dushui River emerges from Baima Lake, runs northeast, and then rejoins the Huai River."[27] Here, the Zhongdu River refers to the Han Canal excavated by King Fuchai of Wu. Building upon this, Li Daoyuan (commentator of the *Commentary on the Water Classic*) asserted that this channel extended all the way to Shanyangkou before merging with the Huai River.[28] While Ban Gu described the Han Canal as flowing north into Sheyang Lake, the Classic of Waterways and Li Daoyuan claimed it diverted the Yangtze into the Huai. Tang Dynasty scholar Li Ao once traveled this waterway and wrote in his *Record of My Journey to the South,* "From the Huai River, we sailed downstream to Gaoyou Lake, then proceeded upstream to reach the Yangtze."[29] Whether moving downstream or upstream, it is evident that the terrain near the Huai and Yangtze sections of the Han Canal was relatively elevated. Shen Kuo, however, argued differently. Citing *Mencius*' statement about "channeling the Huai and Si Rivers into the Yangtze," he maintained that these rivers naturally flowed into the Yangtze—a legacy of Yu the Great's flood control efforts. He further noted that during the Xining era (1068–1077 CE), imperial envoys surveyed the area and found the ancient river courses still clearly traceable.

Due to the deepening of the Yangtze and Huai River beds, the water flow could no longer reach Gaoyou.[30] As for how Shen Kuo determined the so-called 'relics of Yu,' it remains unclear. The claim that northern waters could no longer flow south to Gaoyou was attributed to the deepening of the Yangtze and Huai Rivers. But if the rivers had indeed deepened, why did the water flow only extend north of Gaoyou? Shen Kuo failed to explain this clearly. Zhu Xi disagreed with these arguments. First, he pointed out that the Huai River flowed east into the sea, which was why it was classified as one of the "Four Great Rivers." If it had flowed south into the Yangtze, it would not have fit the definition of a 'great river.' Second, he criticized Li Ao's use of the terms 'downstream' and 'upstream' as inappropriate. He argued: "Throughout history, travel through Huainan has relied solely on the Han Canal, where dams and locks were built to regulate tidal flows for grain transport—not natural currents. Therefore, the journey from the Huai to Gaoyou should not be called 'downstream,' nor should the route from Gaoyou to the Yangtze be called 'upstream.' Li Ao (courtesy name Xizhi) also mentioned entering the canal from the Huai with the tide, which might have given the illusion of going downstream. Presumably, after passing Gaoyou, he encountered the Yangtze tide flowing outward, creating a false impression of going upstream. His careless observation led to this error."[31] He attributed the Huai's failure to reach Gaoyou to deepened riverbeds, though this explanation lacked clarity.

Zhu Xi rejected both views: first, the Huai's status as one of the "Four Great Rivers" required it to flow east to the sea, not south; second, Li Ao's terms "downstream" and

[27] Commentary on the Huai River. In *Water classic.*

[28] Commentary on the Huai River. In *Water classic.*

[29] Cited in *Dream pool essays, vol. 24: miscellaneous records I.* This passage is absent in extant editions of *Record of my journey to the south.*

[30] Miscellaneous records I. In *Dream pool essays*, vol. 24.

[31] Casual reading notes. In *Collected works of Zhu Wengong*, vol. 71.

"upstream" misrepresented tidal lock operations in the Tang-Song canal system. Zhu contended that Fuchai's original Han Canal relied not on tides but natural gradients, as lock technology postdated the Wu era.

The Qing scholar Yan Ruqu believed that the King of Wu State excavated the Han Canal, which connected the Yangtze River to the Huai River. Emperor Wen of the Sui Dynasty opened the Shanyang Ditch, and Emperor Yang opened the Han Canal, both from Shanyang to the Yangtze River. The flow direction of the water was opposite to the previous one.[32] King Wu started digging the canal from the Yangtze River bank, while Emperor Sui began construction from the Huai River bank. Given that the starting locations were different, could the water flow in the canal remain consistent without any changes? This is also a question worth considering.

Hu Wei argued that Ban Gu's statement—that the canal water flows into a lake rather than into the Huai River—was measured and more accurate than the views of many other scholars. He wrote: "The terrain of Gaoyou and Baoying is the lowest, like the bottom of a cauldron. The Han Canal first receives water from the Yangtze River and flows northeastward until it reaches Sheyang Lake. Du Yu stated that from Sheyang, the water continues northwest to Mokou and then enters the Huai River. This merely describes a transportation route for grain from the Yangtze to the Huai, but while the route may reach the Huai River, the canal water itself does not flow into it."[33] Why did Hu describe Gaoyou and Baoying as lying "like the bottom of a cauldron"? His reasoning was based on the observations of Pan Jixun. In his *Treatise on the Two Rivers* (*Lianghe Yi*), Pan noted: "The elevation difference between Gaojiayan and Baoying exceeds one zhang and eight chi; between Gaojiayan and Gaoyou, it exceeds two *zhang* and two *chi*."[34]

This reasoning is in fact quite straightforward: if the canal water were to flow directly from the Yangtze River into the Huai River, the topography would necessarily require that the Yangtze lie at a higher elevation than the Huai. Under such conditions, a lake could not possibly form between the two rivers. Conversely, the same logic applies if water were diverted from the Huai into the Yangtze.

Many scholarly interpretations have become overly entangled in textual debates while neglecting to investigate the local geography and hydrological realities. As a result, their conclusions often fall short of objectivity and accuracy.

The He River Between Shang and Lu

If the sole purpose of King Fuchai of Wu's northern campaign was to conquer Qi, then after excavating a canal from Han City into the Huai River and then proceeding upstream along the Si River, passing through the state of Lu, he could reach the border of Qi. At that time, Lu was under the pressure of Wu, and King Fuchai

[32] Yan Ruoqu. *Explanations of geography in the Four Books.*

[33] *Detailed commentary on the "Tribute of Yu"*, vol. 6.

[34] Yan Ruoqu. *Explanations of geography in the Four Books.*

repeatedly summoned the Marquis of Lu to form an alliance, which the Marquis dared not refuse. It would have been no issue for Wu to pass through Lu to attack Qi, especially since Lu was following behind Wu in the campaign against Qi. However, if King Fuchai intended to meet with the Marquis of Jin, it would have posed a problem. At that time, both the Yellow River and the Ji River flowed northeastward, and the Hong Canal in the Central Plains had not yet been constructed. It would have been difficult for King Fuchai to lead his fleet to the Jin State to meet the Marquis and display his military might. Coincidentally, the area between the Si River and the Ji River was then a marshland, featuring not only the famous Daye Marsh but also Lei Marsh and He Marsh, all relatively close to one another, making it an ideal location for canal construction. Thus, King Fuchai once again employed his proven strategy and excavated a waterway between the Si River and the Ji River, described at the time as connecting the regions of Shang and Lu, with the Yi River to the north and the Ji River to the west.[35] This waterway later became known as the He River. King Fuchai could then lead his fleet from the He River into the Ji River, directly reaching Huangchi, where he would meet and form an alliance with the Marquis of Jin. Huangchi, located south of present-day Fengqiu County in Henan Province, was situated on the banks of the Ji River at that time.

According to later records, this He River connected with the Ji River northeast of Dingtao[36] (present-day Dingtao County, Shandong Province). The He River flowed eastward from Dingtao and joined the Si River at Gutingcheng, about 60 *li* west of Huling County.[37] Huling County was located east of present-day Yutai County, Shandong, with Gutingcheng situated in modern-day Yutai County.

However, the canal constructed by King Fuchai was said to extend north to the Yi River and west to the Ji River. How should this be explained? The Yi River was a tributary of the Si River, originating southeast of Lucheng, which is present-day Qufu County, Shandong) and merging with the Si River southwest of Lucheng.

As Li Daoyuan noted: "Following the waterways, I found only the Si River. Generally, it extended north to the Yi River, passed through the regions of Shang and Lu, and connected with the Ji River. This waterway was merely the result of dredging by the State of Wu, not as some claim—originating from the Yi River in the northeast and flowing southwest into the Ji River."[38]

Thus, the so-called "northern extension to the Yi River" simply meant following the natural course of the Si River upstream. Since the Si River was already a natural

---

[35] *Guoyu·Discourses of Wu* records: "Less than a year after King Fuchai of Wu executed Shen Xu, he launched a military campaign northward. He excavated a deep-water channel, connecting the regions of Shang and Lu. This waterway extended north to the Yi River and west to the Ji River, serving as the route for his meeting with Duke Wu of Jin at Huangchi". Fuchai executed Shen Xu (i.e., Wu Zixu) in 485 BCE and convened with the Jin ruler in 482 BCE. The canal construction between Shang and Lu thus occurred during these three years. At that time, the Lu capital was at Qufu and the Song capital at Shangqiu. As Song descended from the Shang Dynasty remnants, and the He River flowed between these states, the canal was described as "Linking Shang and Lu."

[36] Commentary on the Ji River. In *Water Classic*.

[37] Commentary on the Si River. In *Water Classic*.

[38] Ibid.

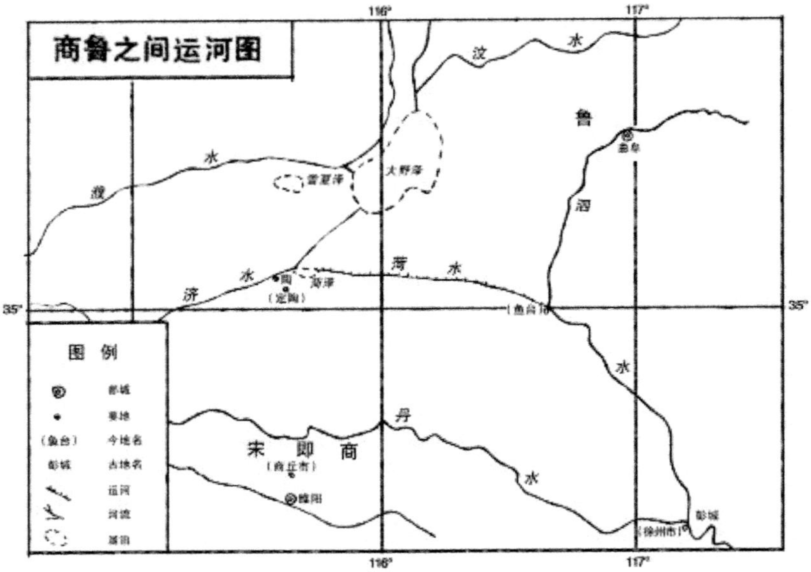

图例 - Legend
都城 – Capital Cities
要地 – Strategic Locations
今地名 – Present Names
古地名 – Historical Names
运河 - Canals
河流 - Rivers
湖泊 - Lakes

Fig. 2.3 The Canal between Shang and Lu

waterway, there was no need to excavate a deep channel—at most, it required dredging and widening. The only true deep-water canal dug was the He River (See Fig. 2.3).

As previously noted, Huangchi—the site where King Fuchai of Wu met with the Marquis of Jin—was located along the Ji River, south of present-day Fengqiu County in Henan Province. Near Huangchi, there was a waterway known as the Huanggou River (黄沟), which branched off from the Ji River and flowed eastward. Its course passed south of ancient Huang County (southeast of modern Lankao County, Henan), south of ancient Chengwu County (now Chengwu County, Shandong), and south of ancient Pingle County (east of present-day Shan County, Shandong), before eventually joining the Si River south of ancient Pei County (east of present-day Pei County, Jiangsu).

The Huanggou River thus formed a connection between the Ji and Si rivers, and its confluence with the Ji River was located near Huangchi. This has led some to speculate that it may have been the canal excavated by King Fuchai. However, the *Water Classic* (*Shuijing*) and its commentary in *The Si River* section describe the Huanggou River in detail without any mention of King Fuchai, suggesting no historical link between the two.

Moreover, although the Huanggou River branched off from the Ji River at Huangchi and flowed eastward, its course was notably winding and indirect—at one point even turning northward through the present-day areas of Chengwu and Shan County in Shandong. Its overall length far exceeded that of the He River, making it highly improbable that it was a man-made waterway. When King Fuchai ordered the excavation of a canal to connect Shang and Lu in order to meet the Marquis of Jin in the west, it would have been neither practical nor strategic for him to choose such a lengthy and circuitous route as the Huanggou River.

To the south of the He River, there existed another artificial waterway called the Huo River. The Huo originated from the Bian River—both being man-made canals situated in the same region between Shang and Lu as the He River. However, the Bian and Huo were branches of the Hong Canal system, not excavated by King Fuchai of Wu.

These two nearly parallel waterways, located in close proximity, could easily be confused. In reality, both the Hong Canal and Ji Rivers diverged from the Yellow River at Shimen—far upstream from Huangchi. Shimen was located northwest of the Xingze County of modern Henan Province. King Fuchai would never have gone to such lengths, taking such a circuitous route and expending excessive labor, just to dig the Bian and Huo Rivers. Thus, although these two waterways were geographically close to the He River, their purposes and origins were entirely different and should not be conflated.

Some may inquire: Wasn't the tribute route described in *Tribute of Yu*—passing from the Huai River to the Si River, then entering the He River before connecting with the Ji River—precisely the same waterway King Fuchai of Wu used when meeting the Marquis of Jin at Huangchi? As previously noted, *Tribute of Yu* was actually composed during the Warring States period by a unification-minded scholar writing under the name of Yu the Great. The inclusion of this He River watercourse provides conclusive evidence that the compiler could not have been a contemporary of King Fuchai or lived earlier. At the very least, by the author's time, the He River had already become a major waterway, leading him to incorporate it into his envisioned transportation network—unwittingly revealing the text's true dating through this very inclusion.

Another compelling piece of evidence suggests that the He River was indeed an artificial canal rather than a natural waterway. The most conspicuous outcome of its successful excavation was the rapid ascension of Tao to become the economic hub of the realm. Tao's strategic location at the juncture where the He River diverged from the Ji River allowed it to serve as a pivotal nexus: to the east, it connected with Qi and Lu; to the west, it linked Qin and Jin; to the south, it extended to Wu and Chu; and to the north, it reached Yan and Dai. This intricate network of waterways facilitated extensive trade and communication, propelling Tao's swift development into the preeminent economic center of its time. If we accept that the renowned Venerable Zhu of Tao was none other than Fan Li, the strategist who assisted King Goujian of Yue in conquering Wu, we can better appreciate the astonishing pace of Tao's transformation. Historical records indicate that when Venerable Zhu settled in Tao, the city had already burgeoned into a metropolis of paramount importance—a remarkable feat achieved within a mere decade following the He River's completion. Such rapid urban ascendancy underscores the transformative impact of this artificial waterway on regional economic development.[39]

During the Spring and Autumn period, Tao was originally the fiefdom of the Earl of Cao. Remarkably, throughout the two centuries of this era, the State of Cao remained insignificant among the various feudal states. This obscurity was not due to its territorial size—after all, the State of Zheng was no more extensive, yet became a focal point of international disputes. During the Battle of Chengpu between Jin and Chu, Duke Wen of Jin first besieged Cao not because of its strategic location, but rather than a personal vengeance—during his exile before ascending the throne, Duke Wen had suffered humiliating treatment from the Earl of Cao.[40] Had the He River been constructed earlier or existed as a natural waterway, the descendants of the Earl of Cao might not have idled through successive generations in obscurity until their state's eventual demise.

Canals Between the Zi and Ji Rivers and Within Chengdu

This section focuses on the canal connecting the Zi River (淄水) and Ji River during the Warring States period. The Ji River flowed through the territory of Qi State, passing northwest of its capital Linzi (modern-day Old Linzi County east of Zibo City, Shandong Province) before emptying into the Bohai Sea. The Zi River originated north of present-day Laiwu County in Shandong, ran east of Linzi city, then joined the Shi River northeastward before also entering the Bohai Sea.[41] At that time, the Ji River flowed just north of the Shi River, with the two waterways remarkably close in proximity. Despite this geographical nearness, precise documentation regarding the exact location of the connecting canal remains elusive. The Zi River's downstream course has been subject to differing historical interpretations. While *The Water Classic* records it as flowing directly into the sea, Du Yu's commentary suggests it merged with the Wen River instead.

However, the Treatise on Geography in *The Book of Han* states regarding Laiwu County in Taishan Commandery: "Mount Yuan, the source of the Zi River, flows east to Bochang where it joins the Ji River." Various annotations differ in their interpretations.[42]

How could the Zi River cross the Mountain Tai to join the Wen River when they are separated by this mountain range? Between the Zi and Ji Rivers flows the Shi River, which itself joins the Zi River, indicating the Zi originally couldn't connect directly with the Ji. *The Book of Han*'s record of the Zi River joining the Ji River

---

[39] Biographies of the money makers. In *Records of the grand historian*, vol. 129.

[40] *Zuo commentary*, the 28th year of Duke Xi.

[41] Commentary on the Zi River. In *Water classic*.

[42] Wang Xianqian. *Collated commentary on the water classic: Zi River*, citing Zhao Yiqing's *Commentary on the water classic*.

may not necessarily be an error, but could indicate the existence of a canal between them. With such a canal, the westward diversion of the Zi River might have caused its original downstream section to gradually dry up, resulting in the Zi River flowing into the Ji rather than the sea.

When was the canal between the Zi and Ji Rivers constructed? This is a question that is not easily answered. When Su Qin lobbied the King of Qi to ally with Zhao State, he mentioned the prosperity of Linzi, which at the time had seventy thousand households, all of whom were quite wealthy.[43] As the capital of Qi, Linzi's wealth was plausible, but such a claim by Su Qin was rare among the states of that era. This prosperity might have been related to the exist of the canal between the Zi and Ji Rivers, as the canal would have facilitated such affluence. However, some further explanation is needed here: as previously mentioned, the construction of the canal between the Zi and Ji Rivers utilized the Shi River, a tributary of the Zi River. The Shi River also had a tributary called the Mian River, which originated east of Ying City, a small town within Linzi. The Mian River flowed into the Shi River northwest of Beiqiu. After the two rivers merged, they were interchangeably referred to, so the Shi River was also called the Mian River.

During the late Warring States period, when Tian Dan was preparing to attack Di, Lu Zhonglian said to him: "Now, General, you enjoy the revenues of Yeyi in the east and the profits along the Zi River in the west. With a golden belt at your waist, you ride freely between the Zi and Mian Rivers."[44] Based on this, one might infer that the Mian River had not yet diminished by the late Warring States period, suggesting that the canal might not have been constructed by then. As mentioned earlier, this canal utilized the course of the Shi River and only began to diverge below Beiqiu to connect with the Ji River. The Mian River merged with the Shi River northwest of Beiqiu, meaning the construction of the canal did not involve the Mian River's course before its confluence with the Shi River. This is the first point.

As early as the Spring and Autumn period, the people of Qi had already valued the Mian River. Duke Jing of Qi once said at a banquet in Jin, "Our wine flows as abundantly as the waters of the Mian River."[45] Lu Zhonglian's pairing of the Mian and Zi Rivers might have been a customary expression of praise at the time, not a literal reference. This is the second point. Based on these two points, it can be concluded that the excavation of the canal between the Zi and Ji Rivers could not have been as late as the end of the Warring States period.

Topographically, this scenario is plausible. The Zi River flowed east of Linzi city, while the Shi River originated 25 *li* northwest of Linzi—their proximity would have allowed interconnection. The Shi River, originating northwest of Linzi, flowed consistently northwest until turning northeast west of Beiqiu. Beiqiu was located west of Bochang County during the Han dynasty (southeast of modern Boxing County, Shandong), approximately 40 *li* southeast of Linzi.[46]

---

[43] Qi strategies I. *In Strategies of the Warring States.*

[44] Qi strategies VI. *In Strategies of the Warring States.*

[45] *Zuo commentary,* 12nd year of Duke Zhao.

[46] Commentary on the Zi River. In *Water classic.*

At that time, the Ji River flowed south of Le'an County and north of Bogu City, which was 50 *li* from Linzi. Bogu City and Le'an County faced each other across the Ji River at close distance.[47] This confirms the Shi and Ji Rivers were separated by less than 10 *li*. Constructing a canal across such a short distance would have been feasible. Such a canal would likely have diverted the Zi River northwest from east of Linzi, utilized the Shi River's course, and then connected to the Ji River via a newly dug channel west of Beiqiu.

As for the excavation of the Lidui Hill in Shu to mitigate the flooding of the Mo River and ensure the smooth flow of the two rivers through Chengdu, Sima Qian explicitly attributed this achievement to Li Bing, a governor of Shu. Li Bing was undoubtedly a historical figure of Qin during the Warring States period. However, there are two differing accounts regarding the time of his appointment as governor of Shu: one claims he was appointed during the reign of King Zhao of Qin,[48] while the other suggests he was sent by King Xiaowen of Qin.[49] After Qin annexed Shu, it initially appointed a marquis to govern the region, and it was not until the 30th year of King Zhao's reign (277 BCE) that the position of governor of Shu was established. At that time, the governor was Zhang Ruo.[50] If Li Bing served as a governor during King Zhao's reign, it would have been toward the end of his rule.

The two rivers flowing through Chengdu, according to *Records of the Lands*, were the Da Jiang, which originated from the boundary of Wenjiang in the southwest, and the Pi River,[51] which flowed from Xinfan County in the northwest. During the Tang dynasty, Wenjiang County and Xinfan County retained the same names as they do today. Li Tai, in his reference to the Da Jiang(大江) flowing from Wenjiang County, also mentioned alternative names such as the Wen Jiang (汶江), Guanqiao Shui (管桥水), Qing Jiang (清江), and Shui Jiang (水江). This river actually flowed west of Wenjiang and Shuangliu (双流) counties and, at the time, did not appear to reach Chengdu. The Pi River flowed through Pi County, which seems somewhat distant from Xinfan County. Du Yu, in his *Records of Yizhou*, identified the two rivers as the Pi River and the Liu River (流江).[52]

Commentary on the Water Classic records that Li Bing constructed a large dam at Du'an County, damming the river to create a diversion dike with two outlets, dividing the river into the Pi River and the Jian River (捡江).[53] This Jian River is likely the same as Liu Jiang. *Commentary on the Water Classic* further states: "The river then flows eastward past Chengdu County, where it splits into two rivers that run side by side. Thus, Yang Ziyun wrote in his *Rhapsody on the Capital of Shu*: 'Two rivers curve before it like ear pendants.'"

---

[47] Commentary on the Ji River. In *Water classic.*

[48] Customs and traditions cited in *Water classic:* Commentary on *the Yangtze River.*

[49] History of Shu. In *History of the lands south of Mount Hua*, vol. 3.

[50] The basic annals of Qin. In *Records of the grand historian,* vol. 5.

[51] Records of the grand historian, vol. 29 cited in *Commentary on 'Book on Rivers and Canals'.*

[52] Ibid.

[53] Commentary on the Yangtze River. In *Water classic.*

When Li Bing served as the governor of Shu, his primary achievement in water management was the irrigation of farmland. The prosperity of the Shu region can largely be attributed to this irrigation system. However, the two rivers flowing through Chengdu were also navigable, much like the canals between the Zi and Ji rivers. Over time, as the irrigation benefits proved far more significant, the navigational function gradually diminished in importance.

The Development of the Hong Canal System

Following King Fuchai of Wu's construction of the Shang-Lu Canal between the states of Song and Lu, the Spring and Autumn Period soon gave way to the Warring States era. During this turbulent age, the Central Plains witnessed extensive hydraulic engineering projects, including the excavation of several major canals. While each of these waterways maintained distinct names and terminal points according to their specific courses, they were in fact interconnected components of an integrated system—collectively known as the Hong Canal. The term "Honggou" has frequently led to historical confusion. Most notably during the Chu-Han contention, the rival leaders Liu Bang and Xiang Yu famously established the Honggou as their temporary boundary line before the final showdown at Gaixia. This presents an apparent paradox: if Honggou referred to an entire canal system with multiple branches bearing different names, which specific segment served as the actual demarcation line? This apparent contradiction stems from the fluid nature of hydrological nomenclature in ancient China. While "Honggou" originally designated the complete network, common usage gradually transferred this name to particular sections—particularly the critical north–south artery known as the Langtang Canal.[54] Thus, when historical records mention the "Honggou boundary" between Chu and Han, they specifically refer to the Langtang Canal segment, not the entire sprawling system.

The configuration of the Hong Canal has long been subject to scholarly discussion. The *Suoyin* commentary to the *Records of the Grand Historian*'s *Annals of Emperor Gaozu* provides detailed annotations, citing Ying Shao's assertion that it was located twenty *li* southeast of Xingyang, while Zhang Hua described it as originally comprising two channels: one flowing eastward past the southern side of Xunyi (modern Kaifeng, Henan) —purportedly excavated by the First Emperor of Qin to flood Daliang—which constituted the true Hong Canal; and another extending east to Yangwu County (southeast of modern Yuanyang, Henan) known as the Guandu River.

These accounts in the *Suoyin* refer specifically to the Hong Canal's main channel. As Li Daoyuan noted in his Commentary on the Water Classic: "The Bian River was also called the Honggou, likely because it served as the demarcation line when Chu and Han divided their territories. As recorded in the *Treatise on Commanderies*

[54] The Langtang Canal is also written as *Langqu*, *Langtangqu*, *Langdangqu*, or *Langqu*. Although the written forms differ, all refer to the same watercourse.

and States: 'The Honggou waterway lies within Xingyang.'"[55] This aligns with Ying Shao's description and refers to the initial segment where the Hong Canal diverged from the Ji River before branching into separate canals.[56] This primary channel of the Hong Canal flowed from west to east.

However, during the Chu-Han demarcation, with Chu occupying the eastern territories and Han the western, it would have been illogical to use an east-flowing waterway as a boundary. Within the Hong Canal system, only the north–south flowing Langtang Canal could have served this purpose. Although Li Daoyuan initially equated the Honggou with the Bian River and located the boundary at Xingyang, he apparently recognized the inadequacy of this explanation, as he later stated: "At this point (near Xunyi County), the channel was known variously as the Yingou or Honggou. When Xiang Yu and Liu Bang divided their realms, they used this waterway to mark their eastern and western boundaries. As Su Qin told King Xiang of Wei: 'Your territory extends south to the Honggou.'"[57] Here, Li not only confirmed this section as the Chu-Han boundary but also refuted alternative theories locating the dividing line at the Honggou Pavilion west of Xiao County (modern Xiao County, Jiangsu) or the Hongkou Pavilion east of Suiyang County (modern Shangqiu, Henan).[58]

The Canals of the Hong Canal System

The excavation of the Hong Canal began at Xingyang, where water was diverted southeastward from the Yellow River. Due to its numerous branches and wide geographic reach, the canal system at the time extended to the states of Song, Zheng,

---

[55] Commentary on the Ji River. In *Water classic*.

[56] Jing Sanlin et al., "Investigation of the Ancient Site of Aocang" (published in *Central Plains Cultural Relics*, No. 1, 1984), discussed the Hong Canal. The report states: "East of the location where the Douwang Stele stands at Niukouyu Dapo, there is a large ditch approximately five *li* in length. Crossing a mountain ridge leads to Liugou, which lies south of Heyin County and appears to have been the headgate of the Hong Canal for channeling water. Extending southward from here is an artificial canal about four *li* long, roughly 300 m wide and 200 m deep. Residential dwellings are situated on earthen platforms about two meters above the canal bed on both sides. Clear traces of water erosion are visible along the canal banks, with snail shells and calcareous nodules (*liaojianshi*) scattered throughout. At the four-*li* mark, the canal turns eastward, traverses Mount Guangwu, reaches Shangren Village, then bends south into the Zhanran River, eventually flowing southeast toward the Kaifeng area. Local residents still refer to this ditch as the 'Honggou' and believe it to be an ancient canal." However, this account conflicts with the *Commentary on the Water Classic*. Mount Guangwu rises steeply along the southern bank of the Yellow River, making it highly improbable that any watercourse from its northern slopes could surmount such a high elevation to flow southward—let alone for Liugou's waters to cross the mountain and merge with the Zhanran River. Moreover, snail shells and calcareous nodules (*liaojianshi*) are unreliable as archaeological evidence, particularly since calcareous nodules are unlikely to form in a perennial riverbed. Thus, the local tradition appears inconsistent with hydrological and geographical realities.

[57] Commentary on the canal. In *Water classic*.

[58] Ibid.

Chen, Cai, Cao, and Wei, eventually merging downstream with the Ji, Ru, Huai, and Si Rivers.[59] The respective capitals of these states were as follows: Song was centered at Shangqiu (modern-day Shangqiu, Henan); Zheng at Xinzheng (modern Xinzheng, Henan); Chen at Huaiyang (modern Huaiyang, Henan); Cai at Fengtai (modern Fengtai, Anhui); Cao at Dingtao (modern Dingtao, Shandong); and Wei at Puyang (modern Puyang, Henan). Geographically, the Hong Canal system occupied the region east of Xingyang, west of the Si River, north of the Huai River, and south of the Ji River. It thus encompassed eastern Henan, southwestern Shandong, northern Anhui, and northwestern Jiangsu. Within the broader Hong Canal system, several major waterways branched off, including the Bian River, Huo River, Langtang Canal, Sui River, Lugou River, and Guo River.[60]

An examination of modern maps reveals that the region traversed by the Hong Canal system is a vast plain, with the overwhelming majority of its elevation lying below fifty meters above sea level. Even in the slightly higher western areas—such as Lankao, Minquan, Qi County, Sui County, Ningling, Taikang, Huaiyang, and further northwest to Kaifeng, Weishi, Tongxu, Zhongmu, and Zhengzhou—the elevation rarely exceeds 200 m. These figures reflect present-day topographical measurements. However, even these elevations are partly the result of long-term sediment accumulation caused by the frequent flooding of the Yellow River. Rich in sediment, the Yellow River slows as it flows into the lower reaches, causing its suspended silt to settle and gradually raise the riverbed. This in turn increases the likelihood of breaches. When such breaches occur, the sediment-laden floodwaters spread across the surrounding land, depositing layers of silt. Since the early Southern Song period, levee failures and flooding in this region have become increasingly frequent. This trend continued—and even intensified—through the Ming and Qing dynasties. The longer this process persisted, the deeper the sediment accumulation became. Entire elevated terrains, hills, and even villages and towns were buried and entombed beneath layers of silt.[61] During the Warring States period, however, such sedimentation had not yet occurred to a significant degree, and the terrain would have been appreciably lower than it is today. Contemporary maps further show that, across this expansive plain, only a few low hills are scattered between present-day Xuzhou in Jiangsu and Su County in Anhui. Aside from these, ancient records scarcely mention any notable mountains in the area. Today, the hills between Xuzhou and Su County stand isolated and disconnected, failing to form continuous ridges or ranges. Given this, it is unlikely that the

---

[59] Sima Qian. Treatise on rivers and canals. In *Records of the Grand Historian (Shiji)*, vol. 29. The text erroneously states: "[They] connected the Hong Canal between the Yangtze and Huai Rivers." This is incorrect—the waterway linking these regions was in fact the Han Canal dug by King Fuchai of Wu, not the Hong Canal. A textual collation with the *Book of Han*'s "Treatise on ditches and canals" reveals that the *Shiji* passage likely contains a scribal error with an extraneous character *hong* (鸿), rendering the statement problematic. This issue has been addressed earlier in our discussion.

[60] The Guoshui (涡水), alternatively written as Guoshui (過水) in historical texts, is the present-day Guo River (涡河).

[61] Shi Nianhai. *Erosion and sedimentation in the yellow river basin through historical periods.*

landscape during the Warring States period would have differed substantially from the general flatness observed now.

The excavation of the transport canals within the Hong Canal system constituted a monumental engineering project of its time. Carving navigable waterways through such an expansive and low-lying plain, largely devoid of prominent hills, represents a remarkable achievement in both harnessing and transforming the natural environment. Today, this region is crisscrossed by a dense network of rivers and streams, resembling a web. It is reasonable to infer that in remote antiquity, this vast plain was already traversed by several natural watercourses. The Bian River, Huo River, Sui River, Lugou River, and Guo River—all components of the Hong Canal system—were long-established rivers, though initially unconnected. The engineers behind the Hong Canal system assessed the terrain and devised ways to link these waterways into an integrated network. For instance, the Bian River, one of the system's primary branches, was already named "Bian" as early as the Spring and Autumn period—a point to be discussed later. Downstream, the Bian connects with the Huo River, another tributary of the Hong Canal system. In fact, the Bian and Huo were originally two distinct rivers. The Bian flowed to Meng County (north of present-day Shangqiu, Henan),[62] where it joined the Huo; part of its remaining flow entered Suiyang (modern Shangqiu County) to the south. This remaining flow, referred to as "residual waters," was likely once the Bian's main current. Later, the Bian was diverted into the Huo, leaving the original main course reduced and thus called residual. The evolution of the name "Huo River" offers further evidence of this transformation. *The Bamboo Annals* records: "The state of Song executed its grandee Huang Yuan on the banks of the Dan River."[63] This incident is also recorded in *Zuo Zhuan*, in the 18th year of Duke Ai of Lu—corresponding to 477 BCE, the 35th year of Duke Ding of Jin. The Dan River was an alternate name for the Huo River. At the time of Huang Yuan's execution, the powerful house of Zhi had not yet been destroyed, and the three families of Han, Zhao, and Wei had not risen as independent states—thus the Hong Canal had certainly not yet been excavated. The Huo River's existence before the canal indicates that it was a natural waterway. Because the Bian and Huo Rivers were geographically proximate, when the Hong Canal was constructed, the Bian was redirected into the Huo. As a result, the original downstream Bian came to be referred to merely as a remnant flow, which eventually disappeared altogether. But why was the Huo also called the Dan River? Li Daoyuan explains: "It is a variant name of the Bian River."[64] Yet such changes in nomenclature are rarely without reason. Li offers no rationale for this name change, making his explanation seem forced. The Sui River provides another example of a pre-existing waterway later incorporated into the Hong Canal system. According to the *Water Classic*, "The Sui River originates in Yan County of Liang Commandery." During the Eastern Han, Yan County was situated between present-day Ningling and Zhecheng in Henan, southwest of Shangqiu County. Li Daoyuan, however, contested this claim, stating:

[62] Commentary on the Mian River. In *Water classic*.

[63] Commentary on the Huo River. In *Water classic*.

[64] Ibid.

"The Sui River arises from the Langdang Canal west of Chenliu County and flows northeast. *The Geography Treatise* states: 'The Sui River begins from the Langtang waters of Xunyi in Chenliu.' Thus, the claim in the *Classic* that it originates in Yan is mistaken."

Within the Hong Canal system, the Sui River flows eastward from Chenliu through Yongqiu, Xiangyi, Ningling, and Suiyang, continuing southeastward. These correspond to today's Qi County, Sui County, Ningling, and Shangqiu, respectively. In the Han period, Yan County lay in the remote southwest of Suiyang, far from the Sui River's course. If the *Water Classic*'s author had been fabricating, it would make little sense to cite a county not along the river's route as its source. This suggests the author may have recorded the Sui River's original natural course—distinct from the one modified in the Hong Canal system. The engineers of the Hong Canal likely only excavated the upper stretch of the Sui River between Chenliu and Suiyang. Downstream, they made use of the river's natural channel. Because this lower section remained the natural Sui River, the artificially constructed upper course also retained the same name.

A similar situation is found with the Guo River (涡水).[65] According to the *Water Classic*, "The Yingou River originates from the Langdang Canal in Langdang, Yangwu County, Henan, and flows southeast to Pei, where it becomes the Guo River (涡水)." Li Daoyuan, in his commentary, offers clarification and raises objections: "The Yingou initially merges with the Langdang Canal, then diverges to become the Sha River, from which the Guo River emerges. The Guo River first receives water from the Sha at Fugou County. Xu Shen also notes: 'The Guo River originates from the Langdang Canal at Fugou County in Huaiyang.' It is incorrect to claim that it only becomes the Guo River upon reaching Pei." Pei County in the Eastern Han period corresponds to present-day Peixian in Jiangsu, which lies far beyond the Guo River's basin and bears no relation to it. However, Pei Commandery in the Eastern Han—with its seat in present-day Huaibei, Anhui—encompassed a vast area. Its westernmost county, Qiao (present-day Bo County, Anhui), lies directly along the present course of the Guo River. Therefore, "Pei" in this context likely refers to Pei Commandery rather than Pei County. Fugou County remains to this day in eastern Henan. The Langdang Canal passed through Fugou and continued southward to Chen County (modern Huaiyang in Henan), where it joined the Ying River. Given the great distance between Fugou and Qiao County, the Langdang Canal could not possibly have extended that far west. Although the *Water Classic* states that the Yin Canal originates in Yangwu County and joins the Langdang Canal, in practice, from Daliang (modern Kaifeng) downstream to Fugou County, the Yingou River and Langdang Canal flow together as a single watercourse. It cannot be the case that the Guo River only begins at Pei. Nonetheless, the *Water Classic* explicitly locates the Guo River in Pei Commandery. This may be similar to the Sui River's attribution to Yan County: both cases suggest the rivers originally existed as natural waterways in these locations. The Hong Canal system was likely formed by cutting channels from

[65] The Guo River (涡水) is also referred to as the Guo River (涡水) in the *Commentary on the Yingou River* from the *Water Classic*.

Fugou County to direct water toward Pei Commandery, where it joined the natural course of the Guo River.

It is now necessary to trace the origins of the Hong Canal system and its various branches. The Hong Canal originally drew its waters from the Yellow River. At that time, the Yellow River had not yet encroached upon the former course of the Ji River, and the points at which the Ji River and the Hong Canal diverged from the Yellow River were located near present-day Xingyang, Henan. Since both the Ji River and the Hong Canal branched off from the Yellow River at this location, they flowed eastward together until reaching the area south of present-day Yuanyang County, where they separated: the Ji River continued eastward, while the Hong Canal veered southeast.[66] Upon reaching the area northwest of present-day Kaifeng, the Hong Canal again divided into two branches—the Bian River, which flowed east, and the Langtang Canal, which flowed south.

After the Bian River and the Langtang Canal diverged, the Bian flowed between present-day Lankao and Shangqiu Counties, where it came to be known as the Zaihuo Canal. Further east, it was renamed the Huo River, which flowed southeast and entered the Si River to the north of Pengcheng (modern Xuzhou, Jiangsu).[67]

The Langtang Canal was known by several names throughout history. It was referred to as the Langdang Canal,[68] the Xunyi Canal,[69] or more simply as "the Canal",[70] with its lower reaches also called the Sha River.[71] The name Xunyi Canal originated from Xunyi County—located near present-day Kaifeng—through which the canal once flowed. The term Canal was naturally a shorthand for the Langtang Canal itself. Scholars have long debated the canal's point of origin. One view holds that it branched directly from the Yellow River north of Xingyang County, diverging in parallel with the Ji River.[72] Another posits that it stemmed from the Ji River, a theory that itself comprises two variants: one placing the divergence near Xingyang County,[73] the other situating it in Yangwu County, east of modern-day Yuanyang County, Henan.[74] In actuality, the eastward flow from Xingyang formed the main channel of the Hong Canal system before it split into subsidiary canals.[75] To assign the Langtang Canal's origin to this early diversion is to retroactively conflate it with the Hong Canal's primary course—an oversimplification that likely projects the canal's

---

[66] The interconnections among the Yellow River, Ji River, and Hong Canal are discussed in detail in the *Commentary on the Yellow River*, *Commentary on the Ji River*, and *Commentary on the Canal (Qu)* in the *Water Classic.*

[67] Treatise on geography. In *Book of Han*, vol. 28; Commentary on the Bian River *and* commentary on the Huo River. In *Water classic.*

[68] Commentary on the Yellow River. In *Water classic.*

[69] Commentary on the Ji River. In *Water classic.*

[70] Commentary on the Canal. In *Water classic.*

[71] Ibid.

[72] Commentary on the Yellow River. In *Water classic.*

[73] Treatise on geography. In *Book of Han*, vol. 28.

[74] Commentary on the Ji River. In *Water classic.*

[75] Book on rivers and canals. In *Records of the grand historian*, vol. 29.

identity farther upstream than historical and geographical evidence supports. As it flowed southward, the Langtang Canal branched northwest of present-day Chenliu Township (Kaifeng) to form the Sui River, and again southwest of Chenliu to produce the Lugou River. Continuing past western Taikang County, it further divided to create the Guo River (涡水). Ultimately, the Langtang Canal emptied into the Ying River south of present-day Huaiyang County.[76] The Ying River itself proceeds southward, joining the Huai River near modern Yingshang County in Anhui Province.

As previously noted, the downstream branches of the Hong Canal connected with the Ji, Ru, Huai, and Si rivers. The Ru River lay west of the Ying River. Given that the Langtang Canal flowed into the Ying with no Hong Canal branches further west, how could it possibly intersect with the Ru River? In fact, this connection was not without basis. Although the Ru and Ying rivers followed independent courses, they were not entirely separate. At the midstream section of the Ru River, near the northwest of Qi'e (modern Yanling County, Henan), a distributary called the Fen River—also known as the Dayin River—branched off eastward and eventually joined the Ying River north of Ruyang County (present-day south of Shangshui County, Henan).[77] Of particular interest is the hydronym Fen River. The *Erya* states: "The Yellow River has its Yong [overflow channels]; the Ru has its Fen." Guo Pu's commentary clarifies that this term refers to a natural distributary formed when a major river overflows into a smaller channel.[78] Thus, the Fen River was a natural watercourse—an eastern-flowing spillway from the Ru into the Ying. This Dayin River bore another name: *Bie Ru* ("Branch of the Ru"). As the term suggests, it originated as a bifurcation of the Ru. During its eastward flow, a subsidiary channel called *Si Ru* ("Dead Ru") further attested to its hydrological ties with the main Ru River.[79] The alternate name Yin River likely arose from phonetic convergence between Fen and Yin.[80] Some might speculate that the Yan Shui was an artificial canal, but such a claim would be overly contrived. *The Book of Songs* during the Western Zhou Period mentions "following the banks of the Ru Fen"[81]—where Fen (坟) denotes the fertile, elevated loess soil along the Ru's distributaries.[82] This confirms the Fen River's natural existence as early as the Western Zhou dynasty, long before the conceptualization of engineered canals. It was precisely this natural distributary that enabled the Hong Canal system—via the Ying River—to establish an indirect link with the Ru River.

Returning now to the Sui River: after diverging from the Langtang Canal, it flowed through present-day Qi County, Sui County, Shangqiu, and Xiayi counties in Henan, as well as Su County and Lingbi counties in Anhui, before entering the Si River

---

[76] Commentary on the canal. In *Water classic*.

[77] Commentary on the Ru River. In *Water classic*.

[78] Explanation on rivers. In *Erya*.

[79] Commentary on the Ying River. In *Water classic*.

[80] Commentary on the Ru River. In *Water classic*.

[81] Classic of Poetry. "Ru Fen." In *Airs of the States – Zhou Nan*.

[82] Zhao, Y. Annotations to the commentary on the water classic (Cited in Wang Xianqian's *Collated edition of the commentary on the water classic)*.

near Suqian in Jiangsu.[83] During this historical period, the Si River itself coursed southward through modern Pei County, Xuzhou, and Suqian County in Jiangsu before merging with the Huai River—thus providing a confluence point for both the Huo and Sui rivers. As for the Lugou River, after branching off from the Langtang Canal, it traversed Qi County before joining the Guo River west of Taikang County.[84]

Among these waterways, the course of the Guo River presents particular complexity. While the Guo River originated as a distributary of the Langtang Canal, its ultimate source traces back to the Yingou River. This channel branched independently from the Yellow River near the site of the former Yuanwu County in Henan. After diverging from the Yellow River, the Yingou River flowed southeast, crossing the Ji River before converging with the Langtang Canal near present-day Kaifeng. The subsequent distributary from this merged channel formed what became known as the Guo River. Notably, the segment between Yuanwu and Kaifeng later disappeared from historical records—likely due to siltation or abandonment. From its divergence at the Langtang Canal, the Guo River coursed through modern Taikang and Luyi counties in Henan, then through Bozhou, Guoyang, and Mengcheng counties in Anhui, before finally joining the Huai River at Huaiyuan.[85] Today, only the lower reaches of the Guo River remain visible. Another surviving waterway, the Sui River, retains little of its original form within the Hong Canal system. All other branches have since vanished (See Fig. 2.4).

The Construction Period of the Hong Canal

Given the Hong Canal's extensive reach and complex network, when exactly was it excavated? This question has long attracted scholarly attention, with numerous interpretations offered over time. A brief review of some key views is presented here.

Li Daoyuan attributed the canal's construction to Yu the Great.[86] However, Yan Ruoqu strongly disagreed. Citing the *Records of the Grand Historian*, "Yu's achievements extended across the Three Dynasties. Thereafter, the river was diverted at Xingyang to the southeast, becoming the Hong Canal," he argued that the canal was the work of Yu's descendants. Yan advanced three conjectures in support of a later date: (1) Su Qin told King Xiang of Wei, "To the south of Your Majesty's realm lies the Hong Canal," implying its existence before the Warring States period; (2) the Battle of Bi, where Chu fought Jin, occurred at Bi—identified with the Bian River—suggesting the canal existed before the Spring and Autumn period; and (3)

[83] Commentary on the Sui River. In *Water classic*.

[84] Treatise on geography. In *Book of Han*, vol. 28.

[85] Commentary on the Yingou River. In *Water classic*.

[86] Commentary on the Yellow River. In *Water classic*.

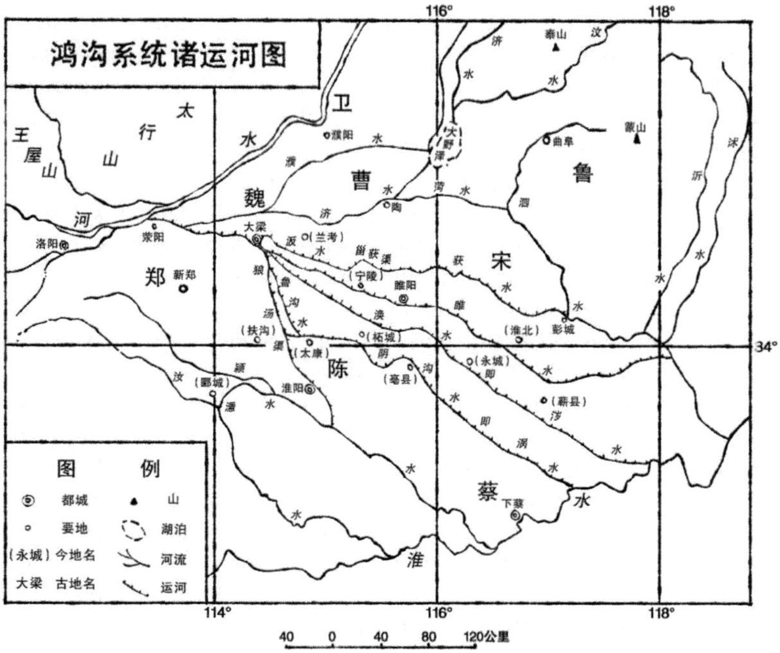

图例 Legend
都城 Capital Cities
要地 Strategic Locations
(永城) Yongcheng
大梁 Daliang
今地名 Present Names
古地名 Historical Names
山 Mountains
湖泊 Lakes
河流 Rivers
运河 Canals
王屋山 Wangwu Mountain
太行山 Taihang Mountain
河水 Yellow River
大野泽 Daye Marsh
鲁沟水 Lugou River
狼汤渠 Langtang Canal
汳水 Bian River
甾获渠 Zihuo Canal
涣水即濊水 Huan River (Hui River)
阴沟水即涡水 Yingou River (Guo River)
蔡水 Cai River
淮水 Huai River
菏水 He River
济水 Ji River
汶水 Wen River
获水 Huo River
睢水 Sui River

Fig. 2.4 The Hong Canal network

the *Erya* states that "the river flowing from the Yellow River is called Yong," and as Yong was originally the name for the Bian River, this too hints at an early date.[87]

While Yan Ruoqu rightly rejected the notion that Yu the Great excavated the Hong Canal, his own evidence does not conclusively establish that it existed prior to the Spring and Autumn period. The exact date of the *Erya*'s compilation remains uncertain; many scholars now believe it was a product of the Warring States period. As for Bi, the battlefield was located near Mount Ao in the vicinity of Xingyang. If Bi did refer to the Bian River, it would have been a local natural watercourse rather than a man-made canal segment. Moreover, King Xiang of Wei lived well after the Spring and Autumn period, making it untenable to date the canal to before the Warring States based on Su Qin's mention alone.

Quan Zuwang, drawing on Liu Chengguo's *Geographical Records of Xuzhou* (as cited in the *Commentary on the Water Classic, Jishui section*), referred to a passage stating, "King Xu Yan governed with benevolence and righteousness, and, desiring to travel by boat to the upper states, opened a canal between Chen and Cai." From this, Quan inferred that the Hong Canal—specifically its section around the Sha River—was excavated during the reign of King Xu Yan.[88] According to traditional accounts, Xu Yan lived well before King Fuchai of Wu. However, during his supposed era, the lands between Chen and Cai were already under the control of the state of Chu. How could Xu Yan, whose own domain was far removed, have undertaken canal works within Chu territory? Furthermore, if King Xu Yan had indeed excavated a canal connecting the Huai and Ji Rivers, why would King Fuchai of Wu later go to great lengths to dig another canal between Shang and Lu? Modern scholar Qian Mu identified King Xu Yan with King Yan of Song, a hypothesis that holds some plausibility. However, Qian too relied on the *Geographical Records of Xuzhou* to argue that King Yan of Song constructed a canal between Chen and Cai.[89] Yet by that time, the Song state had been reduced to a small enclave centered on Pengcheng (modern Xuzhou), besieged on all sides—especially by its archenemy Wei. Under such political and military duress, it seems highly unlikely that King Yan of Song had the means or authority to excavate a canal reaching into Chu-controlled territory, let alone pass through land dominated by Wei. Moreover, as King Xiang of Wei's reign predates that of King Yan of Song by several years, and the Hong Canal already existed during King Xiang's time, attributing the canal's construction to King Yan remains untenable.

Although Li Daoyuan stated that the Hong Canal was excavated by Yu the Great, in the *Water Classic*, he cited two entries from the *Bamboo Annals* that suggest otherwise. The first states: "In the tenth year of King Hui of Liang (360 BCE),[90]

---

[87] Cited in *Detailed commentary on the tribute of Yu.*

[88] Quan Zuwang. Questions and answers on the classics and history (Jingshi wenda). In *Collected works of Jieqi pavilion (Jieqi ting ji)*, vol. 8.

[89] Qian Mu. A study on Pengcheng as the capital of Song State during the Warring States period. *Yugong Bimonthly*, 3(3).

[90] Regarding the chronology of King Hui of Liang, there are inaccuracies in the *Records of the Grand Historian: Chronological Table of the Six States*. The dates adopted here follow Chen Mengjia's *Chronology of the Six States*. The same applies below.

river water was brought into Futian, and a great canal was also dug to divert the waters of the Fu River." After quoting this, Li Daoyuan added: "There is another watercourse that receives the Yellow River from Suanzao, diverted from the Pudu, passing Suanzao and running south of Yangwu County. It is commonly known as the Shizi Canal and is connected to the main canal. Some claim this watercourse was opened during the reign of King Hui of Liang, though the details are unclear." The second entry reads: "In the thirty-first year of King Hui of Liang (339 BCE), a great canal was constructed at the northern outskirts to carry the waters of Futian." This "great canal" likely refers to the Hong Canal, and "King Hui of Liang" is the same as King Hui of Wei.

Further supporting this identification, the *Strategies of the Warring States: Strategies of Chu* records: "During the Handan crisis, the State of Chu dispatched general Jing She to lead troops to rescue Zhao. After Handan fell, Chu seized the territory between the Sui and Hui rivers." Here, the Sui refers to the Sui River within the Honggou Canal system, while the Hui denotes the Huan River. This Sui-Hui interfluvial region corresponds to present-day Shangqiu, Ningling, and Suixian counties—constituting the southeastern frontier of Wei and the northeastern border of Chu.[91] The so-called "crisis at Handan" refers to Wei's surrender of the city of Handan, which occurred in the eighteenth year of King Hui of Liang (352 BCE). From this passage, it can be inferred that the waterways of the Sui and the Hui Rivers had already been completed by that time.

The Sui River was a tributary of the Hong Canal, and the Hui River a tributary of the Sui. That Chu was able to seize the region between the two suggests that the canal system had largely been established by then. After all, any functioning water system requires both a source upstream and a discharge point downstream. If King Hui of Liang had not already excavated the main canal, from where would the Sui and the Hui Rivers have flowed? Without a source, they would have been nothing more than water without origin. Conversely, without completed downstream branches, where would the water in the great canal have drained? What's more, since the canal drew from the Yellow River, any mismanagement could have easily led to disastrous flooding. Thus, it is implausible that King Hui would have undertaken such a project without first drafting a comprehensive plan. For these reasons, it is reasonable to conclude that the excavation of the Hong Canal began sometime between the tenth and eighteenth years of King Hui's reign. Additional construction likely continued afterward, as evidenced by the aforementioned works in the thirty-first year of his reign.

In his early years, King Hui of Liang was an ambitious and enterprising ruler. The capital of the state of Liang had originally been located in Anyi (present-day northwest Xia County, Shanxi), but during King Hui's reign, it was moved to Daliang (present-day Kaifeng, Henan). Although the state had formerly been known as Wei, from this time onward it came to be referred to as Liang. There are differing accounts regarding the exact year of King Hui's relocation of the capital—some sources claim it occurred in the sixth year of his reign (364 BCE), while others place it in the ninth

[91] Cheng Enze. *Geographical investigation of the strategies.*

year (361 BCE).[92] Both dates are recorded in the *Bamboo Annals*, though given the text's fragmentary state and the inconsistencies in various citations, some degree of error is to be expected. Nevertheless, both dates precede the commencement of the Hong Canal's excavation. Therefore, it is unnecessary to dwell on whether the relocation took place in the sixth or ninth year of his reign.

Following the move to Daliang, King Hui launched military campaigns against the states of Zhao and Han, dispatching his chariots on multiple expeditions, each yielding notable success. It is said that during the Fengze Assembly, King Hui rode in a Xia-style chariot and styled himself the "King of Xia,"[93] an act that vividly reveals his soaring ambitions. At the time, he not only ruled a territory spanning a thousand *li* but also commanded the allegiance of twelve feudal lords.[94] During the reign of King Hui, although the situation of the Seven Powers standing in rivalry had long taken shape, there still remained several smaller states interspersed among them, which King Hui could thus harness and direct. These twelve vassal states, as described by Wei Yang when he met King Hui, were "none other than Song, Wei, Zou, Lu, Chen, and Cai."[95] The states of Song, Wei, Chen, and Cai have already been discussed. The state of Lu was centered in Qufu, in present-day Qufu County, Shandong, while the former territory of Zou lay in present-day Zouxian, also in Shandong—both within the Si River basin. Although Lord Shang mentions only six states, falling short of the full tally of twelve, his list already accounts for half. Even considering just these six, they all lay within the regions accessible by the Hong Canal system and its associated waterways—namely, the Ji, Ru, Huai, and Si Rivers. That King Hui achieved such political and military successes cannot be considered unrelated to the development and use of the Hong Canal and its transport network.

The Rise of Economic Centers

King Hui of Liang's excavation of the Hong Canal, though primarily driven by military and political motivations and yielding notable achievements in those realms, had its most far-reaching and enduring impact in the economic sphere. The capital of the Wei state, Daliang (present-day Kaifeng, Henan), was naturally the first to flourish under the influence of the canal. While historical records offer few direct accounts of Daliang's prosperity at the time, indirect evidence strongly suggests that it was indeed a thriving metropolis. On one occasion, Prince Xinling of Wei, in conversation with the King of Wei, remarked that Wei commanded "hundreds of large counties and dozens of renowned cities across the regions north and south of

---

[92] Hereditary house of Wei *with Pei Yin's c*ollected explanations. In *Records of the Grand Historia,* citing the *Bamboo Annals*; Zhao Qi's *Correct Meanings of Mencius* likewise references this text.

[93] Qin strategies IV. In *Strategies of the Warring States.*

[94] Qin strategies V. In *Strategies of the Warring States.*

[95] Qi strategies V. In *Strategies of the Warring States.*

the Yellow River"[96]—a testament to the state's affluence. These areas—referred to as Hexi and Henei—surrounded Daliang along both banks of the Yellow River[97] and owed much of their prosperity to the presence of the Hong Canal. If even Wei's secondary cities had reached such a level of wealth, it follows that Daliang, as the capital, must have been even more prosperous. The destruction of Daliang occurred when the Qin forces, during their conquest of Wei, diverted the waters of the Hong Canal to flood the city.[98] Its former splendor was thus obliterated almost overnight. The devastation was so complete that the city remained in ruins for a long time; when Sima Qian passed through the area during the reign of Emperor Wu of Han, he still found it desolate.[99] Although Daliang was reduced to rubble, the transport functions of the Hong Canal were not interrupted. A replacement for such an important city was inevitable. In the wake of Daliang's fall, the upstream city of Xingyang (modern Xingyang County, Henan) and the downstream city of Suiyang (modern Shangqiu County, Henan) emerged to fill the vacuum. Both cities inherited Daliang's legacy and began to prosper in turn. Xingyang, located at the point where the Hong Canal branched off from the Yellow River, had remained overshadowed during the Wei period due to its proximity to Daliang. After the latter's destruction, however, the Qin state—seeking to facilitate canal transport—established the Ao Granary near Xingyang to store grain tribute shipped from various parts of Shandong. As a key node in the canal network, Xingyang gradually rose to economic prominence. Suiyang, while somewhat less prosperous than Xingyang, also absorbed a portion of Daliang's former vitality. Unlike Xingyang, it did not benefit from government investment but developed naturally through its favorable geography. Though it was not situated at the main canal mouth like Daliang or Xingyang, Suiyang was served by the Sui River to its south and the Huo River not far to the north, enabling the flow of goods from the Jianghuai region. During the Spring and Autumn and Warring States periods, Suiyang had served as the capital of the State of Song. Its later designation as "Liang" in the Han Dynasty reflects its inheritance of Daliang's former status.

The numerous downstream branches of the Hong Canal gave rise to many flourishing urban centers. Among them was Chen at the confluence of the Langtang Canal (a branch of the Hong Canal) and the Ying River. Chen functioned not only as a transit port for goods from the Jianghuai region bound for Daliang, but also as a hub for distributing supplies needed in the upper reaches of the Ying River. As previously mentioned, the Yan River provided a navigable link between the Ying and Ru Rivers, thereby spurring economic development in the Ru River basin. The mercantile orientation of Chen's inhabitants was undoubtedly shaped by this favorable geographical

[96] Wei strategies III. *In Strategies of the Warring States*. In the text, Lord Xinling is written as Zhu Yi, which is a scribal error for Wuji.

[97] During the Warring States period, the Yellow River flowed northeast through present-day Xun County and Puyang in Henan Province and emptied into the Bohai Sea. As a result, the area southeast of the Taihang Mountains and north of the Yellow River was referred to as "within the River" (*Henei*), while the region south of the Yellow River was correspondingly called "beyond the River" (*Hewai*).

[98] Hereditary house of Wei. In *Records of the grand historian*, vol. 44.

[99] Ibid.

environment.[100] In the late Warring States period, the Chu state relocated its capital from Yingdu to Chen[101]—not only to evade Qin encroachment, but also, crucially, to take advantage of Chen's economically strategic position.

At the time, Chu repeatedly shifted its capital. While external threats were certainly a factor, economic motivations were arguably even more decisive. When Chen became untenable, the capital was moved again, this time to Shouchun (modern-day Shou County, Anhui), another major city within the Hong Canal network. Situated at the confluence of the Ying and Huai Rivers, Shouchun lay just north of Hefei, where two rivers—the Fei and Shi—originated close to one another. These waterways together formed another navigational route between the Yangtze and Huai regions, outside of the Han Canal. Because of its position at the junction where the Fei entered the Huai, Shouchun held strategic importance for transport, forming a mutually reinforcing relationship with Hefei. With the successful excavation of the Hong Canal, this regional network was greatly enhanced. Shouchun and Hefei became further integrated with the Central Plains, developing into prominent metropolitan centers of the time.

Another prominent economic center along the lower reaches of the Hong Canal was Pengcheng, corresponding to present-day Xuzhou. To its north, the Huo River flows into the Si, and just to its south, the Sui River also joins the Si. Pengcheng thus sat not only between the Huo and Sui Rivers but also at their confluence with the Si River. If one accepts that Suiyang rose to prominence after Daliang precisely because of its location between the Huo and Sui Rivers, then there is no reason to deny that Pengcheng, with similar geographic advantages, had the potential for equal prosperity. Moreover, Pengcheng lay along the banks of the Si River, whose upper reaches connected with the He River. Along the He River stood Tao, known in ancient times as the "center of the world." Whether traveling from Tao to the Jianghuai region or from Daliang and Suiyang southward, one invariably passed through Pengcheng. Its strategic position along these key transportation arteries made its emergence as a commercial metropolis all but inevitable. When Xiang Yu, the Hegemon-King of Western Chu, overthrew the Qin dynasty, he stubbornly refused to establish his capital in Guanzhong—the so-called land of heavenly bounty—and instead chose Pengcheng. His decision, while mocked by some as the act of "a macaque donning a crown",[102] in fact reveals a keen sensitivity to the economic realities of the time.

During the Warring States period, the He River and the Hong Canal formed two vital waterway systems. While they jointly enabled the rise of Pengcheng in their lower reaches, their upper courses connected with other natural rivers, giving rise to several additional economic centers. Chief among them was Linzi. As the capital of the state of Qi, Linzi was located along the Zi River. As previously noted, by the time Su Qin visited Qi during his eastern travels, Linzi had already developed into a large city with a population of 70,000 households.[103] Although Su Qin had

[100] Biographies of the money-makers. In *Records of the grand historian*, vol. 129.

[101] Hereditary house of Chu. In *Records of the grand historian*, vol. 40.

[102] Basic annals of Xiang Yu. In *Records of the grand historian*, vol. 7.

[103] Qi strategies I. In *Strategies of the Warring States*.

journeyed across many states, he reserved his praise for Linzi—clearly signaling its superiority over other contemporary capitals. Even by modern standards, a city of such size would be considered substantial. Linzi's prosperity was intimately tied to the affluence of Qi. The region was naturally well-suited for agriculture and sericulture, with fertile lands that encouraged the flourishing of both. Qi's silk textiles were long renowned among the various states, and its grain surpluses were still widely admired well into the Qin and Han dynasties. This favorable natural environment spurred Linzi's continued economic expansion.[104] Likely during the early Warring States period, Qi also excavated a canal connecting the Zi River to the Ji River—a development discussed earlier. This canal enabled goods from Linzi to be transported via the Zi River into the Ji River system and further distributed to other regions. Situated as the eastern terminus of this Tao-centered water transport network, Linzi's strategic location played a decisive role in its prosperity. A modern analogy can be drawn from the construction of the Longhai Railway. As with railways, the terminus of a transport network often enjoys greater development than points along the route. The Longhai Railway began as the Bian-Luo Railway, with its western terminus initially at Luoyang. Later, as construction proceeded westward in fits and starts, towns like Guanyintang and Shanxian in Henan experienced brief periods of economic flourishing when they temporarily served as endpoints. However, these towns soon declined as the railway extended further. When the line eventually reached Baoji in Shaanxi, that city temporarily emerged as a major economic center in the northwest. By comparison, ancient Linzi, as the eastern terminus of a major waterborne trade network, owed its prosperity not to chance but to its critical role within this broader system of regional transport.

Luoyang, serving as a counterpart to Linzi in the east, functioned as the western terminus of ancient China's waterborne trade routes. Whether goods moved westward via the He River or the Hong Canal, they invariably passed through Luoyang for transshipment and redistribution. This centrality stemmed from navigational constraints on the Yellow River: only the southern stretch of the Yellow River in Henan allowed reliable navigation.[105] Beyond this segment, both upstream and downstream travel was fraught with difficulty. Consequently, commodities from Tao and Daliang had to be transferred into the Luo River and shipped upstream to Luoyang. These geographic imperatives fundamentally shaped Luoyang's mercantile activity during the Warring States period.[106] In the early part of this era, the need to traverse Han territory when transporting goods westward to Qin stimulated the rise of Yiyang (located northeast of present-day Xihancheng in Yiyang County) as a secondary hub. Originally a minor county, Yiyang rapidly expanded into a de facto commandery-level urban center[107] due to its strategic location on alternative routes that bypassed Han control. Two main routes led west from Luoyang: a northern route through Mianchi to Shanxian,

[104] Biographies of the money-makers. In *Records of the grand historian*, vol. 129.

[105] The term "southern stretch of Yellow River" (*Nan He*) first appears in the *Tribute of Yu* and refers to the section of the Yellow River lying south of the regions of Hedong and Henei.

[106] Biographies of the money-makers. In *Records of the grand historian*, vol. 129.

[107] Qin strategies II. In *Strategies of the Warring States.*

following the Jian River valley—a path famously taken by Qin general Meng Mingshi during a surprise attack on Zheng in the Spring and Autumn period, when his army passed between the southern and northern ridges of Mount Xiao.[108] By the Warring States period, the southern route through Yiyang had become dominant. This is reflected in King Wu of Qin's campaign to seize Yiyang,[109] led by Gan Mao, as well as in King Zhao of Qin's diplomatic meeting with the King of Wei at Yiyang.[110] Though geographically distant from the Hong Canal, Yiyang's prosperity—second only to Luoyang—was nonetheless indirectly sustained by the broader economic network shaped by canal-based transportation.

During the Warring States period, Tao and Wei were frequently mentioned together[111]—an indication that they held comparable status in terms of economic importance. Tao was described as being located "at the center of the world," a phrase that not only reflected its economic prominence but also its strategic position as a nexus of transportation routes. As previously noted, from Tao one could travel downstream along the He River to reach the region between the Yangtze and Huai Rivers; eastward along the Ji River to arrive at Linzi; or upstream via the Ji River into the Yellow River and then into the Luo River, eventually reaching as far as Luoyang. These were the transportation routes to the east, west, and south. But what of the north? Northward travel from Tao led directly to Wei—the very state so often paired with Tao. Wei referred specifically to Puyang, which was the capital of the State of Wei. The name "Puyang" (*Pu-yang*, "north of the Pu") derives from its location on the northern bank of the Pu River. This river branches off from the Ji River, flowing through present-day Fengqiu County and Puyang County in Henan Province, before rejoining the Ji River in Dingtao County.[112] It functioned as a distributary of the Ji River. Although the Pu River was relatively short, it was sufficient to maintain transportation and communication between Tao and Wei. Wei's prominence in the eyes of contemporaries was due not only to its riverine connection with Tao, but also to other geographical factors. After the Yellow River bends northeastward in the Central Plains, that section of the river became virtually useless for transportation. Accordingly, movement between the northern states of Yan and Zhao (in present-day Hebei) and the Central Plains no longer relied on the Yellow River, but rather on overland routes running along the eastern slopes of the Taihang Mountains—a route that roughly corresponds to the present-day Beijing–Guangzhou railway line[1]. For these regions to access the economic center of Tao, they had no choice but to pass through Wei. In this context, Wei naturally prospered and gradually came to be regarded as equal to Tao. By the end of the Warring States period, however, the State of Wei had become severely weakened, though it continued to maintain its capital at Puyang. While Qin had little interest in Wei's narrow territory, it had no intention

[108] *Zuo commentary*, 32nd year of Duke Xi.

[109] Qin strategies II. In *Strategies of the Warring States.*

[110] Basic annals of Qin. In *Records of the grand historian*, vol. 5.

[111] Both *Strategies of the Warring States* and *Han Feizi* contain frequent parallel references to Tao and Wei.

[112] Li Daoyuan. Commentary on the Ji River. In *Commentary on the Water Classic.*

of allowing Wei to continue monopolizing control of Puyang, a key economic hub. Thus, in the sixth year of the reign of the First Emperor of Qin (241 BCE)[2], the ruler of Wei was ordered to relocate the ancestral altars and capital to Yewang (in present-day Qinyang County, Henan), and Puyang County was formally annexed by Qin.[113]

The artificial waterway systems of the He River and the Hong Canal, centered around Tao and Daliang, together with several related natural waterways, formed a network that linked more than half of China east of Hangu Pass. The economies across this extensive region were closely interconnected and mutually dependent. Significantly, this economic network did not include the state of Qin—a fact clearly reflected in Tao's designation as the "center of the world." Qin's subsequent aggressive expansion eastward was, in essence, an attempt to forcibly incorporate itself into this preexisting economic system (See Fig. 2.5).

Cultural Integration Facilitated by Canals

The construction of these canals was of immense economic significance, but its impact on cultural integration was equally profound. During the Spring and Autumn and Warring States periods, China's natural waterway system consisted of four major rivers—the Yellow River (*He*), the Ji River (*Ji*), the Yangtze River (*Jiang*), and the Huai River (*Huai*)—each flowing independently into the sea, mostly in a parallel fashion. Each river had its own network of tributaries and thus formed a distinct system. Although the lower courses of the Ji and Yellow Rivers were relatively far apart, their upper reaches were connected, since the Ji River branched off from the Yellow River. These hydrological configurations had clear cultural implications.

The proximity and interconnection between the Ji and Yellow Rivers meant that their basin regions shared a common culture. During the Spring and Autumn period, the area was divided into many feudal states, which, despite political fragmentation, maintained a consistent cultural identity. These states referred to themselves collectively as *Zhuxia* or *Huaxia*, signaling their shared cultural heritage. In contrast, the so-called *Rong* and *Di* peoples beyond the *Zhuxia* sphere had distinct customs and cultures, and relations between the Huaxia and the Rong-Di were marked by sharp cultural divisions.

When Guan Zhong assisted Duke Huan of Qi in establishing hegemony, he put forward the call to "honor the king and repel the barbarians," a slogan that achieved some success and was later praised by Confucius. Confucius famously remarked, "Without Guan Zhong, we would be wearing our hair loose and fastening our garments on the left,"[114] referring to the sartorial customs of nomadic peoples—practices wholly unrelated to those of the Jiang-Huai region. Yet even the state of

---

[113] Table of the six states. In *Records of the grand historian*, vol. 15; Hereditary house of Wei Kangshu. In *Records of the grand historian*, vol. 37.

[114] Confucius. Xian Wen. In *The analects*.

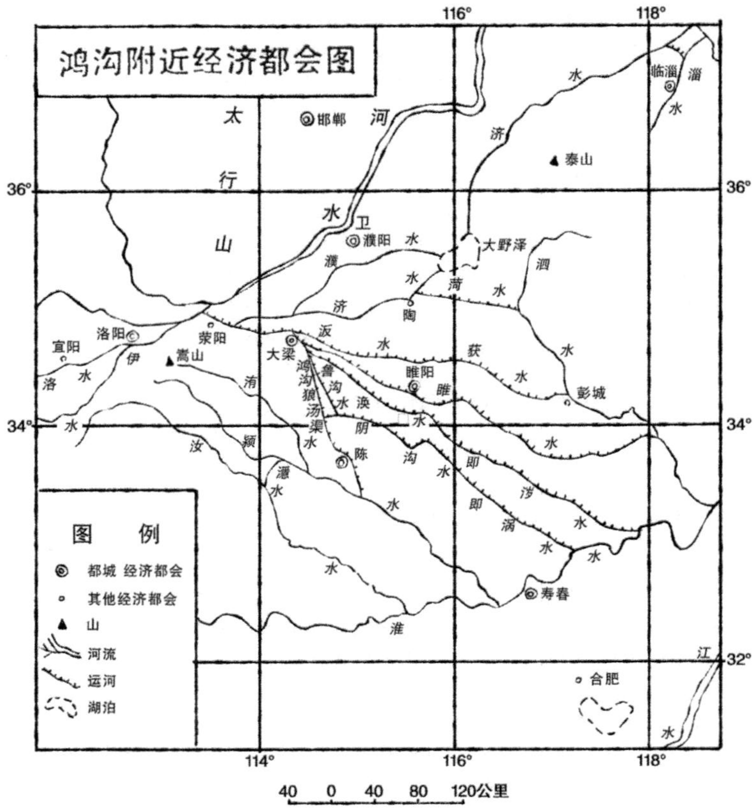

图例 Legend
都城经济都会 Economic Centers in Capital Cities
其他经济都会 Other Economic Centers
山 Mountains
河流 Rivers
运河 Canals
湖泊 Lakes

Fig. 2.5 Major economic centers near the Hong Canal

Chu, located in the middle reaches of the Yangtze River, was regarded at the time as culturally alien, as expressed in the saying: "With the southern *Man* and northern *Di* crossing paths, China remains hanging by a thread."[115] This aptly illustrates the deep cultural anxieties of the time. Indeed, Chu itself did not conceal its identity and openly acknowledged its status as a non-Huaxia state.[116]

Because of the cultural isolation between regions, significant differences emerged—in daily life, cuisine, dress, and even state institutions. Chu and Wu were

115 *Gongyang commentary on the Spring and Autumn annals*: The fourth year of Duke Xi.

116 Hereditary house of Chu. In *Records of the grand historian,* vol. 40.

respectively situated in the middle and lower reaches of the Yangtze River. In theory, their cultures should have been similar, but due to the vast and difficult waters of areas like Yunmeng and Jiujiang, effective transportation between them was lacking. As a result, despite both being along the Yangtze, Chu and Wu developed divergent cultural traits. Early on, Wu was even referred to as *Jingman* (southern barbarians), and customs such as tattooing and cutting the hair—common in Wu—were rarely seen in Chu.[117]

These cultural differences gradually began to dissolve as interactions between regions increased. The growing frequency of such exchanges was greatly aided by the development of waterborne transportation. Once the canals enabled smooth passage, the barriers to cultural integration diminished and, in time, were effectively erased.

The Formation of Unification

During the pre-Qin period, the continuous excavation of canals gradually linked previously unconnected waterways into an integrated system. As transportation became more convenient, interactions among populations in different regions grew more frequent, facilitating cultural exchange, stimulating commerce, and promoting economic development. These changes laid the groundwork for transforming the fragmented landscape of feudal states and paved the way for eventual political unification. As discussed earlier, the transportation system envisioned by the author of the *Tribute of Yu* already reflected a strong aspiration for unification. At the time of its composition, the Hong Canal had not yet been fully excavated, yet such a comprehensive plan had already been conceived. Once the Hong Canal was completed, waterborne transportation became even more efficient, further advancing the cause of unification. Mencius, in his advocacy of benevolent governance, proposed policies such as "establishing border checkpoints without collecting tariffs,"[118] which aimed at easing internal barriers within individual states. However, such ideas remained largely confined to a single feudal domain. By that time, prominent economic centers had already emerged across various states, each of which established customs checkpoints and levied taxes, thereby impeding broader economic integration. Xunzi advanced the idea even further, asserting that "within the Four Seas, all should be as one family."[119] This statement expressed a clearer and more comprehensive vision of unification, reflecting the growing intellectual and ideological momentum toward ending the fractured state of the realm. Such notions of unity not only inspired efforts to overcome the feudal divisions of the Warring States period but also created

117 Hereditary house of Wu Taibo. In *Records of the grand historian,* vol. 31.

118 Gongsun Chou I. In *Mencius.*

119 Royal regulations. In *Xunzi.*

the necessary ideological conditions for realizing a centralized state. Although the Qin ultimately unified China through military conquest, the form of unification it achieved fulfilled a long-cherished aspiration shared by thinkers and policymakers alike—an aspiration that the construction of canal networks had both embodied and helped bring to fruition.

Chapter 3
The Rehabilitation of the Canal Transport Network During the Qin and Han

Political and Economic Centers

After the Qin dynasty unified China, it soon became evident that there was a significant geographical separation between the political and economic centers of the empire. The political center was located in Xianyang, while the economic hub was situated in Tao. Bridging the considerable distance between these two vital regions posed a serious challenge. To understand the necessity of linking them, one must first appreciate the extent to which the political center depended on the economic one. Xianyang was located in the Guanzhong region, historically praised as the "Inland Sea" and the "Heavenly Granary,"[1] expressions that reflect its agricultural abundance. However, the Guanzhong Plain was relatively narrow, encompassing only the lower reaches of the Jing and Wei Rivers. Though rich in natural resources and worthy of its laudatory titles, the plain's limited size ultimately placed constraints on its capacity to support a large population. Once this capacity was exceeded, fears of food shortages would inevitably arise. In fact, this terrifying prospect had been apparent long before the Qin had unified the country. At that time, the Qin state had no choice but to rely on the grain supplies from the Ba and Shu regions (modern-day Sichuan and Chongqing) to address this issue. While these areas produced grain in abundance, transporting it to Guanzhong was no easy task since the formidable Qinling and Bashan mountain ranges lay between them, creating a substantial natural barrier. Even with today's vastly improved transportation infrastructure, traversing these mountains on foot remains arduous. One can imagine how difficult and labor-intensive it must have been to transport grain along this route during the Qin and Han periods.

In the late Warring States period, the Qin state had constructed the Zhengguo Canal in the Guanzhong region. Drawing water from the Jing River at Hukou in western Zhongshan, the canal flowed eastward along the northern foothills and eventually

[1] Strategies of Qin I. In *Strategies of the warring states*; Biography of Liu Jing. In *Records of the grand historian*, vol. 99; Rhapsody on the western capital. In *Selections of refined literature*, vol. 1.

N. Shi, *A Historical Geography Research of Canals in China*,
https://doi.org/10.1007/978-981-95-3517-0_3

emptied into the Luo River. Stretching over 300 *li* (approximately 150 km), it irrigated more than 40,000 hectares of saline-alkaline land, yielding up to one *zhong* (a large grain measure) per mu.[2] Prior to unification, this major irrigation project had effectively solved the food crisis in Qin. When Xianyang came to serve as the imperial capital after unification, however, the food supply problem reappeared.

The Capital of the Qin Dynasty

Given this problem, why didn't Qin consider a more permanent solution after unification, relocating its capital to an economic center, or somewhere in close proximity? Qin did not adopt such a strategy due to the very grandness of the Qin state. The state did not pursue immediate, short-term benefits at the expense of addressing its long-term challenges. At this time, what concerned Qin the most was internal instability and the looming threat of the Xiongnu in the northwest. Xianyang was the old capital of Qin, strategically situated in a terrain fortified by mountains on all sides. Particularly important was Hangu Pass to the east (located in present-day Wangduo Village, Lingbao County, Henan), which could block military advances from that direction.

In the late Warring States period, Qin had also achieved multiple successes in dealing with the Xiongnu to the north, one result of which was the establishment of the Beidi Commandery (around present-day Qingyang County, Gansu). However, despite suffering successive defeats, the Xiongnu remained a constant threat and continued to be the Qin state's greatest concern. Keeping the capital at Xianyang was optimal for directing military campaigns against the Xiongnu. Moreover, Xianyang had been the capital since the reign of Duke Xiao of Qin and could not easily be relocated. The Qin state therefore chose to endure economic difficulties rather than sacrifice a critical move in its grand strategy. Hence, it did not prioritize relocating its capital to an economic hub, choosing instead to maintain Xianyang as the political and military center of the state.

The Need for Water Transport

Although the economic and political centers of Qin were situated in locations that were geographically distant from each other, they were fortunately located along an east–west axis allowing for them to be connected via waterways. Xianyang was located on the banks of the Wei River, while Tao was situated on the banks of the Ji River. The Ji flows into the Yellow River, from whence one could navigate upstream into the Wei. These three waterways were therefore interconnected, offering considerable convenience for the Qin people. The primary means of transporting

[2] Book of canals and waterways. In *Records of the grand historian*, vol. 29.

grain into the Guanzhong region was by river. A portion of this grain came from the He River Basin and the area between the Yangtze and Huai River regions, while another portion originated from the Hong Canal Basin and Huainan. Qin established a large-scale storage depot known as the Aocang Granary near Xingyang, strategically located where the Hong Canal and Ji River diverged from the Yellow River. The reasons for placing the Aocang Granary at this juncture was likely twofold: first, it served as a central hub for grain transported by both the Ji River and Hong Canal; and second, since the vessels traveling on the Ji River and Hong Canal were unsuitable for navigating the Yellow River and their grain had to be transferred to different types of ships, the Aocang Granary functioned as a storage site during ship transfers.

How much grain did the Qin transport from the eastern regions to Guanzhong each year? The historical records do not provide an exact figure, so we can only speculate. However, it was undoubtedly a large quantity. Grain-producing areas in the east included places near Linzi, the region between the Ji and Si rivers, the Hong Canal Basin, and the lower reaches of the Yangtze and Huai rivers. All of these areas were, of course, subject to taxation and grain collection by the Qin state via a requisitioning system that extended as far as the eastern coastline, including the easternmost tip of the Shandong Peninsula, which was encompassed by the commanderies of Huang, Chui, and Langya.[3] Huang was located in present-day Huang County, Shandong, and Chui in present-day Wendeng County, Shandong. Langya was the name for both a county and a commandery at that time. Langya lay mostly within the present-day Shandong peninsula and was known as the "coastal commandery."

The eastern tip of the Shandong Peninsula was a prosperous area. Notably, Langya County, the seat of Langya Commandery (in present-day northwestern Jiaonan, Shandong), had once served as the capital of King Goujian of Yue.[4] Qin Shi Huang also had a particular fondness for this region and stayed there for three months during his royal tours.[5] Despite its prosperity, transportation was highly inconvenient. It is likely that the quantity of grain that could be drawn from regions near the Zhengguo Canal in Guanzhong was insufficient to meet demand, which led to the collection of grain from such distant and difficult to reach coastal areas. This also suggests that the amount of grain transported from these eastern regions to Guanzhong was certainly not a small quantity.

How large was the transport capacity of these canals? This is not an easy question to answer with precision. Nonetheless, it can be reasonably surmised that their combined capacity far surpassed that of the southern course of the Yellow River. Though each canal derived its flow from the Yellow River[6] and was individually smaller in volume, their cumulative transport power exceeded that of the main river itself. This conclusion is supported by the example of the Aocang Granary, situated

---

[3] Biography of Zhufu Yan. In *Book of Han*, vol. 64, part 1.

[4] Treatise on geography. In *Book of Han*, vol. 28.

[5] Annals of the first Emperor of Qin. In *Records of the grand historian*, vol. 6.

[6] Among these canals, although the He River does not have a direct connection to the Yellow River, it is influenced by the Ji River, which itself is a tributary of the Yellow River. Therefore, the He River can also be considered a branch of the Yellow River.

at the point where these canals emptied into the Yellow River. If the total capacity of the canals had been equal to or less than that of the Yellow River, the grain they conveyed could have been transferred directly onward without the need for long-term storage. In reality, however, the volume of grain accumulated at Aocang was enormous. The Yellow River lacked sufficient transport capacity to dispatch it in a timely manner, and even after the fall of the Qin dynasty, large quantities of grain remained stored there. The strategic importance of Aocang became evident during the civil war between Liu Bang and Xiang Yu. The advisor Li Shiqi urged Liu Bang to advance swiftly and seize Xingyang in order to gain control of the grain at Aocang, thereby securing a reliable supply of provisions. He also criticized Xiang Yu for failing to hold Aocang after capturing Xingyang, instead diverting his forces to defend Chenggao—a miscalculation that contributed to his eventual defeat. Liu Bang accepted Li Shiqi's counsel, took control of Aocang, and thereby altered the course of the war.[7] This episode clearly illustrates the vast scale of the grain reserves at Aocang. Without such abundance, it would not have been a decisive factor in the conflict.

The Allure of the Economic Center

After the fall of the Qin, both Xiang Yu and Liu Bang considered regions with developed canal networks—such as the Hong Canal and He River basins—when determining the location of their respective capitals. Xiang Yu established his capital in Pengcheng, which was situated at the confluence of the Si and Huo rivers. Although Liu Bang ultimately chose Chang'an as his capital, he initially took the throne in the area near the Si River.[8] Given the Si River's proximity to Tao, this suggests that Liu Bang may have originally intended to establish his capital in Tao. In the end, however, Liu Bang resisted the temptation of Tao's wealth and commercial vitality. Prioritizing long-term political stability and strategic advantage, he chose Guanzhong—specifically Chang'an—as the seat of imperial power.

Grain Transport in the Early Han Dynasty

In the aftermath of the Qin dynasty's collapse, the early Han government embraced the principles of Huang-Lao governance, focusing on restoring peace and allowing the people to recover from the devastation. All administrative and infrastructural measures were aimed at basic functionality rather than grandeur, resulting in a streamlined and modest governmental structure. Owing to the widespread chaos and destruction, the population of the Guanzhong region suffered greatly, with many

---

[7] Biography of Li Shiqi. In *Records of the grand historian*, vol. 97.

[8] Annals of Gaozu. In *Records of the grand historian*, vol. 8.

people displaced or perishing during the turmoil. This naturally led to a decline in household numbers. Consequently, Guanzhong was not excessively dependent on grain shipments from the eastern regions, particularly Shandong.[9] Historical records indicate that during this time, the amount of grain transported from Shandong each year was only a few hundred thousand *dan* (石, a traditional Chinese unit of ship capacity, where 1 *dan* $\approx$ 60 kg).[10] Compared to the Qin era, this was a dramatic decrease. This policy of recovery and frugality was carried out by Emperor Gaozu and continued under Emperor Jing. The period of around sixty to seventy years following the establishment of the Han Dynasty was marked by peace and prosperity. In the twelfth year of Emperor Wen's reign (168 BCE), a decree was issued that halved the land tax across the empire. The following year, even the remaining half was waived entirely—an extraordinary measure virtually unparalleled in Chinese history. This remission of taxes continued for over a decade, and it was not until Emperor Jing ascended the throne that the land tax was reinstated—though even then, it only returned to the reduced rate of Emperor Wen's twelfth year.[11] Given the state's minimal grain requirements and the long-standing suspension of land taxes, the dedicated canal systems that had once played a vital role in the logistics of state grain transport were rendered largely dormant during this period.

The Decline of Transport Along the Hong Canal

Although the name "Hong Canal" attracted considerable attention from the Warring States period through the Qin and early Han dynasties, references to it gradually fade from the historical record after the early Han. This decline is closely related to the reduced demand for grain transport, as previously discussed, as well as the widespread tax remissions that eased burdens on the populace during the early Han Dynasty. Another significant factor contributing to the canal's declining role was the political configuration of the time, particularly the decentralized structure of the feudal kingdoms.

At the outset of the Han Dynasty, the territories of the feudal kings were spread across the Guandong region, and several of them were located in or near the Hong Canal basin. Among them, the Kingdom of Liang was located in the upper reaches of the Hong Canal, while the Kingdoms of Huaiyang and Chu were situated along

[9] In the Qin and Han periods, when people referred to "Shandong," they were specifically referring to the area east of the Xiao Mountains. This region was called "Shandong" at the time, but it is different from the present-day Shandong Province. The Hangu Pass, which was located on Mount Xiao, marked the boundary between the "Guandong" region (east of the pass) and what was referred to as Shandong. Therefore, in this historical context, "Guandong" and "Shandong" were synonymous terms.

[10] Treatise on equable taxation. In *Records of the grand historian*, vol. 30.

[11] Annals of Xiaowen. In *Records of the grand historian*, vol. 10. Sima Qian. Annals of Xiaojing. In *Records of the grand historian*, vol. 11. Annals of Emperor Wen. In *Book of Han*, vol. 4. Annals of Emperor Jing. In *Book of Han*, vol. 5.

its lower reaches. The capital of Liang was Suiyang, located on the middle reaches of the Sui River in present-day Shangqiu, Henan. The capital of Huaiyang was Chen County, located on the lower reaches of the Langtang Canal, in present-day Huaiyang County, Henan. The capital of the Kingdom of Chu was Pengcheng, located at the confluence of the Huo and Si rivers. Further south, the Kingdoms of Huainan and Wu were located along the Yangtze and Huai River basins. The capital of Huainan was Shouchun, on the southern bank of the Huai River in present-day Shou County, Anhui. The capital of Wu was Wuxian, on the eastern shore of Taihu Lake, in present-day Suzhou, Jiangsu. Despite their distance from the Hong Canal basin, the capitals of Huainan and Wu still required transportation through the Hong Canal basin to reach the capital, Chang'an. These kingdoms were largely fiscally autonomous.

From the founding of the Western Han, it was established that "the revenues from land taxes—from the emperor down to the feudal lords and their fiefs—were to be privately retained and not included in the central government's treasury."[12] In other words, the revenues of the feudal domains were not necessarily remitted to the imperial court. The Kingdom of Wu, in particular, stood out. Wu had access to the copper mines at Zhang County (modern-day Anji County, Zhejiang), where copper could be smelted to mint coins, and being located near the East China Sea, it also produced salt from sea water. As a result, the Kingdom of Wu abolished domestic taxes.[13] In fact, during the reign of Emperor Wen, the policy of exempting land taxes was already in effect in Wu. The Kingdom of Wu was also remarkably prosperous. It was said that the grain supplies from the various regions, whether transported by land or water, could not compare to the wealth stored in the sea granary of Hailing (modern-day Taizhou, Jiangsu) in Wu.[14] Similarly, the Kingdom of Liang, located in one of the most fertile regions of the empire, was also prosperous, with its treasury reportedly containing vast amounts of gold and silver.[15] These feudal kingdoms maintained differing relations with the Han court. Liang and Huaiyang were closely tied to the imperial family across generations, while Huainan and Wu often stood

---

[12] Treatise on equable taxation. In *Records of the grand historian*, vol. 30.

[13] Biography of King Bi of Wu (Wuwangbi Zhuan). In *Records of the grand historian*, vol. 106.

[14] Biography of Mei Cheng. In *Book of Han*, vol. 51. According to the *History of Song*, vol. 96, *Treatise on canals and waterways*, an edict from the fourth year of Emperor Huizong's Xuanhe reign (1112 CE) traces the origins of the Han Canal. The edict states: "The water transport system in the Jiang-Huai region dates back to ancient times. During the Spring and Autumn period, Wu dug the Han Canal, linking to Sheyang Lake in the northeast and reaching Mokou in the northwest. Later, the King of Wu, Liu Bi, expanded the Han Canal to connect with Hailing." However, the Han Canal in the Spring and Autumn period did not connect with Hailing. If King Wu Bi did open a canal, it must have been another waterway that simply took the name of the Han Canal. Liu Bi, the King of Wu, established his capital at Guangling (present-day Yangzhou, Jiangsu Province). Hailing lay to the east of Guangling. If such a waterway existed, its course would have differed significantly from that of the Han Gou Canal dug during the Spring and Autumn period. However, the claim that Liu Bi, the King of Wu, expanded the Han Gou Canal is only recorded in this edict, with no prior historical evidence. It might have been based on hearsay, and the drafters of the edict could have incorporated it without verification.

[15] Annals of King Xiao of Liang (Liangxiaowang Shijia). In *Records of the grand historian*, vol. 58.

in implicit or open antagonism toward the central government. At a time when the Han government itself had reduced reliance on grain shipments from the east, and when the feudal kingdoms had little incentive to offer tribute to the capital, the decline in transport along the Hong Canal was further exacerbated. That said, the canal was not abandoned entirely. The diminishing use of the Hong Canal for state logistics did not mean that boat traffic ceased altogether or that the waterway became completely impassable. Commercial activity and local trade among the populace likely continued, albeit on a reduced scale.

The Initial Misfortune of the Canals

The prosperity of the canal system, however, was short-lived. In the third year of Emperor Wu's reign (132 BCE), a sudden disaster struck. The Yellow River breached its banks near Puyang, flowing southeastwards into the Daye Lake. From there, it overflowed into the He River, eventually joining the Si River and then continuing towards the Huai River.[16] Not only were the He, Si, and Huai rivers affected by this flood, but even parts of the Hong Canal system suffered damage. As these canals were artificially constructed, their channels were relatively shallow and thus particularly susceptible to siltation and obstruction. Though their primary water source was the Yellow River, the canals relied in part on the Ji River, which in turn derived its flow from the Yellow River before entering the Yingze, located in what is now eastern Xingyang County, Henan. The flow of water through the Ying Marsh was notably slow, allowing sediment to settle, and as a result, the Ji River remained clearer than the Yellow River itself. Since their construction, these canals had not experienced any significant blockages.

This breach of the Yellow River differed markedly from previous incidents. The rupture was both extensive and deep, resulting in the near-total overflow of the river's waters. Given the river's notorious sediment load, such floodwaters could have been carried away in regions with sufficient gradient. However, the breach occurred across a broad and level plain, where the sluggish flow caused sediment to accumulate gradually. Had the breach been swiftly repaired, the damage might have remained manageable. Unfortunately, the rupture remained unsealed for over twenty years. During this prolonged period, the canal network suffered progressive siltation. Although some sections—being located at a greater distance—were relatively spared, the overall connectivity between the canals deteriorated significantly. Consequently, their transport capacity declined, and the system never regained the efficiency it once possessed.

[16] Book of canals and waterways. In *Records of the grand historian*, vol. 29.

The Decline of Economic Hubs

With the canals becoming silted and obstructed, several once-prominent economic centers also fell into decline. During the Warring States period, Tao and Wei were frequently mentioned together as flourishing metropolises. However, following the drastic upheaval caused by the Yellow River's course change, Puyang (the ancient Wei) faded from prominence. Although Dingtao (known in earlier times simply as Tao) managed to retain a semblance of urban status during the Han period, it no longer resembled its former glory as the "center of the realm." If one counts from the time when King Fuchai of Wu ordered the canal to be opened between the states of Shang and Lu, to the breaching of the Yellow River near Puyang during the Yuanguang reign of Emperor Wu, this period spans approximately 350 years. These three and a half centuries marked the height of prosperity for Tao and Wei. Afterward, they never again experienced such flourishing. Cities like Suiyang and Pengcheng shared a similar fate. Though they were also affected by the breach of the Yellow River, the damage they suffered was less severe than that endured by Tao and Wei. Nonetheless, both cities experienced decline. Unlike Tao and Wei, however, Suiyang and Pengcheng were eventually able to revive and regain some of their former vitality. Although the breach of the Yellow River was ultimately repaired, the economic configuration of the eastern region had already undergone a fundamental transformation.

The Increase in River Transport

It was precisely during the period when these canals were in decline that the political landscape of the Han Dynasty began to change. In the early years of the Han, both the imperial court and the people focused on recovery and frugality, avoiding unnecessary extravagance. By the time of Emperor Wu, however, the foundation of the state had become stable. Emperor Wu, a ruler who was active and unwilling to remain idle, took a proactive approach to internal development while also engaging in continuous military campaigns. As a result, the state's expenditures increased significantly, and the demand for grain from the Guandong region rose substantially. In the early Han period, only a few hundred thousand *dan* of grain were transported from the Guandong region each year. This number gradually increased to over one million *dan*, and later reached over six million *dan*.[17] This continual increase in demand surpassed not only the transport capacity of the aging canal systems but also the carrying capacity of both the Yellow and Wei Rivers. The Yellow River, near the Hangu Pass, was notoriously treacherous, with natural hazards such as the Dizhu Rapids making upstream navigation particularly difficult. Grain boats, when navigating this stretch, faced great risk of capsizing or being submerged if not handled with extreme

[17] Book of canals and waterways. In *Records of the grand historian* (Shiji), vol. 29; Treatise on equable taxation. In *Records of the grand historian* (Shiji), vol. 30.

caution. The lower reaches of the Wei River were also difficult to navigate, with shallow waters and sandy banks that caused further obstacles. The river's course was winding, stretching over 900 *li*,[18] which made the westward journey of the grain boats particularly cumbersome. When the river transport was less critical in the early Han period, these obstacles were not much of a concern. However, during Emperor Wu's reign, with demand for transported grain rising sharply and navigation hazards becoming more evident, the urgent need to resolve these logistical challenges became a shared concern across both the court and the broader administration.

The Excavation of the Canal South of the Wei River

At the time, the solutions to these problems boiled down to two approaches: one was improving the transport routes to make shipping more efficient; the other was to fundamentally reduce the dependence on grain from the Guandong region. Regarding the first point, many plans were suggested by officials. However, as the state's construction projects and military campaigns increased, reducing the consumption of eastern grain seemed practically impossible. Thus, the only viable solution was to increase grain production in the Guanzhong region. To achieve this, the only and most effective method was improving water management. Guanzhong had long been renowned for its fertile soil, and with proper irrigation, its agricultural output could be significantly increased, thereby alleviating the burden on the eastern grain transport. Interestingly, the first major irrigation projects were not started in Guanzhong, but in Hedong, the region east of Guanzhong. In Hedong, canals were constructed to divert water from the Fen River, irrigating farmland in Pishi (modern-day Hejin County, Shanxi) and Fenyin (modern-day Wanrong County, Shanxi). Additionally, water from the Yellow River was used to irrigate the farmland in Fenyin and Puban (modern-day Yongji County, Shanxi). These two irrigation projects were capable of irrigating around 5000 *qing* of farmland. If the farmland in Hedong yielded a good harvest, the grain could yield over two million *dan,* which could then be transported westward via the Wei River, eliminating the need to contend with the hazards of the Dizhu Rapids. Unfortunately, these irrigation projects were hindered by the local westward shift of the Yellow River, and the canals were not successfully completed.[19] Subsequently, efforts were directed towards the lower reaches of the Luo River. The Luo River here refers to the river north of the Wei River in present-day Shaanxi, not the Luo River in Henan. The project involved creating a canal from Zhi County (modern-day Chengcheng County, Shaanxi) to divert water for irrigation, extending to the fields east of Chongquan (modern-day Pucheng County, Shaanxi) and reaching over 10,000 *qing* of farmland in Linjin (modern-day Dali County, Shaanxi). It was projected that each mu of irrigated land could yield ten *dan* of grain. However, this canal had to pass through the steep and unstable Shangyan Mountain, which required

[18] Treatise on ditches and irrigation works. In *Book of Han*, vol. 29.

[19] Book of canals and waterways. In *Records of the grand historian*, vol. 29.

the excavation of tunnels to allow water to flow through. This was the famous Long-shou Canal. Although the canal was eventually completed, its results were not as significant.[20]

At the time, the most successful irrigation projects were those that diverted water from the Wei River and Jing River. The canals that drew from the Wei River included the Chengguo Canal, Menglong Canal, and Lingzhi Canal. The Chengguo Canal was excavated from the Wei River in Mei County (modern-day Meixian, Shaanxi), flowing eastward to the northern part of Zhouzhi County (modern-day Zhouzhi, Shaanxi), where it entered the Menglong Canal. The Lingzhi Canal was also in Zhouzhi, but its water source is unclear; it was likely an alternative name for the Menglong Canal, rather than a separate canal.[21] Another notable project was the Wei Canal, which likely drew from the Wei Stream (水), a tributary of the Wei River (渭水).[22] Prior to the Qin unification of the six states, the Jing River had already been diverted into the Guanzhong region via the construction of the Zhengguo Canal. Later, six auxiliary canals—known collectively as the "Six Auxiliary Canals"—were excavated alongside the Zhengguo Canal to support its function. Also worth mentioning is the later construction of the Bai Canal (White Canal), which likewise drew water from the Jing River. Like the Zhengguo Canal, it originated at Guokou and flowed southeast to Liyang (modern-day northern Lintong County, Shaanxi), where it emptied into the Wei River. The Bai Canal extended over 200 *li* and irrigated more than 4500 *qing* of farmland,[23] earning it a reputation equal to that of the Zhengguo Canal.

The excavation of these irrigation canals met with mixed outcomes—some were successful, others less so—but on the whole, they contributed significantly to increasing agricultural productivity. Although grain output in the Guanzhong region improved, it was only sufficient to offset part of the demand for grain shipments from the Guandong region; the remainder still needed to be transported westward as before. As a result, improving the waterways used for transport became an urgent necessity. The Dizhu Rapids in the Yellow River, located near Hangu Pass, posed a serious navigational hazard. Given the limitations of contemporary engineering, there was no viable solution for improving this treacherous section, and alternatives had to be sought elsewhere. At the time, the Grand Minister of Agriculture, Zheng Dangshi, recognized this challenge and proposed a solution: to improve the water transport system by focusing on the Wei River. He noted that the Wei River's course was excessively winding, shallow, and heavily silted—conditions that severely hindered grain transport. He therefore proposed constructing a new canal parallel to the Wei River. As a man-made waterway, it could be engineered to run straighter and narrower, thereby deepening the channel and minimizing silt accumulation that would obstruct navigation. If such a canal could be completed, the main course of the Wei River could be largely avoided for transport purposes. Emperor Wu accepted

[20] Ibid.

[21] Ibid.

[22] Treatise on ditches and irrigation works. In *Book of Han*, vol. 29.

[23] Ibid.

Zheng's proposal and ordered the mobilization of laborers to begin construction. The canal diverted water from the Wei River near Chang'an and ran along the foothills of the southern mountains (Zhongnan Mountains), eventually rejoining the Yellow River near its confluence with the Wei. The project took three years to complete and proved a success. Compared to the earlier route along the Wei River, the new canal made upstream grain transport significantly easier. Moreover, when not in use for transport, the canal water could be redirected for irrigation purposes. This dual-purpose infrastructure delivered notable benefits and stood as a remarkable hydraulic achievement of its time.[24]

The use of the Wei River as the primary water source for the canal was unquestionable. However, there was also an instance where the canal drew water from Kunming Pool. Kunming Pool was located to the southwest of the Han capital, Chang'an, near the present-day Doumen Town in Xi'an. Li Daoyuan once mentioned: "The canal that runs through the old canal of the Ba River and eastward through Xinfeng County joins the canal constructed by Zheng Dangshi, the Grand Minister of Agriculture of Han. Due to the difficulty of using the Wei River, Xu Bobiao, a hydraulic engineer of the Qi State, was tasked with digging a canal to divert water. The canal starts from Kunming Pool, runs south along the mountain plains, and heads east to the river."[25] He identified the old Kunming Canal passing through the south of Chang'an and outside the Qingmen Gate as the canal that linked Kunming Pool with the transport system.[26] The canal was dug in the sixth year of Emperor Wu's Yuan Guang reign (129 BCE),[27] while Kunming Pool itself was constructed in the third year of the Yuan Shou reign (120 BCE).[28] The diversion of water from Kunming Pool to the canal likely occurred after the canal had already been established, intended to supplement the water supply for the canal when it became insufficient.

There are different opinions on the exact location where the Wei River was first diverted for the canal. In the mid-Tang period, Han Liao, who was responsible for reopening the canal, mentioned: "The old canal was 18 *li* west of Xi'an… it had been dredged and excavated since the Qin and Han periods, but later became silted and abandoned."[29] The canal that Han Liao reopened was the Xingcheng Weir, which followed the old course of the Sui Dynasty's Yongtong Canal. The Qin period had no records of canal excavation. The Han canal, often referred to as the "transport canal," followed the plans established by the Sui and Tang dynasties, based on earlier practices. The "18 *li* west of Xi'an" likely refers to the area near Diaoyutai in present-day Xianyang. The Wei River in this region is relatively narrow, unlike the broader expanse near Chang'an, making it easier to construct dams and divert the water.

[24] Ibid.

[25] Commentary on the Wei River. In *Water classic*. Xinfeng County corresponds to present-day Xinfeng Town in Lintong County, Shaanxi Province.

[26] Commentary on the Wei River. In *Water classic*. Qingmen Gate was the third gate from the north on the eastern side of the Han Dynasty Chang'an city.

[27] Annals of Emperor Wu. In *Book of Han*, vol. 6.

[28] Ibid.

[29] Biography of Li Shi. In *Old book of Tang*, vol. 172.

The ancient canal passed north of Hao Pool[30] and south of Cishimen Gate,[31] both of which are located northwest of Kunming Pool, which is also southwest of the ancient Qin capital, Xianyang.[32] This suggests that Kunming Pool was not the main source of water for the canal. Some argue that the canal's diversion of the Wei River took place in the northwest part of present-day Xi'an. However, this view contradicts historical records and is geographically implausible based on the topography of the area.

To the northeast of Xi'an, east of the Ba River, there is a relatively low-lying area starting from the vicinity of Xinzhu, extending eastward through Wanshengbu, Taojia, Tianjia, Xinhe, Chunshuzhuang, Tangjiacun, and Zhoujiawan. This region is about one meter lower than the surrounding areas and is prone to water accumulation. In certain sections, only reeds can be grown. The local residents refer to this area as the "Canal." From Xinzhu, if one crosses the Ba River and moves slightly southwest, it directly reaches the ruins of Han-era Chang'an. This low-lying area, referred to as the Canal, is likely the remains of the canal system from the Han Dynasty.

From that point, the canal continued eastward, skirting the foot of the southern mountains and ultimately discharging into the Yellow River. Its confluence with the river is generally believed to have been near the old city of Tongguan in present-day Shaanxi, just west of the modern suspension bridge. Some scholars have suggested that the canal may have instead flowed into the Wei River east of present-day Huayin County. However, this theory does not align with the actual topography. The terrain in the northwest of Huayin County rises considerably and slopes eastward, only beginning to descend near the area of the suspension bridge. This topographical pattern can be verified not only through visual inspection but also through recent 1:50,000 scale topographic maps, which clearly indicate this gradient. Moreover, no remnants of the ancient canal have been found in the elevated areas east of Huayin, making it unlikely that the canal could have entered the Wei River from that direction. Since the terrain begins to lower near the suspension bridge, it is more plausible that the canal flowed northward from this point to merge with the Yellow River. This area lies near the confluence of the Yellow, Wei, and Luo rivers, where the Yellow River's course has historically been highly unstable, shifting across a wide swath of territory. At various times, the Yellow River has moved westward as far as Sanhekou, located to the west of the current suspension bridge. The Luo River, which originally flowed into the Wei near Sanhekou, has at times—due to the Yellow River's oscillations—been diverted eastward into the Yellow River instead of flowing south into the Wei. When the canal was constructed, Fan Xi, Governor of Hedong, was diverting Yellow

[30] Annals of the first Emperor of Qin. In *Records of the grand historian*, vol. 6, citing *Kuo Di Zhi*: "The source of the He River originates from the Hechi in the northwest of Chang'an, Yongzhou. Li Yuan's annotation in *Water classic* states: 'The He River flows from Hechi, heading north into the Wei River.' However, according to recent investigation, the Hechi water flows into the Yongtong Canal, indicating that Li Yuan's interpretation is erroneous." The source of the Yongtong Canal water is the same as the supply of the Cao Canal. It flows eastward along the ancient course of the Han Canal, passing through the same locations as the Han Canal did in the past.

[31] Song Minqiu. Chang'an. In *Gazetteer of Chang'an*, vol. 12.

[32] Yang Shoujing. *Illustrated commentary on the Water classic*.

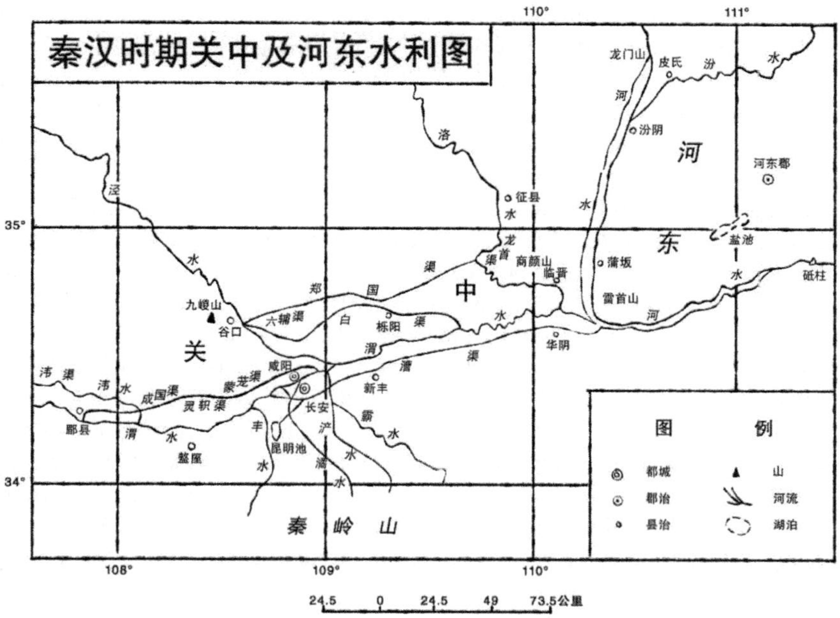

图例 - Legend
都城 Capital Cities
郡治 Commandery Seats
县治 County Seats
山 Mountains
河流 Rivers
湖泊 Lakes

Fig. 3.1 Water conservancy map of Guanzhong and Hedong during the Qin and Han dynasties

River water to irrigate the lowlands along the river in Fenyin (modern-day Wanrong County, Shanxi) and Puzuo (modern-day Yongji County, Shanxi). Due to the shifting of the river, the canal was left without water to draw from.[33] Clearly, the Yellow River was shifting westward at the time, which is why the canal emptied into the Yellow River rather than the Wei River (See Fig. 3.1).

The Attempted Excavation and Failure of the Baoxie Road (褒斜道)

After Zheng Dangshi had overseen the excavation of the transportation canal, a man submitted a memorial to Emperor Wu proposing the construction of a new route known as the Baoxie Road—a passage connecting the Bao and Xie rivers. Both

[33] Book of canals and waterways. In *Records of the grand historian*, vol. 29.

rivers originate in the Qinling Mountains: the Bao River flows southward into the Mian River (modern Han River), while the Xie River flows northward into the Wei River. Though the two rivers are not directly connected, they lie just over one hundred *li* apart. The proposal suggested that if both waterways could be dredged and made navigable, and if the land between their sources could be made passable, then grain boats from the Shandong region would no longer need to pass through the perilous Dizhu Rapids of the Yellow River. Instead, grain could be transported from the vicinity of Nanyang to the Mian River, then upstream to the Bao River. From the headwaters of the Bao, it could be hauled by carts over the Qinling ridge, descend along the Xie River, and finally be carried via the Wei River to Chang'an. While this route was more circuitous, it would avoid the treacherous waters of the Yellow River.

Emperor Wu, intrigued but skeptical, instructed the Censor-in-Chief Zhang Tang to evaluate the proposal. Zhang, an energetic and ambitious official, conducted an investigation and reported favorably. He noted that if the route could be completed, it would not only facilitate grain transport from Shandong but also enable large-scale exploitation of timber and bamboo arrow shafts abundant in the region between the Bao and Xie rivers. Furthermore, the existing overland route from Ba and Shu to Chang'an had to detour west of these rivers through what is now Feng County, Shaanxi—a path both inconvenient and unnecessarily long. Zhang concluded that opening the Baoxie route would be a superior alternative. Emperor Wu accepted the recommendation and appointed Zhang Tang's son, Zhang Ang, to oversee the project. The plan involved both dredging the rivers and constructing the overland road. While the road was successfully completed, the water route could not be properly cleared: the current remained too swift, and the riverbeds were strewn with rocks, making navigation impossible. Nevertheless, the undertaking was not entirely in vain—travelers from Ba and Shu still benefited from a shorter route than before.[34]

The Canal Near Nanyang

The person who submitted the petition to Emperor Wu, advocating for the transport of grain from Shandong to the vicinity of Mian River near Nanyang, and then up the Xie River, had his reasons for this proposal. At first glance, it might seem that transporting grain via Nanyang would require the use of carts. However, the person was actually proposing a completely water-based transport system, with no reliance on land transport. To the east of Nanyang, there is Tanghe County, and near it flows the Tang River. In ancient times, this river was called the Zi River. The upper reaches of the Zi River were not far from the Wu River and the Wei Stream to the east, and it seemed that these rivers were once interconnected. The lower courses of the Wu and Qin rivers flowed into the Ru River, which then emptied into the Huai River. Additionally, the Yin River provided a connection to the Ying River. Both the Ying and Huai rivers were originally part of the Hong Canal system. Given this

[34] Ibid.

interconnected network of rivers, the grain from Shandong could travel along the Hong Canal, Huai, Ru, Wu (or Qin), and Zi rivers all the way to Nanyang. The lower course of the Zi River flows into the Yu River (modern-day Bai River), which then empties into the Mian River. As a result, the grain could be transported from the vicinity of Nanyang down the Zi River to the Mian River. While this route may seem somewhat circuitous, the interconnection of these water routes would allow the grain boats to follow the flow of the water, making the journey not particularly difficult despite the apparent distance.

To return to the topic—how exactly were the Zi River, Wu River, and Qin River connected, and when did this connection occur? The Qing scholar Quan Zuwang posits that these waterways were originally unconnected and were likely linked during the Spring and Autumn period by the people of the State of Chu. His argument is drawn from insights found in the *Commentary on the Water Classic*, particularly the annotations on the Wu River (潕水), Zi River (泚水), and Qin River (瀙水). The annotation on the Qin River states that it merges with the Wu River; the Wu River annotation similarly notes its confluence with the Zi River; and the Zi River annotation further claims that the Zi River merges with the Wu River before entering the Huai River. All three rivers are tributaries of the Huai River. However, the Zi River annotation also states that the Zi merges with the Du River and Qin River, as well as the Wu River, before flowing into the Yu River. Both the Du and Yu rivers are tributaries of the Han River. Nanyang, located in a region where the Huai and Han river systems run in parallel, had areas—such as modern-day Xinye and Xinyang in Henan—where their tributaries intermingled. This suggests that before the Yangtze and Huai Rivers were fully connected, the Huai and Han river systems were already hydrologically linked. At the time, the power of the State of Wu had not reached this region, so if the water routes were not connected by the people of Chu, then who else could have done so?[35] Quan Zuwang's reasoning is compelling. During the Spring and Autumn and Warring States periods, the Chu's northward route passed through what is now Xiangyang, then to Nanyang, and from there through Fangcheng and Ye County into the Central Plains. This route was relatively flat and accessible. The Zi, Wu (also known as Wu Shui), and Qin rivers ran close to this path. The Chu people were renowned for their skill in water management and navigation; given the proximity of these rivers' upper courses, it is entirely plausible that the Chu undertook the effort to link them for transport. Even if the people of Chu did not complete the connection of these rivers, it is likely that someone had undertaken the task before Emperor Wu of Han. Otherwise, how could the person who submitted the petition to the emperor have so confidently proposed such a plan? Upon hearing the suggestion, Emperor Wu had doubts about the feasibility of the Baoxie Road but had no doubts about how grain from Shandong could be transported to the vicinity of Mian River via Nanyang. As an astute and capable monarch, Emperor Wu would not have easily accepted such a plan unless he had good reason to believe in the route's viability. Nevertheless, we know little about the actual function and usage

35 Quan Zuwang. Questions and answers on the classics and history (Jingshi wenda). In *Collected works of Jieqi pavilion (Jieqi ting ji)*, vol. 8.

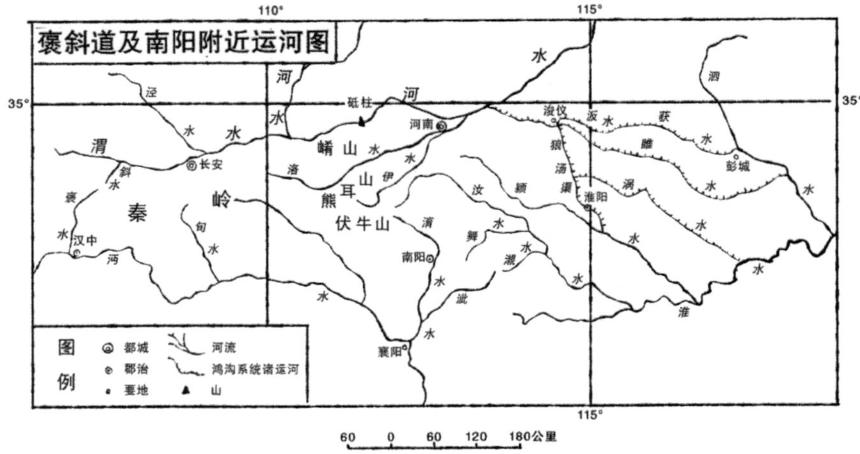

图例	- Legend	河流	Rivers
都城	Capital Cities	鸿沟系统诸运河	Honggou System and Related Canals
郡治	Commandery Seats	山	Mountains
要地	Strategic Locations		

Fig. 3.2 The Baoxie Road and the Canal near Nanyang

of this water route at the time. Historical records are sparse, and since the Baoxie watercourse was ultimately never successfully dredged, the plan to transport grain southward from Shandong was never realized. Thus, this water route bloomed briefly like an ephemeral flower, only to fade into obscurity, never to be mentioned again (See Fig. 3.2).

The Destruction and Restoration of the Hong Canal

Since Emperor Wu's closure of the Yellow River breach near Puyang, the river continued to flood intermittently, often affecting the Hong Canal. The most severe flooding occurred during the reign of Emperor Ping. During this flood, the Yellow River surged into the Hong Canal, flowing southeastward into the Huai River, while another branch diverted into the Ji River and eventually reached the sea from the northeast.[36] The floodwaters completely ravaged the Hong Canal. By the time of Emperor Ping, the Western Han Dynasty was in its final years, and the government no longer had the time to repair the damage caused by this disaster. After Wang

[36] Supplementary discussion on the Yellow River's historical course changes (Fulun [huanghe] lidai tuliu). In *Detailed commentary on the "Tribute of Yu"* (*Yugong zhuizhi*), vol. 13, part 2.

Mang usurped the throne, the Yellow River again breached its banks in Weijun (located southwest of present-day Linzhang County, Hebei). Despite Wang Mang and his ministers frequently proclaiming their commitment to restoring the Yellow River, their words were empty, and no meaningful action was taken.[37] When Emperor Guangwu of the Eastern Han restored order within the empire, he sought ways to seal the breach. However, following the chaos of the civil war, the people were exhausted, and there were no immediate resources to address the issue. It wasn't until the reign of Emperor Ming, over sixty years later, that the breach continued to deepen and worsen. The Hong Canal became completely unusable, and the surrounding lands in Yuzhou (southeastern Henan) and Yanzhou (southwestern Shandong) had lost all hope of being cultivated again. By this time, not only could the grain transport be delayed, but the people also voiced their dissatisfaction, lamenting that the government was preoccupied with other matters while failing to consider the suffering of the people. As a result, they watched helplessly as the disaster continued to spread day by day.

During the reign of Emperor Ming, society had gradually stabilized, making large-scale projects feasible. At this time, there was an expert in water management named Wang Jing from Rangnang-gun (in present-day northern Korea). He advocated for two main measures: first, that the breaches of the Yellow River should be repaired, and second, that the original course of the Hong Canal should be restored. By this period, the Yellow River had multiple branches, with one of them still flowing into Hong Canal, while another had largely become a permanent watercourse. The branch that flowed into Hong Canal, however, only used the old channels of the Bian and Huo Rivers within the Hong Canal system. Other tributaries of the Hong Canal system had already been blocked. Wang Jing realized that it was no longer feasible to return the Yellow River to its original course during the Western Han period. The only viable solution, he determined, was to separate the Yellow River from the Hong Canal system, redirecting it through another route so it would no longer disrupt the Hong Canal. His proposal won the support of Emperor Ming, and Wang was given free rein to carry out the plan. He began by repairing the sluice gates at the point where the Yellow River branched into the Hong Canal, thereby regulating the volume of water entering the canal and preventing the kind of uncontrolled flooding that had occurred in the past. Remarkably, his rehabilitation proved successful.[38]

After Wang Jing successfully mitigated the disasters caused by the Yellow River and brought its flow under control, the regions that had suffered from over sixty years of flooding began to recover. However, the landscape had changed in many ways. The He River was once again silted up, and the small marshes near Dingtao were transformed into dry land. These changes are referenced in the imperial edict issued in the 13th year of Emperor Ming's Yongping reign (AD 70), which states: "The Yellow River and the Bian River are now separated, restored to their former courses;

[37] Treatise on ditches and irrigation works. In *Book of Han*, vol. 29.

[38] Annals of Emperor Ming (Mingdi ji). In *The later book of Han* (Houhan shu), vol. 2; Biography of Wang Jing, section on exemplary officials (Xunli: wangjing zhuan). In *Later book of Han*, vol. 76.

to the north of Taoqiu, the land gradually merges with grave mounds."[39] Moreover, the transport canals of the Hong Canal system also underwent major transformation. Previously, one branch of the Yellow River had surged into the Bian River; after the Yellow River was redirected to a new course, the Bian River became an independent waterway. It remained the sole surviving vestige of the Hong Canal system after the flood. Originally, "Hong Canal" referred broadly to the network of man-made waterways in the Central Plains, but from this point forward, the term gradually faded from use. The only remaining branch came to be known as the Bian Canal. In fact, "Bian" (汴) is simply a variant of "Bian" (汳); the Bian Canal and the Bian River referred to essentially the same watercourse, with little distinction between their paths before and after.

The Bian Canal During the Eastern Han Dynasty

During the Eastern Han dynasty, when the capital was moved to Luoyang, it was strategically positioned in the fertile eastern regions, which necessitated special attention to the construction and maintenance of the Bian Canal. The primary goal was to prevent the Yellow River from flooding into the canal, and also to ensure that no other water sources could infiltrate it, nor could the canal itself overflow. In the reign of Emperor Shun (Yangjia period, 132–135 AD), large-scale construction work was initiated along the canal. From its estuary (Biankou) to the Huai River mouth, stone embankments were built along the banks as a comprehensive flood prevention measure. Later, during the reign of Emperor Ling (Jianning period, 168–171 AD), additional stone gates were constructed at Biankou to further restrict the flow of water into the canal.[40] These measures demonstrate that the Eastern Han government took great care in maintaining the Bian Canal, constantly reinforcing and safeguarding it to prevent any neglect or lapse in its protection.

The Yang Canal

The head of the Bian Canal reached only as far as the vicinity of Xingyang; beyond Biankou, vessels had to enter the Yellow River. The capital of the Eastern Han Dynasty, Luoyang, was located on the northern bank of the Luo River. Thus, after passing from the Bian Canal into the Yellow River, transport boats still needed to navigate upstream along the Luo River to reach the capital. However, the natural course of the Luo River was not easily navigable without significant improvement—much like the need to improve the Wei River during the Western Han, when the capital was at Chang'an, to ensure smooth water transport. The earliest efforts to

[39] Annals of Emperor Ming. In *Later book of Han*, vol. 2.

[40] Commentary on the Yellow River. In *Water classic*.

improve the Luo River waterway were undertaken by Wang Liang during the reign of Emperor Guangwu. In the fifth year of the Jianwu era (AD 29), Wang Liang was appointed Intendant of Henan, and the Luo River lay within his jurisdiction. Observing that the river was difficult for boats to navigate, he proposed to dig a canal at the foot of Luoyang to divert the Gu River into the Luo River. Although the canal was successfully excavated, the water failed to enter it, and the project ended in failure. Wang Liang, taking responsibility for the setback, resigned from his post.[41] Twenty years later, in the twenty-fourth year of the Jianwu era (AD 48), the Minister of Works Zhang Chun revived the initiative to improve the Luo River's navigability. This time, he adopted a new strategy: a canal, later known as the Yang Canal, was excavated to divert water from the Luo River southwest of Luoyang. The canal extended eastward to Yanshi County, where it rejoined the Luo River. This newly constructed channel greatly facilitated the upstream passage of transport boats.[42] As Luoyang continued to serve as the capital in later dynasties, the Luo River waterway was repeatedly maintained and improved, though subsequent modifications largely followed the framework established by Zhang Chun (See Fig. 3.3).

The Ling Canal

Following the excavation of the Hong Canal, several other canals were gradually constructed across China up until the Eastern Han dynasty. These canals were not part of a unified system; each served its own distinct function. The first among them that deserves mention is the Ling Canal. Located between the Xiang and Li Rivers, this canal was first excavated during the reign of Qin Shi Huang as part of his military campaign against the Nanyue Kingdom. The name "Ling Canal" appears to be a later addition; at the time of its construction, it was simply referred to as a canal without any specific designation. The records regarding the construction of the Ling Canal are sparse and unclear. During the reign of Emperor Wu of the Han Dynasty, historical figures such as Zhufu Yan, Yan Zhu, and Yan An mentioned Qin Shi Huang's deployment of General Tu Sui (under Zhao Tuo's command) with naval forces against the Baiyue tribes, while Supervising Censor Lu was ordered to construct the canal to transport grain.[43] Commentators throughout history have offered little elaboration on these statements, with the exception of Wei Zhao, who briefly noted: "Jian refers to an imperial censor named Lu." Even so, this provides little concrete clarity. Slightly earlier, Liu An had recorded a similar account in his Huainanzi, echoing the statements attributed to Yan An and others. Gao You,

[41] Biography of Wang Liang (Wangliang zhuan). In *Later book of Han*, vol. 22. According to the biography, Liang served as governor of Henan in the fifth year of Emperor Guangwu's Jianwu reign (29 CE).

[42] Biography of Zhang Chun (Zhangchun zhuan). In *Later book of Han*, vol. 35.

[43] Biography of Zhufu Yan. In *Records of the grand historian*, vol. 112; Biographies of Yan Zhu and Yan An (Yanzhu zhuan, Yanan zhuan). In *Book of Han*, vol. 64.

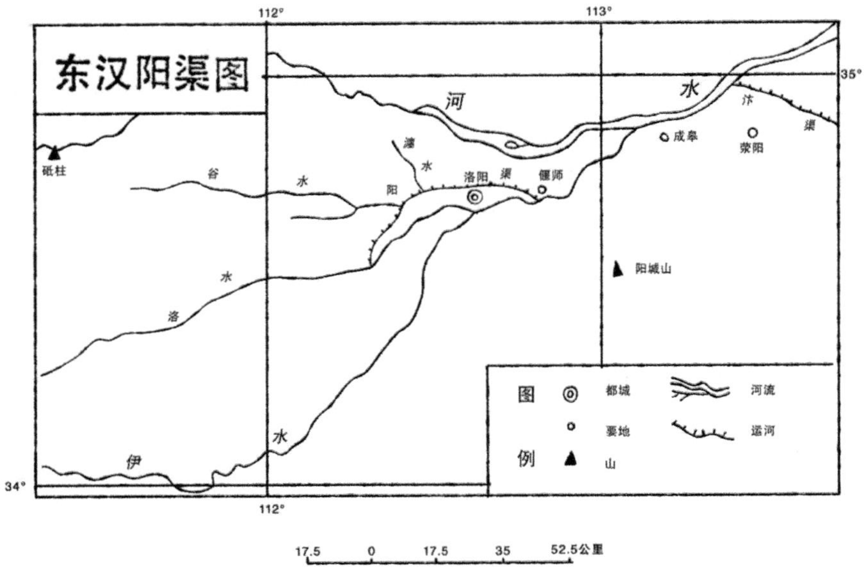

图例	Legend	都城	Capital Cities
要地	Strategic Locations	河流	Rivers
山	Mountains	运河	Canals

Fig. 3.3 The Yang Canal during the Eastern Han dynasty

in his commentary on Huainanzi, explained that the canal overseen by Censor Lu ran between the Xiang and Li Rivers.[44] While Gao You's sources are unknown, his account is generally considered reliable. Among the water systems on either side of the Five Ridges, the headwaters of the Xiang and Li Rivers are geographically the closest. It is said that both rivers originate from the same mountain, but diverge at the source to flow in separate directions. At the point of greatest proximity, the land between them is barely more than a hundred paces wide—an area known as Shi'an Qiao, also called Yuecheng Qiao.[45] Given this close geographical relationship, constructing a canal would have been relatively straightforward. It is unlikely that Qin Shi Huang would have chosen a more difficult location when a simpler solution was available. Subsequent records provide more detailed accounts of the canal's construction. For example, the *Illustrated Gazetteer of Lingui* states that Emperor Qin ordered Censor Shi Lu to oversee the digging of a canal from Lingling, which then emptied into the Li River.[46] This seems to be based on Gao You's explanation.

---

[44] Teachings of the human realm (Renjian xun). In *Huainanzi*, vol. 18.

[45] Commentary on Li River. In *Water classic*.

[46] Excerpt from the geography section (Dibu) in *Taiping imperial encyclopedia (Taiping yulan)*, vol. 65.

However, Lingling, located in present-day Lingling County, Hunan, is far from the Li River. How could a canal be constructed there to divert water southward? *The Guihai Yuheng Gazettee*r clarifies, saying: "The Ling Canal is located in Xing'an County, Guangxi. During Qin Shi Huang's southern expedition, Censor Shi Lu dug this canal for grain transportation, and its remnants remain today."[47] Xing'an County, once known as Quanyi County in the Tang Dynasty,[48] is now modern-day Xing'an County. This account places both the name and location of the Ling Canal in a more precise context.

When Emperor Qin Shi Huang commissioned the excavation of the Ling Canal, his original intent was merely to facilitate the transport of grain and military provisions for a limited period. Yet over time, the canal evolved into the sole shortcut connecting the Central Plains with Lingnan. The term "Lingnan" refers to the region lying south of the Five Ridges—a mountain range that stretches east to west and forms a formidable natural barrier, obstructing travel and communication between Lingnan and northern China. Traveling from the Central Plains to Lingnan required either scaling these steep mountain paths or taking a long detour by sea. The mountainous routes were treacherous and difficult, while the maritime path was often plagued by storms and unpredictable conditions, making passage perilous and unreliable. Since the construction of the Ling Canal during the reign of Qin Shi Huang, the Xiang River was connected to the Li River. From there, boats could travel downstream along the Xiang River to enter the Yangtze River, or down the Li River to reach the West River. Although the Ling Canal itself is relatively short in distance, its strategic location has made it an important route throughout history. Due to its critical position, it has always been of significant importance, bridging the gap between the northern and southern regions.

How was the Ling Canal constructed? The Song dynasty scholar Fan Chengda provided an explanation. He states, "To construct the canal, stones were piled in the sandy bed of the Xiang River to form a sharp wedge-shaped structure known as a *hua zi* (plowshare dam), with its pointed end facing upstream. This split the Xiang River into two branches, diverting part of the flow into a canal that ran sixty *li* before joining the Rong River, continuing southward. The canal wound through the Xing'an region. It was only a few feet deep and slightly over ten feet wide. Along its sixty-li length, thirty-six sluice gates were installed, which locals simply

[47] Tang annals, part 66 (Tangji liuliu), In *Comprehensive mirror to aid in government* (Zizhi tongjian), vol. 250: "In the fourth year of Emperor Yizong's Xiantong reign, Hu's commentary quotes Fan Chengda's *Guihai Yuheng Gazetteer* (Guihai yuheng zhi)." Ouyang Xiu and Song Qi. Treatise on geography, In *New book of Tang*, vol. 43: "Guizhou's Liding was originally named Xing'an… In the second year of the Zhide reign, the name was changed. Ten li to the west lies the Ling Canal, which diverts water from the Li River, originally dug by Lu of Qin." Although Liding County was renamed from Xing'an County, this Xing'an County is not the same as the later Xing'an County. *Treatise on Geography* mistakenly attributes the Ling Canal to this county. *Treatise on Geography* further states: "Quanyi, originally named Linyuan, was established in the fourth year of Wude (621 AD) by partitioning from Shi'an, and was renamed in the third year of Dali (768 AD)." The Quanyi and Linyuan mentioned here are, in fact, the Tang-era names of what is now Xing'an County.

[48] Treatise on geography VI. In *History of Song*, vol. 90.

called *dou*. As a boat entered a sluice, the gate would be closed behind it, and water allowed to accumulate until the boat could rise and move forward. In this way, vessels could ascend along the cliffs and descend like water from a tipped jar. Even ships carrying a thousand hu of cargo could navigate the passage." Due to this clever engineering, Fan Chengda praised the Ling Canal, calling it "the most skillful achievement in waterworks."[49] Zhou Quifei also believed that this described the method used by Shi Lu in constructing the canal, asserting that "Lu [Shi Lu] was truly an exceptional man."[50] However, Ouyang Xiu disagreed, stating that during the early Baoli era of the Tang Dynasty's Jingzong reign (825 CE), the official Li Bo oversaw the construction of 18 sluice gates along the Ling Canal to facilitate grain transport. Unfortunately, these sluice gates were not preserved for long. More than forty years later, in the ninth year of the reign of Tang Emperor Yizong (868 CE), the Governor of Guizhou, Yu Mengwei, constructed a 40-li-long embankment using stone and installed large wooden beams to create sluice gates, allowing for the passage of large boats.[51] Yu Mengwei later wrote a "*Record of the Ling Canal*", in which he stated: "At the beginning of the Baoli era, Grand Master Li Bo came here on an official inspection. Observing the longstanding disrepair, he re-excavated and extended the original canal routes to facilitate navigation. He constructed *hua di* embankments to block side currents and designed stepped sluice gates to enable direct flow, allowing boats to travel upstream without resistance. Li Bo's efforts were truly innovative." After Li Bo's departure, the canal once again deteriorated. In 868 CE, Yu Mengwei undertook comprehensive repairs. As his record continues: The hua di embankments were rebuilt entirely with massive stones, stretching forty *li*, with strict prohibitions against using inferior materials. The sluice gates were constructed with sturdy timber, expanded to eighteen layers, and no miscellaneous debris was allowed between them. Gravel beds were dredged, channels redirected, and hazards cleared. The canal soon roared with water again—so much so that even boats carrying a hundred *hu* of grain could pass with ease.[52] From this record, it is evident that neither the *hua zi* embankment nor the sluice gate designs originated with Yu Mengwei. He merely repaired and improved the Ling Canal, increasing the number of sluice gates. Ouyang Xiu's claim appears to contain inaccuracies. According to Yu Mengwei's account, the *hua di* design actually dates back to Li Bo. However, the very concept of redirecting water through engineered embankments must have been part of the original design, and likely not invented as late as the Tang dynasty. Historical records

[49] Annals of Tang, part 66. In *Comprehensive mirror to aid in government*, vol. 250. In the fourth year of Emperor Yizong's Xiantong reign, Hu's commentary quotes Fan Chengda's *Guihai Yuheng Gazetteer*. Fan Chengda mentions the Rong River here. Zhou Qufei's Geography in *Lingwai Daida*, vol. 1, states: "The Rong River in fact originates from the Yao tribal valleys. When Emperor Wu of the Han pacified Nanyue, he dispatched troops from Lingling, descended the Li River—likely traveling upstream along the Xiang River, then downstream along a branch canal—entered the Rong River, and continued southward. The Li River flows from Gui, through Zhao, to Cangwu."

[50] Geography. In *Lingwai daida*, vol. 1.

[51] Treatise on geography. In *New book of Tang* I, vol. 43.

[52] Treatise on canals and waterways. In *Guangxi general chronicle* during the Jiajing era, vol. 16. The treatise on arts and literature. In *Xing'an county chronicle* during the Qianlong era, vol. 9.

indicate that in the 17th year of Emperor Guangwu's reign (41 CE), the general Ma Yuan, while campaigning against the rebel leaders Trưng Trắc and Trưng Nhị in what is now northern Vietnam, utilized Shi Lu's old canal, widening sixty *li* of the Xiang River to facilitate the transport of military provisions.[53] This account is also mentioned in Yu Mengwei's Ling Canal Record and is likely based on historical facts. It indicates that a foundation for the canal was already in place during the Eastern Han Dynasty, and as Fan Chengda and Zhou Quifei suggested, the original construction likely dates back to the time of Shi Lu.

Due to the Ling Canal's pivotal role in facilitating transportation between the northern and southern regions of the Five Ridges, it underwent continuous dredging and maintenance throughout successive dynasties. In the third year of the Jiayou era (1058 CE) during the reign of Emperor Renzong of the Northern Song Dynasty, Li Shizhong, then serving as Judicial Commissioner of Guangxi, observed that stones in the canal were obstructing navigation. He employed fire-setting techniques to crack the rocks before excavating them, thereby restoring smooth passage for boats.[54] Later, in the twenty-eighth year of the Hongwu era (1395 CE) under the reign of Ming Taizu, the canal had fallen into disrepair. To address this, the emperor dispatched the Censor Yan Zhenzhi to oversee its repair and dredging. Yan Zhen cleared over 5,000 *zhang* of the canal, built more than 150 *zhang* of embankments, constructed 36 flash locks, and removed the rocks that obstructed boat traffic. As a result, the transportation of goods along the canal was restored.[55] These records of the Ling Canal's restoration represent some of the more substantial and well-documented efforts. In addition to these, countless smaller-scale dredging and maintenance projects were undertaken over the centuries. Taken together, they testify to the strategic importance of the Ling Canal in connecting the regions north and south of the Five Ridges, as well as to the enduring attention and resources devoted to its upkeep by successive dynasties (See Fig. 3.4).

The Dabai Canal in the Taihang Mountains to the East

In the south, the Five Ridges form a natural barrier between the north and south; in the north, the Taihang Mountains serve as an obstacle between east and west. The crucial difference lies in the fact that some rivers west of the Taihang Mountains cut through the range and flow into the Shandong Plain. Among them, the central one is the Hutuo River, the southern one is the Zhang River, and the northern one is the Sanggan River. The Taihang Mountains are steep and rugged, and the rivers within them are rapid and turbulent, rendering them entirely unsuitable for transportation.

---

[53] Guangxi I. In *Essentials of historical geography* (*Dushi fangyu jiyao*), vol. 106.

[54] Biography of Li Shizhong. In *History of Song*, vol. 332.

[55] Biography of Yan Zhenzhi. In *History of Ming*, vol. 151; Quotation from Chen Lian's *Record of the renovation of the Ling Canal*. Treatise on canals and waterways. In *Guangxi general chronicle* in Jiajing, vol. 16.

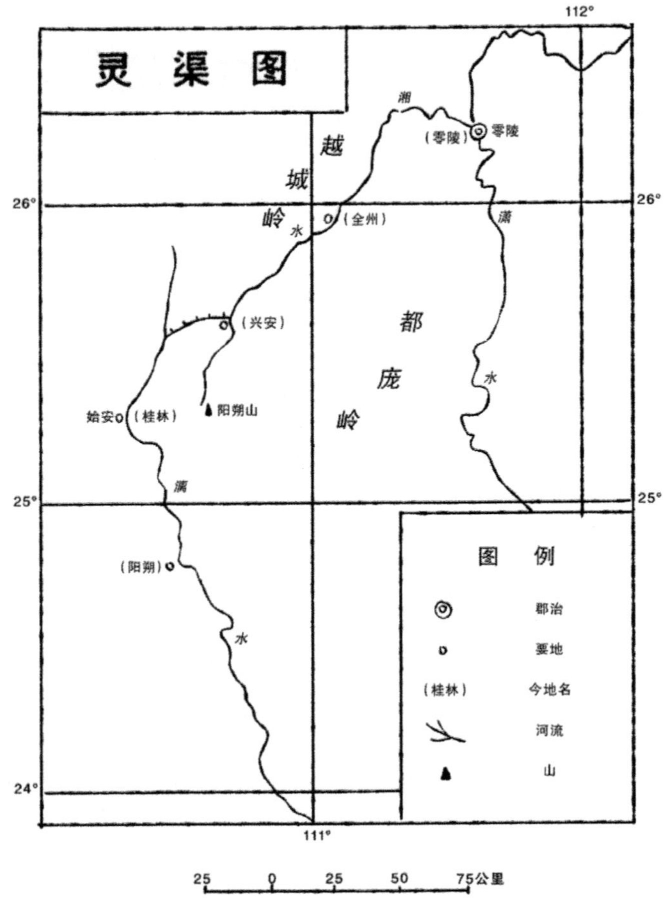

图例 - Legend
郡治 Commandery Seats
要地 Strategic Locations
今地名 Present Names
河流 Rivers
山 Mountains

Fig. 3.4 The Ling Canal

Among these, the upper reaches of the Hutuo River extend the farthest, which is why people once attempted to construct a canal utilizing its waters. It is believed that during the Western Han period, a canal was excavated near this river. This canal was known as the Dabai Canal. The Dabai Canal is first mentioned in the *Book of Han: Treatise on Geography*, which indicates that the canal could not have been constructed earlier than the Western Han period. To be precise, the construction of the Dabai Canal involved the diversion of the Mianman River. Today known as the Ye River, the Mianman River also originates from the western side of the Taihang Mountains and flows into the Hutuo River at Pingshan County in Hebei Province. The canal was constructed near the Hutuo River, closely linked to it. The Dabai Canal in ancient times was located in Puwu County, where it diverted water from the Mianman River. The site of Puwu County's ancient capital is now in the southeastern part of Pingshan County, Hebei. The Dabai Canal then flowed into the Sixiao River at Xiaquyang County. During the Han period, Xiaquyang County was situated in what is now Jin County, Hebei.[56] Where did the Sixiao River originate? This question warrants further investigation. In fact, the Sixiao River was also an artificial canal, not a natural watercourse. It was formed by the diversion of water from the Dabai Canal at Mianman County. During the Han period, Mianman County was located north of present-day Huolu County in Hebei. This area is not far from Puwu County, where the Dabai Canal was fed by the Mianman River. After the Dabai Canal's water joined with the Sixiao River downstream, it flowed into the ancient Zhang River.[57] The construction of this canal clearly aimed to connect the Hutuo River and the Zhang River. Both the Hutuo and Zhang Rivers are two major rivers east of the Taihang Mountains. While their lower reaches eventually converge, their upper reaches are quite distant from each other. This canal effectively bridged the gap between them, linking the two rivers. Why do we say that the Dabai Canal connected the Hutuo and Zhang Rivers? Because the upper source of the Dabai Canal was the Mianman River, a tributary of the Hutuo. In this way, by diverting the Mianman River, the Dabai Canal in effect linked to the Hutuo River as well (See Fig. 3.5).

During the Eastern Han period, the Dabai Canal and the Sixiao River were no longer mentioned in historical records, but another canal—the Puwu Canal—was. This Canal was managed in the 10th year of Emperor Ming's Yongping reign (67 CE)

[56] Treatise on geography. In *Book of Han*, vol. 28: "In Puwu County of Changshan Commandery, the Dabai Canal originates from the Mianman River and flows southeast to enter the Sixiao River at Quyang Lower District."

[57] Treatise on geography, In *Book of Han*, vol. 28. It states: "In Mianman County of the Zhengding Kingdom, the Sixiao River first receives water from the Dabai Canal and flows east to join the river at Qiao." The phrase "flows east to join the river" here actually means it merged into the Zhang River, specifically what was historically called the Hengzhang. Li Daoyuan. Commentary on the Zhuozhang River. In *Water classic*: "The Hengzhang (Transverse Zhang) flows northeast past the ancient city of Taoxian… where it converges with the old course of the Sixiao River. The Sixiao River first receives water from the Dabai Canal, which in turn draws from the Mianman River." This clearly demonstrates that the Dabai Canal and the Sixiao River were originally interconnected. Taoxian was located in present-day Hengshui County, Hebei Province, with the Han-era Qiao County situated to its north.

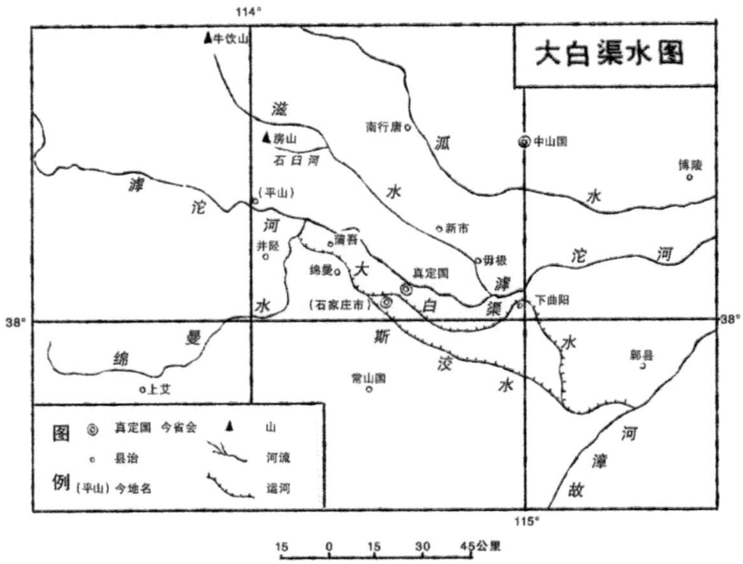

图例 - Legend
真定国 Zhendingguo
今省会 Present-day Capital Cities
县治 County Seats
（平山） (Pingshan)
今地名 Present Names
山 Mountains
河流 Rivers
运河 Canals

Fig. 3.5 The Dabai Canal

as part of the water management efforts within the territory of Changshan Commandery at the time.[58] This canal was named after Puwu, a subordinate county of Changshan, which had also been the site where the Dabai Canal had originally diverted water from the Mianman River during the Western Han. It is clear that the Eastern Han's Puwu Canal was essentially the same waterway as the Dabai Canal of the Western Han. The name may have changed, but the canal's function remained largely unaltered—aside from likely additional dredging and maintenance. Was the Dabai Canal constructed during the Western Han period intended to transport grain from the plains east of the Taihang Mountains to the foothills and then further on to Shanxi? This is not recorded in ancient texts, so it remains unknown. However, the Puwu Canal, which was renovated during the Eastern Han period, clearly had this purpose. The Eastern Han not only sought to transport grain from the Taihang Mountains to

[58] Liu Zhao, annotation. Quotation from *Notes on the present and the past* (*Gujin zhu*). Treatise on commanderies and kingdoms (Junguo zhi). In *Continued book of Han* (*Xu hanshu*).

Shanxi, but also aimed to use the waterways to facilitate this transportation.[59] To achieve this goal, in the 10th year of Emperor Ming's Yongping reign (67 CE), the riverbed of the Hutuo River was restructured. At that time, the Shijiu River was also dredged. The purpose of these efforts was to use water transport to move grain from Dulü to the Yangchang Granary.[60] Dulü (also recorded as Dulu[61] in some sources) cannot be definitively located today, but Yangchang Granary was situated in the ancient Fengyang County,[62] which is now Jingle County in Shanxi. As for the Shijiu River, some sources suggest it was located to the northeast of Tang County in the Tang Dynasty's Dingzhou.[63] Tang County, which still retains the name today, is far removed from the Hutuo River. It is likely that Tang County refers to Tangchang County by mistake. To the northeast of Tangchang County, there is indeed a river called the Shijiu River.[64] Tangchang County was located in what is now the northeast part of Wuji County, Hebei Province. The Shijiu River was originally called Shui (Water) and was also known as the Lu River. It originated about fifty li northwest of Songping County, flowed through Xingtang County, and entered Boling, where it was called the Mudao Ditch. It then flowed southward into the Hutuo River.[65] Both Songping and Xingtang Counties still retain their historical names today. Although there was no Boling County during the Song dynasty, the Mudao Ditch did pass through areas southeast of Tang-era Xinle County and south of Yifeng County.[66] Xinle County of the Tang corresponds to today's Xinle County, and Yifeng County to present-day Anguo County. Therefore, the Boling mentioned here likely refers to Boling County of the Eastern Han period.[67] During the Eastern Han period, Boling County was located south of present-day Li County. It was at this point that the Shijiu River entered the Hutuo River. Here, the Shijiu River became connected with the Zi River. The Zi River was first recorded during the Han dynasty; it originated from Niuyin Mountain in Nanxingtang County and flowed to Xinshi, where it joined the Hutuo River.[68] During the Han dynasty, Nanxingtang County was located to the north of present-day Xingtang, while Xinshi County was situated to the northwest of present-day Wuji County. Today, the Zi River flows northeast from Wuji and joins the Sha River. This suggests that the source of the Zi lies far to the northwest of the Shijiu River, while its lower course lies to the south of the Shijiu River, forming a crisscross pattern. This configuration appears to be the result of later changes and likely did not exist in the Han period. If a river such as the Shijiu did exist during the Han, it would

---

59 Ibid.

60 Biography of Deng Xun. In *Later book of Han*, vol. 16.

61 Yuan Hong. *Annals of later Han.*

62 Commentary on Fen River. In *Water classic.*

63 Annals of Emperor Zhang, with commentary by Li Xian. In *Later book of Han*, vol. 3.

64 Administrative divisions, part eight. In *Comprehensive institutions* (*Tongdian*), vol. 178.

65 Zhenzhou. In *Taiping huanyu ji*, vol. 61.

66 Dingzhou. In *Yuanhe maps and records of prefectures and counties* (*Yuanhe junxian tuzhi*), vol. 18.

67 Commentary on Kou River. In *Water classic.*

68 Treatise on geography. In *Book of Han*, vol. 28.

have been a tributary of the Zi River. It is likely that in later times, the Zi overflowed and flowed southeast, thereby forming what became known as the Mudao Ditch—a development unrelated to the Han period. In the Han period, the Zi River remained confined to the modern regions of Xingtang and Wuji Counties. Even though this was a relatively short river, its course ran nearly parallel to the Hutuo. The Shijiu, being a tributary of the Zi River and not directly connected to the Hutuo, must have been separated from it by a considerable distance. Hu Sanxing, in his commentary on the *Comprehensive Mirror to Aid in Government*, writes: "According to the *Commentary on the Water Classic*, Sima Biao's *Treatise on Commanderies and Kingdoms* states: 'Nanxingtang County in Changshan Commandery had a place called Shijiu Valley. It seems the intention was to harness the Hutuo River, transporting grain from east of the Taihang Mountains, from Dulü to the Yangchang Granary, and thence to Taiyuan via water transport.'"[69] Shijiu Valley must refer to the area through which the Shijiu River flows. Since it lay within the jurisdiction of Nanxingtang County, it was relatively far from the Hutuo River. How could such a small tributary have borne the burden of transporting grain by boat? If a canal was indeed dredged in this area at the time, it would have had little practical value. According to the *Notes on the Present and the Past*: "In the 10th year of the Yongping reign, the Puwu Canal on the Hutuo River in Changshan was constructed to facilitate grain transport by boat."[70] This indicates that the canal connected only the Hutuo River and the Puwu Canal—not the Shijiu River. Most likely, the proximity between the Puwu Canal and the Shijiu River led to historical confusion or misattribution.

Regarding the function of this waterway, Hu Sanxing offers further commentary. He writes: "Consulting Ban Gu's *Treatise on Geography*, the annotation for Shang'ai County in Taiyuan Commandery states: 'The Mianman River flows east to Puwu, where it enters the Hutuo River.' The annotation for Puwu County reads: 'The Dabai Canal draws its source from the Mianman River and flows southeast to Xiaquyang, where it enters the Sixiao River.' From this, it can be inferred that the grain transport route ran from the Dabai Canal into the Mianman River, and from the Mianman River into the Fen River, ultimately reaching the Yangchang Granary."[71] In light of geographical evidence, however, Hu's interpretation is untenable. The actual canal project aimed to facilitate transport to Taiyuan.[72] The administrative seat of Taiyuan Commandery was located southwest of present-day Taiyuan City, considerably north of the Yangchang Granary. If grain boats could have traveled upstream from Puwu along the Mianman River, they would have reached Taiyuan directly, without the need to detour through Yangchang Granary. In reality, the transport route at the time must have involved navigating upstream along the Hutuo River, followed by land conveyance to Yangchang Granary, and then continuing along the Fen River to reach Taiyuan. It is said that the entire route of grain transshipment involved a

[69] Chronicle of the Han, part thirty-eight. In *Comprehensive mirror to aid in government*, vol. 46.

[70] Liu Zhao, annotation. Quotation from *Continued book of Han*, Treatise on commanderies and kingdoms.

[71] Chronicle of the Han, part thirty-eight. In *Comprehensive mirror to aid in government*, vol. 46.

[72] Yuan Hong. *Annals of the later Han.*

total of 389 checkpoints or difficult passes.[73] The course of the Mianman River is relatively short and could not have involved such a large number of obstacles; these difficulties must have arisen along the Hutuo River. Precisely because of the sheer number of obstructions and hazards, countless boats were lost and the death toll became incalculable. As a result, in the third year of Emperor Zhang's Jianchu reign (78 CE), the government abandoned water transport in favor of land conveyance.[74]

The First Rerouting of the Han Canal

In the mid-Eastern Han period, the original course of the Han Canal underwent renovation. The canal, originally excavated by King Fuchai of Wu, drew water northward the foot of Han City and then turned eastward into Bozhi Lake and Sheyang Lake. Sheyang Lake, being a large body of water, naturally facilitated waterborne transport and reduced the labor required for canal excavation. At the time, the transport route proceeded into Sheyang Lake, then continued northwestward by cutting a canal from the lake's northern edge until it reached Mokou and flowed into the Huai River. While this route did save some of the effort needed for digging, it significantly increased the overall distance traveled. Moreover, because Sheyang Lake was quite expansive and its surface often swept by strong winds, it posed considerable hazards to navigation. During the reign of Emperor Shun of the Eastern Han (Yonghe period, 136–141 AD), Chen Min excavated a section of waterway from the north of Fanliang to connect with Jin Lake. From there, the canal was further dug into Baima Lake, and then proceeded to Jiaxie, where it met the transport route that emerged from the northern end of Sheyang Lake.[75] Later, when Emperor Wen of Wei traveled south and north, he passed through Jin Lake. Due to the low water levels in the lake, navigation became difficult. Initially, he planned to abandon the boats and set them on fire. However, thanks to the efforts of Jiang Ji, who employed various methods, the journey was safely completed, and they eventually entered the Huai River.[76] In his work Discourse on Three Provinces, Jiang Ji particularly praised the achievements of Chen Min, mentioning his excavation of Baima Lake. He noted, "He further dug the Malai, a hundred-li passage across the lake." Upon exiting Baima Lake, one would reach the old city of Sheyang, which was also the location of Shanyang County, the administrative center of Shanyang Commandery during the Eastern Jin period, and is now known as Huaian in Jiangsu. Chen Min's rerouting created a direct route that extended from Yangzhou all the way north to the Huai River mouth, a much more convenient path compared to the previous efforts of King Fuchai of Wu (See Fig. 3.6).

[73] Biography of Deng Xun. In *Later book of Han*, vol. 16.

[74] Ibid.

[75] Commentary on Huai River. In *Water classic*.

[76] Book of Wei: Biography of Jiang Ji. In *Records of the three kingdoms* (*Sanguo zhi*), vol. 14. The text records "Jing Lake" (精湖) in place of "Jin Lake" (津湖).

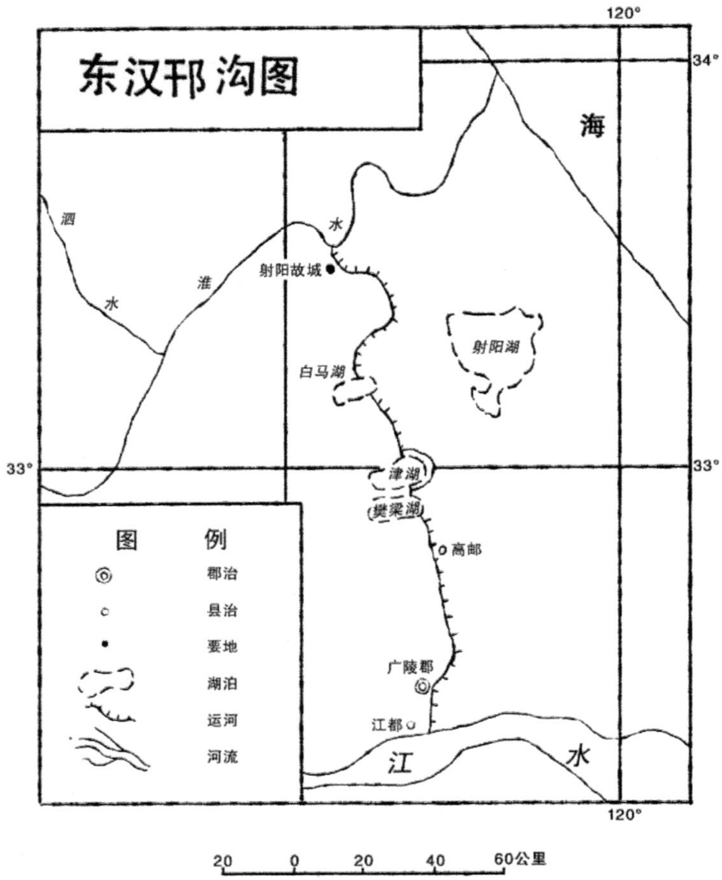

图例 - Legend
郡治 Commandery Seats
县治 County Seats
要地 Strategic Locations
湖泊 Lakes
运河 Canals
河流 Rivers

Fig. 3.6 Eastern Han dynasty Han Canal

During periods of relative peace in the Qin and Han dynasties, considerable effort was devoted to the restoration and improvement of the grain transport network. However, little attention was given to areas outside the Hong Canal system. The reason for this was clear: the regions served by the Hong Canal were generally prosperous. The grain produced in these fertile areas was likely sufficient to meet the needs of the time. Supplementing this with the use of certain natural waterways for transportation rendered the system more or less adequate. The enduring stability of the Han dynasty was closely linked to the efficient integration of these transportation networks. Once these networks fell into disrepair, that long-standing stability likewise proved unsustainable.

Chapter 4
The Destruction and Restoration of the Canal Transport Network After the Eastern Han Dynasty

The Great Turmoil at the End of the Eastern Han Dynasty

Artificially constructed canals required regular maintenance and dredging to prevent silting and ensure continued operation. However, the great turmoil at the end of the Eastern Han Dynasty deprived the Bian Canal of the opportunity for necessary upkeep. During this period, the imperial court moved erratically, from Luoyang to Chang'an, then back to Luoyang, and eventually to Xu County (present-day Xuchang in Henan). Amid such disorder and haste, the court struggled even to sustain its own survival, let alone manage routine state affairs. Meanwhile, control over the regions along the Bian Canal fell successively into the hands of competing warlords. These self-proclaimed "heroes," each entrenched in their own territories, were preoccupied with mutual conflict and had neither the capacity nor the inclination to restore a canal that had already lost its function. Moreover, the entire basin of the Bian Canal was divided among rival factions, lacking any unified administration. Once the vital hub of the transportation network, the canal's dysfunction led inevitably to the collapse of the entire system.

When Emperor Xian of Han returned to Luoyang from Chang'an, the city had already been reduced to ruins by the chaotic soldiers. The prosperity that had flourished for over a century was now gone in an instant. Upon returning to this broken capital, the court found not only a lack of palaces to provide shelter, but also a severe shortage of grain to meet the daily needs of officials and citizens. The surrounding areas of Luoyang could not produce enough food, and the transport vessels on the Bian Canal had long disappeared. In such circumstances, any hope for a peaceful and stable life was unattainable. At that time, Cao Cao was summoned to the capital, and even he found it impossible to maintain order in this shattered state. At that time, the only strategy was to abandon the already broken capital and relocate to Xu County. From a strategic perspective, Xu County was not an ideal place to establish a capital, but given the current state of chaos, it had not yet been severely disturbed. As a temporary solution to the immediate crisis, it was considered a viable option.

N. Shi, *A Historical Geography Research of Canals in China*,
https://doi.org/10.1007/978-981-95-3517-0_4

Fortunately, Xu County lay not far from the Ying River, and its land was still fairly fertile. After the widespread turmoil, much of the population had either perished or scattered, leaving vast tracts of unused land. Reclaiming these lands through state-organized agriculture (*tuntian*) became the most effective way to resolve the grain shortage following the move to Xu County.

Restoration of the Bian Canal's Old Channel

Through Cao Cao's military campaigns to the east and west, the Central Plains were gradually reunified, and society began to stabilize. With this newfound peace, efforts were once again directed toward restoring the old course of the Bian Canal. In the seventh year of Emperor Xian's Jian'an reign (202 CE), Cao Cao initiated the dredging of the canal's upper reaches. This round of work extended only as far as present-day Shangqiu County, which was then known as Suiyang County. Accordingly, the restored section was renamed the Suiyang Canal.[1]

What was the effectiveness of the Suiyang Canal? There are no contemporary records indicating that it played a major role, suggesting that its impact was likely limited. For the Bian Canal to function effectively, its full course needed to be navigable. If the Han Canal section had been open to traffic, it would have further enhanced the canal's utility. However, at the time, the Jiangnan region was under the control of Sun Wu, rendering the Han Canal effectively unusable. In addition, after several major battles, Xuzhou had been devastated. Even if the Bian Canal were restored, it would no longer see the constant flow of vessels it once did. At this point, the capital had been moved to Xu County, which was quite far from the Suiyang Canal. It is unlikely that the grain transported via the Suiyang Canal would be sufficient to supply Xu County. Given these circumstances, why did Cao Cao invest in restoring this section of the canal when the population was already exhausted from war? I suspect the restoration of this canal was likely not aimed at benefiting Xu County, but rather a strategic move to prepare for the management of Hebei. After the Battle of Guandu between Cao Cao and Yuan Shao, although Cao Cao emerged victorious, the bulk of Yuan Shao's forces remained intact. With a powerful adversary still "snoring beside his bed," how could Cao Cao rest easy? At this point, his campaign to secure Hebei was already imminent—like an arrow drawn on the bowstring, ready to be released. During the Battle of Guandu, Cao Cao nearly suffered a defeat due to a shortage of military supplies, which brought him to the brink of collapse.[2] If Cao Cao were to continue advancing, the preparation of sufficient military supplies was essential. The timing of the dredging of the Suiyang Canal, which occurred immediately after the Battle of Guandu and before further

[1] Book of Wei: annals of Emperor Wu. In *Records of the three kingdoms*, vol. 1.

[2] Book of Wei: Annals of Emperor Wu; Book of Wei: Biography of Yuan Shao; In *Records of the three kingdoms*, vols. 1, 6.

consolidation of Hebei, was clearly aimed at ensuring a steady supply of grain for his army.

Cao Cao's Construction of the Bai Canal

In the third year after Cao Cao restored the Suiyang Canal—namely, the ninth year of Emperor Xian's Jian'an reign (204 CE)—he launched his campaign to consolidate power in Hebei. At the time, the Yuan clan's base of operations was in Ye County (present-day southwest of Linzhang County in Hebei), and Cao Cao's initial offensive targeted this strategic location. In conjunction with the military advance, he ordered the excavation of a new canal. This canal drew water from the Yellow River at the banks near present-day Jun County, Henan, diverted the flow of the Qi River into the old course of the Bai Canal, and passed through Neihuang, eventually connecting with the Huan River near Ye County. This stretch of the Bai Canal was originally part of the Qi River's ancient course. The Qi River originates in the southeastern reaches of the Taihang Mountains, near present-day Huguan County in Shanxi. From there, it flows southeast through Qi County in Henan, then turns northeast at a bend not far from the Yellow River, where it was diverted into the new canal.

It was around the late Spring and Autumn period that the Yellow River shifted northward, encroaching upon the Qi River's course and causing the Qi River to redirect its flow southward into the Yellow River. Because a large number of white clam shells had accumulated in the riverbed, the dried-up lower reaches of the river came to be known as the Bai Canal (literally, "White Ditch").[3] This long-buried ancient channel was not restored until this period (See Fig. 4.1).[4]

Some argue that the name "Bai Canal" (White Ditch) is unrelated to the abundance of white clam shells in the riverbed, citing two main reasons. First, according to Shen Kuo's *Mengxi Bitan* (*Dream Pool Essays*), the phenomenon of snail and clam shells forming bands along cliffs in the eastern foothills of Mount Taihang is not uncommon, and may not have been unique to the Qi River. Second, historical records show that rivers named "Bai Canal" include not only the canal constructed by Cao Cao but also others, such as the Bai Canal during the Tang Dynasty, which diverted the Bian River to connect Cao and Yan, and the Bai Canal in the Song Dynasty, which referred to a tributary of the Sui River. There was also a Bai Canal that emerged from the Dukan Reservoir and flowed into the Juma River, among others. A common feature among these rivers is that their sources were reservoirs or lakes designed to store water. These lakes functioned to settle sand and sediment, and the water that flowed out was clear and pristine, which is likely the reason why these rivers were all named "Bai Canal." Thus, it is likely that Cao Cao's Bai Canal was named for

[3] Shi Nianhai. (1984). A study of ancient river courses west of Mount Dapi in Xun County, Henan. In *Historical Research*, (02), 50–71.

[4] Book of Wei: annals of Emperor Wu. In *Records of the three kingdoms*, vol. 1.

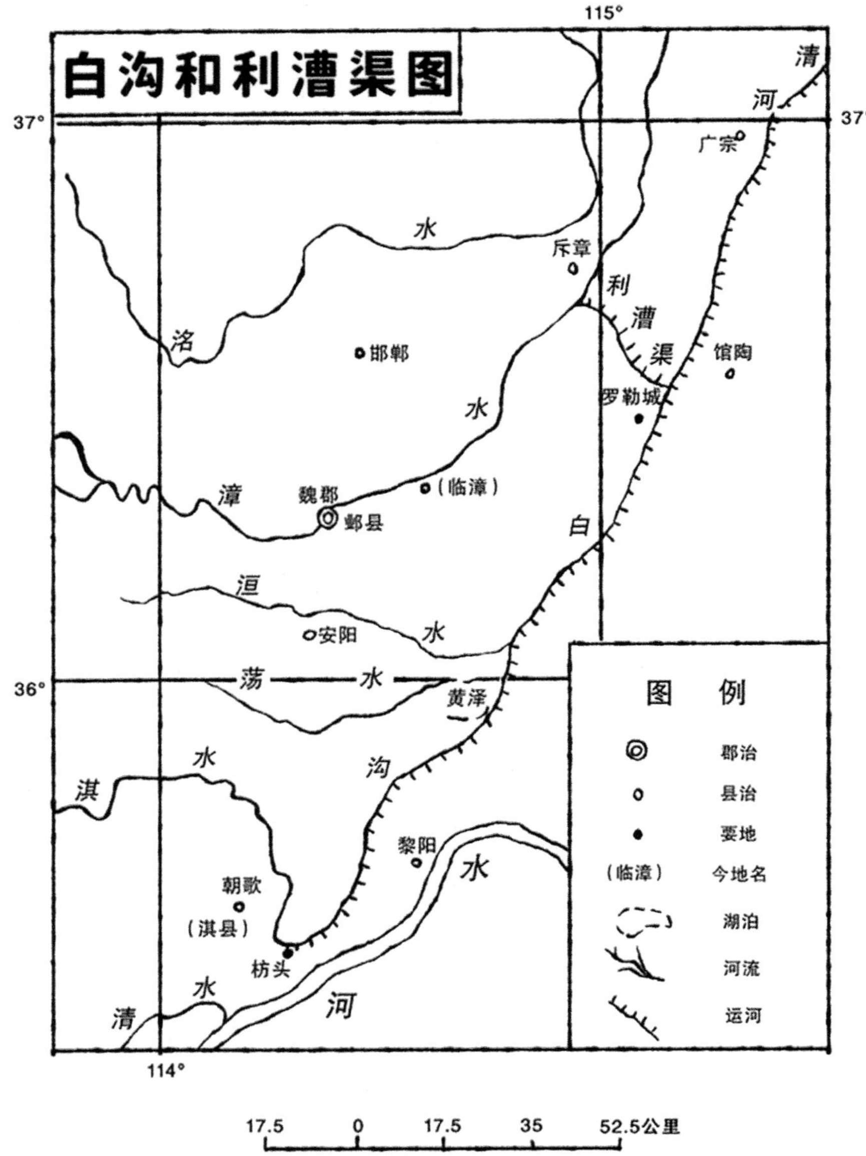

图例 - Legend

都城 Capital Cities

郡治 Commandery Seats

要地 Strategic Locations

今地名 Present Names

湖泊 Lakes

运河 Canals

河流 Rivers

Fig. 4.1 The Bai Canal and the Li Canal

this characteristic, rather than the presence of white clam shells in the riverbed.[5] Both of these reasons are difficult to fully accept. As Shen Kuo mentioned in his *Mengxi Bitan*, while the phenomenon of snail and clam shells being found along the cliffs of Mount Taihang is not uncommon, it is not the case that all rivers flowing from the foothills of Taihang contain such shells. For example, to the north of the Qi River lies the Huan River, and to the south of the Qi River is the Qing River, both of which also flow down from the Taihang Mountains—but neither contains snail or clam shells. Although historical records mention several rivers named Bai Canal, with some shared characteristics, this does not necessarily rule out the distinctive feature of the Bai Canal that Cao Cao restored. The presence of white clam shells in its riverbed could indeed be a relevant factor in its naming, despite the existence of other rivers with similar names. Therefore, we cannot entirely dismiss this possibility. If one insists that all rivers named Bai Canal must originate from reservoirs, then where, exactly, is the reservoir at the source of the Qi River? From its headwaters onward, the Qi River is notably clear and sediment-free, making a sedimentation reservoir unnecessary. Moreover, geological surveys of the river's upper reaches reveal no conditions suitable for forming such a reservoir, nor do historical records mention any. These two explanations, then, appear insufficient. A field investigation would quickly reveal how far removed these arguments are from local realities. Without on-site research, textual analysis alone is unlikely to yield a comprehensive understanding of the issue.

Pinglu Canal and Quanzhou Canal

After launching his northern campaign, Cao Cao swiftly captured Ye, the stronghold of the Yuan family. Unable to hold their ground, Yuan Shao's sons—Yuan Xi and Yuan Shang—fled northward to seek refuge with the Wuhuan, a tribal group originally residing beyond the northeastern frontier. By this time, the Wuhuan had grown into a formidable force, with Tadun, the Chanyu of Liaoxi, wielding particularly strong influence. Tadun had once received favor from Yuan Shao, and so, upon the brothers' arrival, he took them in and offered them protection. When word of this reached Cao Cao, he was unwilling to let the matter rest. Determined to wipe out the last remnants of the Yuan clan, he began preparing for a far-reaching expedition into these distant territories.

During this northern expedition, two additional canals were constructed: the Pinglu Canal and the Quanzhou Canal. The Pinglu Canal diverted water from the Hutuo River (also known as the Huto River) into the Gu River. The Quanzhou Canal was cut from the mouth of the Ju River into the Lu River, ultimately connecting to

[5] Fei Yu. (1983). A critical discussion on the naming of Bai opened by Cao Cao. In *Historical Geography*, (03), 62.

the sea.[6] These two canals effectively traversed the present-day Hebei Province from north to south.

Let us begin with the Pinglu Canal. During Cao Cao's northern campaign against the Wuhuan, Ye served as a major strategic center in Jizhou. Military logistics and the transportation of provisions were all coordinated from Ye. Just north of Ye County lay the Zhang River. Today, the Zhang River flows directly eastward into the Wei River, but at that time, it flowed northeastward from the vicinity of present-day Linzhang, passing through what are now Ji County and Hengshui, and continued northeast to merge with the Qing River before ultimately reaching the sea via the present canal system. At that time, the Hutuo River emerged from the Taihang Mountains and flowed southeastward through counties such as present-day Raoyang and Wuqiang. Northeast of Wuqiang, it merged with the Zhang River, and then branched off again, flowing further northeast into the sea. As for the Gu River, it corresponds to the present-day Sha River. The lower course of today's Sha River is known as the Zhulong River. In the Han period, however, the Gu River ran somewhat south and east of the modern Zhulong, passing north of Raoyang County and west of Hejian County, then veering eastward near the northwest of Wen'an County to empty into the Bohai Sea. Cao Cao's Pinglu Canal diverted the waters of the Hutuo River into the Gu River. Because the Hutuo and Zhang Rivers were interconnected, vessels could navigate from the Zhang River into the Hutuo, and from there enter the Gu River via the Pinglu Canal. Where exactly was the Pinglu Canal excavated? This has long been a matter of scholarly debate. One view holds that it was located two hundred paces south of Qingchi County in Cangzhou during the Song dynasty. It is also recorded that when the Pinglu Canal was completed, a fortified town was constructed on the left bank of the canal.[7] Qingchi County of the Song period was located southeast of present-day Cangzhou in Hebei, corresponding to Fuyang County during the Eastern Han. In the Han era, the Zhang River lay far to the north of Fuyang, not to its south. Moreover, the Pinglu Canal merely served to connect the Hutuo River and the Gu River and had no direct relation to the Zhang River. Therefore, identifying the canal's course in this area seems inconsistent with contemporary historical records. Another view holds that the Pinglu Canal lay within the territory of Lucheng County during the Tang dynasty.[8] Lucheng County of the Tang dynasty was located northeast of present-day Cangzhou, directly east of Qing County and no less than fifty or sixty li west of it. This area corresponds precisely to the course the Hutuo River followed during the Han dynasty. At that time, the Gu River flowed north of what is now Jinghai County.

[6] Book of Wei: Annals of Emperor Wu. In *Records of the three kingdoms*, vol. 1. Book of Wei: the biography of Dong Shao. In *Records of the three kingdoms*, vol. 14. Fan Ye. *Treatise on Commanderies and Kingdoms, In The book of the later Han*: Liu's annotation cites the *Shangdang Records* stating, "In the eleventh year of Jian'an, a canal was dug from the mouth of the Xiong (Ju) River into the Lu River, called the Quanzhou Liang Canal, to connect to the sea." This annotation is erroneous. The Quanzhou Canal actually utilized the Lu River, which flows through present-day Tong County, historically also called Lu County. Because of the same county name, a mistake was made. If the Quanzhou Canal were indeed drawn from Shangdang, how could it possibly connect to the sea?

[7] Cangzhou. In *Universal geographic record of the Taiping era*, vol. 65.

[8] Cangzhou. In *Yuanhe maps and records of prefectures and counties*, vol. 18.

Since modern Jinghai lies not far south of the former site of Lucheng County east of Qing County, it is quite plausible that the Pinglu Canal was excavated in this region. Some have suggested that the so-called Pinglu Canal in this context was the precursor to the present-day southern section of the Grand Canal between Qing County and Duliu Town. However, the modern South Canal lies at some distance from the old site of Lucheng and should not be conflated with it. Others may invoke the Shuijing Zhu, which states that the stretch of the South Canal between Qing County and Duliu Town follows the former course of the Zhang and Qing Rivers. If that were indeed the case, identifying it with the Pinglu Canal would not be unreasonable. However, if the old course still carried water at the time, there would have been no need to excavate a new canal. The very fact that the Pinglu Canal was constructed suggests that, at the very least, the old route was no longer suitable for transportation.

There is yet another theory regarding the location of the Pinglu Canal, which holds that it was situated north of Raoyang County during the Tang dynasty. Historically, the Hutuo River had flowed south of Raoyang, but due to the presence of the old Raoyang River channel, Cao Cao is said to have ordered the diversion of the Xingou River northward, thereby shifting the river's course to the north of Raoyang.[9] During the Tang period, Raoyang County was situated thirty *li* southwest of Shenzhou (whose administrative seat was in present-day Shenxian, Hebei).[10] Furthermore, Tang-era Raoyang County lay to the south of its Song counterpart.[11] The Raoyang County of the Song dynasty corresponds to modern-day Raoyang County in Hebei. Therefore, the Pinglu Canal in question should still be located to the north of present-day Raoyang County. The precise course of the so-called Xingou River remains uncertain. However, since the Pinglu Canal constructed by Cao Cao connected the Hutuo River with the Gu River—and given that the Gu River at that time flowed past the northwest of present-day Raoyang—the Xingou River could not have been far from the western part of Raoyang County. In the *Universal Geographic Record of the Taiping Era (Taiping huanyu ji)* compiled by the Song historian Le Shi, there is the following entry: "In Raoyang County, Shenzhou, the withered Baima Canal (Baima Qu) lies south of the county seat. It was also known as the Yellow River and is now called Baima Gou. It draws from the Hutuo River and flows eastward into Xiabo County. According to Zhao Ji by Li Gongxu, this Baima Canal was excavated by Biao, Prince of Baima, during the Wei dynasty."[12] The Prince of Baima referred to here is Cao Biao, a son of Cao Cao. Cao Biao was enfeoffed as Marquis of Shouchun ten years after the completion of the Pinglu Canal, and his title as Prince of Baima was granted only after the establishment of the Wei dynasty.[13] If Cao Biao had indeed been involved in canal construction in Raoyang County in his early years, it

---

[9] Annals of Emperor Guangwu I. In *Later book of Han I*, vol. 1.

[10] Shen Prefecture. In *Yuanhe maps and records of prefectures and counties*, vol. 17.

[11] Hebei Circuit. In *Treatise on the nine territories in the Yuanfeng reign (Yuanfeng jiuyu zhi)*, vol. 2: Raoyang County of Shenzhou lies 90 li north of the state capital.

[12] Shenzhou. In *Taiping huanyu ji*, vol. 63.

[13] Book of Wei: Biographies of Princes of the Martial and Civil Lineage. In *Records of the three kingdoms*, vol. 20.

must have been as a participant in Cao Cao's project to excavate the Pinglu Canal, rather than as an independent undertaking. Moreover, the Baima County for which he received his princely title belonged to the Eastern Commandery of Yanzhou, located east of present-day Huaxian in Henan—far removed from Raoyang. Given the Wei dynasty's strict restrictions on the activities of its princely vassals, it is highly unlikely that Cao Biao would have been permitted to cross territorial boundaries to direct waterworks in Raoyang. According to Le Shi, the withered Baima Canal lay one li south of Raoyang County. No records suggest that Raoyang's administrative center was relocated in the early Song, indicating that it likely remained where it had been during the Tang. In the Tang period, the river south of Raoyang was in fact the Hutuo River, not the Baima Canal. At that time, the channel was also referred to as the Yellow River. Given that the Hutuo River originates in the western Taihang Mountains and carries a heavy silt load, referring to it as the Yellow River was not without justification. At this point, it would also be important to mention another related river canal. *Yuanhe Maps and Records of Prefectures and Counties* contains the following record: "The administrative city of Raoyang County in Shenzhou is the site of the Jin and Lu Kou City. When Gongsun Quan rebelled, Sima Xuanwang (Sima Yi) was sent to suppress him, digging the Hutuo River into the Gu River to transport grain, and built this city as a result. The Hutuo River was also known as the Lutuo River, and hence the city was called Lukou."[14] This canal mentioned, which was excavated by Sima Yi, is not the same as the Pinglu Canal dug by Cao Cao. It is possible that, after Sima Yi's canal was dug, the name Lukou City became associated with it, leading to the misidentification of this location as the site of the Pinglu Canal (See Fig. 4.2).

Another canal, the Quanzhou Canal, was dug from the mouth of the Ji River into the Lu River. The canal runs alongside Quanzhou County, so it is named after Quanzhou Canal, which is located to the south of Wuqing County in Tianjin, to the northwest of the central district of Tianjin. The meeting point of the Pinglu Canal and the Gu River is in present-day Tianjin. The Quanzhou Canal was excavated northward from the vicinity where the Pinglu Canal and the Qishui meet, reaching Baoqiu River. The mouth of the Ji River is where the Ji River meets Baoqiu River, located in the southeastern part of Baodi County, Tianjin. The confluence of the Quanzhou Canal and Baoqiu River lies near the mouth of the Ji River, which is why the canal is often associated with this location. Since Baoqiu River flows westward through Lu County, it also became known as the Lu River.[15] The reason the Quanzhou Canal was directed into the Lu River is because of this geographical relationship.

The Quanzhou Canal was located 120 *li* west of the then Yongnu County. Yongnu County is located to the southeast of present-day Tongzhou District in Beijing, situated along the banks of the Lu River. In fact, the Lu River in Yongnu County at

[14] Shen Prefecture. In *Yuanhe maps and records of prefectures and counties*, vol. 17. It should be noted that the administrative center of Shenzhou is not in Raoyang County. The term "state administrative city" here should be a mistake for "county administrative city." "Gongsun Quan" should be corrected to "Gongsun Yuan." The Tang dynasty's taboo on the name "Yuan" led to the substitution of "Yuan" with "Quan."

[15] Commentary on the Baoqiu River. In *Water classic*.

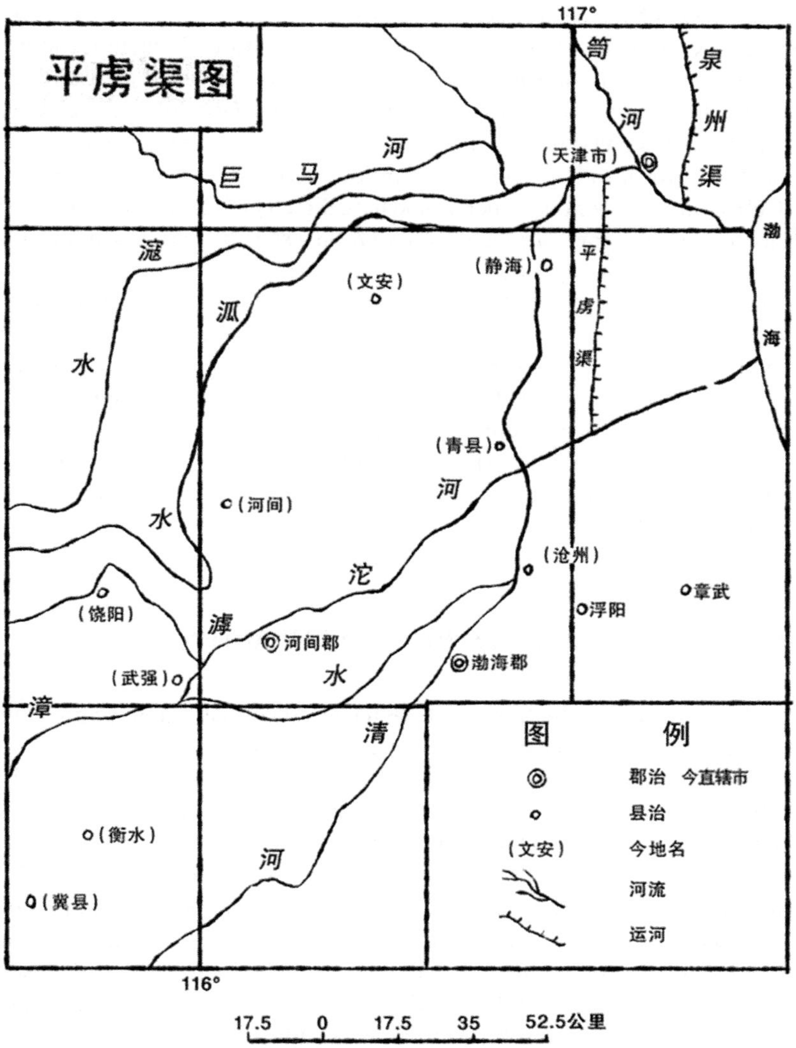

图例 - Legend

郡治 Commandery Seats

今直辖市 Direct-Administered Municipalities (Present Day)

县治 County Seats

今地名 Present Names

河流 Rivers

运河 Canals

Fig. 4.2 The Pinglu Canal

that time should have been called Si Gou (Si Canal), and the Lu River is the same as Baoqiu River, which flows eastward from just north of Yongnu County.[16] The waterway near the city of Yongnu County, whether it was called the Lu River or Si Gou, was always in existence. It essentially became the later North Canal. Given the existence of this waterway, why did Cao Cao still choose to excavate the Quanzhou Canal? Moreover, Cao Cao's original intention was to open a route from the present-day Gubeikou to the frontier, which would have directly utilized the river near Yongnu County to transport grain. There must have been a specific reason why Cao Cao did not use this existing waterway.

Later, during the Ming and Qing dynasties, the North Canal faced several sections with rapid and turbulent water, often leading to dangerous situations. Cao Cao's decision not to use this existing waterway at the time may have been based on similar concerns. As mentioned earlier, the Pinglu Canal, flowing from Qing County to the east of Duliu Town, did not utilize the old course of the Zhang River, which also makes sense. Otherwise, why would it have been necessary to undertake the laborious task of excavating new channels at such a critical moment during military operations? (See Fig. 4.3).

The Canal Between the Lu River and Ru River

After Cao Cao reached the frontier, the Wuhuan forces held strategic positions that prevented his army from advancing further. As a result, Cao Cao altered his original plan and chose to exit the frontier through the areas around present-day Lulong and Qian'an.[17] These regions, along the coastline, have historically been swampy and marshy. During the summer and autumn, they turn into vast waterlogged areas of varying depths, making them inconvenient for both vehicular and naval travel. When Cao Cao's army moved through this area, he also took the opportunity to dig a new canal. This canal was constructed from Yongnu County, leading eastward through Baoqiu River, and continued all the way to the southern part of present-day Luan County, where it joined the Ru River.[18] The Ru River is the modern-day Luan River. This canal traversed several other waterways, including the Geng River, the Fengda River, the Huanxu River, the Su River, and the Qing River, before eventually connecting to the Ru River. The Geng River roughly corresponds to today's Jiyun River; the Fengda River is now the Dou River; the Su River retains its original name; and the Qing River is known today as the Qinghe.[19] Given that this canal cut across multiple water systems, its construction must have been more challenging than that of either the Pinglu Canal or the Quanzhou Canal. At the time, it bore no formal name and was simply referred to as the "New Canal" (Xinhe). Perhaps due to its

---

[16] Ibid.

[17] Book of Wei: biography of Tian Chou. In *Records of the three kingdoms*, vol. 11.

[18] Commentary on the Ru River. In *Water classic*.

[19] Yang Shoujing. *Illustrated commentary on the Water Classic*.

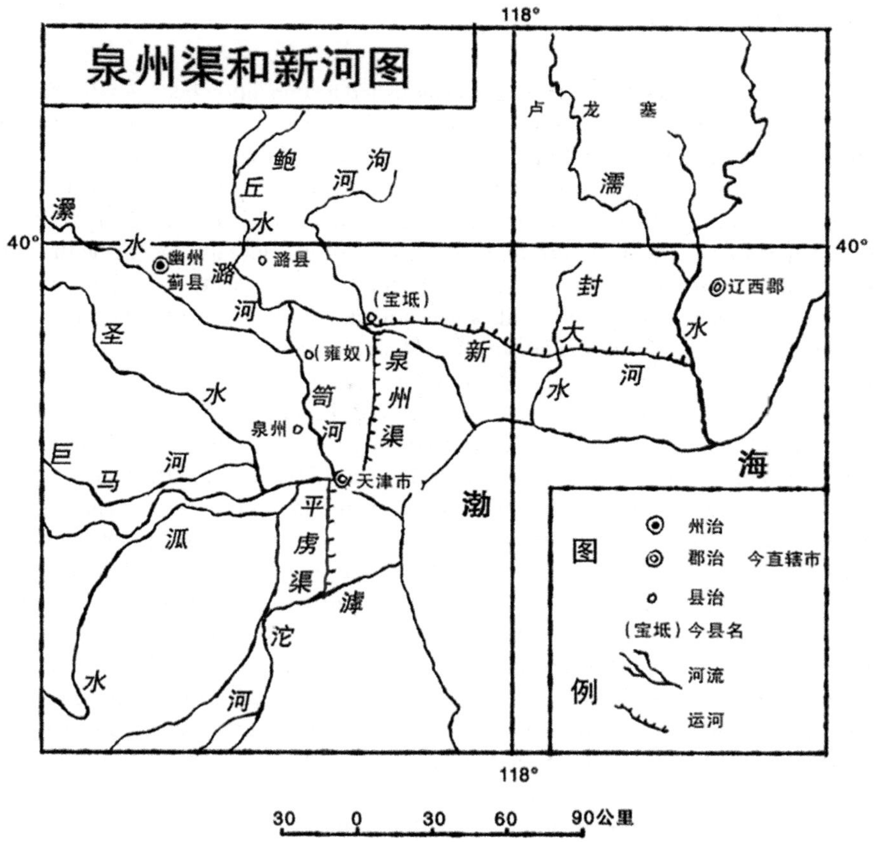

图例 - Legend
州治 Prefecture Seats
郡治 Commandery Seats
今直辖市 Present-Day Direct-Administered Municipalities
县治 County Seats
今县名 Present County Names
河流 Rivers
运河 Canals

Fig. 4.3 The Quanzhou Canal and the Xinhe Canal

limited utility and sparse traffic, the canal gradually faded from public attention and was eventually forgotten.

The Licao Canal

Cao Cao's excavation of the Bai Canal established a connection between the Yellow River and the Huan River; similarly, the Pinglu Canal he ordered linked the Gu River and the Hutuo River. Since the lower reaches of the Hutuo merge with the Zhang River, the Pinglu Canal can also be considered as linking the Gu and Zhang Rivers. The only section left unconnected was between the Huan and Zhang Rivers. Given that these two rivers are relatively close to one another and lie within the same flat plain, it is puzzling why Cao Cao neglected to excavate this final segment. The question is not easily answered. It was not until seven years after the successful completion of the Pinglu Canal, in the 18th year of the Jian'an reign (213 CE) under Emperor Xian of Han, that work on this final section began -nine years after the opening of Bai. This new waterway came to be known as the Licao Canal.[20] The Licao Canal connected the Huan and Zhang Rivers. In fact, the "Huan River" referred to here is the section formed after the Huan River and the Bai Canal merge; since the two waterways join together, either name may be used interchangeably. This stretch of Bai may thus also be called the Huan River. Flowing northeast, the Huan River reaches the east of Luole City, where it arrives at Licao Estuary—the point at which the Licao Canal branches off from the Huan.[21] Luole City is located southeast of present-day Guangping County and northwest of Daming County in Hebei Province.[22] The other end of the Licao Canal joins the Zhang River just south of the Han administrative center of Chizhang County,[23] which corresponds to the southeast of modern Quzhou County, Hebei. It is evident that the Licao Canal was excavated to facilitate transportation and communication with Ye County. However, rather than being dug directly east of Ye, the canal was routed much farther to the northeast. This may have been due to the proximity between the Huan and Zhang Rivers in the region between Luole City and Chizhang County, making excavation easier. As a result, transport vessels—particularly those coming from the south—were compelled to take a longer and more circuitous route.

The Prosperity of Ye County

Following the successful excavation of the Licao Canal, north–south connectivity within what is now Hebei Province could be achieved via waterways. This transformation in transportation routes brought about a shift in the prominence of regional economic centers. Since the Qin and Han dynasties, the economic hubs in the region had been Ji County (modern Beijing), Zhuo County, and Handan County. Their prosperity stemmed from their position along a major north–south land route to the

[20] Book of Wei: Annals of Emperor Wu. In *Records of the three kingdoms*, vol. 1.

[21] Commentary on the Qi River. In *Water classic*.

[22] Yang Shoujing. *Illustrated commentary on the Water Classic*.

[23] Commentary on the Zhuozhang River. In *Water classic*.

east of the Taihang Mountains. However, as this route was terrestrial rather than fluvial, the completion of the Licao Canal—and the resulting integration of north–south waterways—gradually reduced the importance of these locations, which lay too far from the new waterborne artery. In their place, Ye County rose to prominence. Situated southwest of modern Linzhang County in Hebei, Ye was located near the Zhang River. The area's fields were irrigated by the Zhang, and the region gained renown for its agricultural fertility.[24] A vivid depiction of this agrarian abundance is captured in the lines: "Millet and sorghum cultivated on dry land, thick groves of mulberry and zhe trees, lush fields of hemp and ramie, regularly divided plots and well-delineated fields, scattered clusters of farmhouses, abundant ginger and taro, and peach and plum trees providing generous shade."[25] This passage serves as a detailed portrait of Ye County's flourishing agricultural economy. The canal known as the Licao Canal, excavated under the direction of Cao Cao, passed along the eastern outskirts of Ye County. With its fertile lands providing a strong economic foundation, Ye flourished following the successful completion of this waterway. The canal not only facilitated transport but also drew a wide array of valuable local products from surrounding regions into Ye's thriving markets, further enhancing the county's prosperity. Among these sought-after goods were the renowned pears from Zhending (southeast of present-day Zhengding County in Hebei), and the prized chestnuts from Gu'an (southeast of present-day Yixian County, Hebei). Ye also received mellow wines of Zhongshan, whose capital lay in present-day Ding County, Hebei. From the regions along the Qishui and Huanshui rivers—corresponding to modern Qixian and Anyang counties in Henan—came fresh bamboo shoots. Xindu (modern-day Jixian, Hebei) supplied high-quality jujubes, while Yongqiu (present-day Qixian, Henan) contributed its millet. Rice was brought in from Qingliu, and brocades and embroideries came from Xiangyi (modern Suixian, Henan). The region of Chaoge (today's Qixian, Henan) was known for its silks and gauzes, whereas Fangzi (southwest of modern Gaoyi County, Hebei) produced soft cotton and silk floss. Finally, fine silk textiles, known as *jian zong*, were supplied by Qinghe, the commandery seat located northeast of present-day Linqing in Shandong.[26] This abundance of regional specialties greatly enriched Ye County. As a result, the city became a vibrant commercial hub, described as follows: "The clatter of carts echoed through hundreds of alleys, their axles crowded and laden with goods worth thousands. Men leaned on carriage rails, striking their steeds; sleeves and curtains fluttered in the wind like a moving tapestry. People and commodities from all corners of the realm converged here, presenting a spectacle of diverse customs and appearances. Transactions were fair and smooth, and trade in knives and cloth flowed in incalculable volume."[27] Such

---

[24] Book of canals and waterways. In *Records of the grand historian*, vol. 29.

[25] Zuo Si. Rhapsody on the capital of Wei. In *Selections of refined literature*, vol. 6.

[26] Zuo Si. Rhapsody on the capital of Wei. In *Selections of refined literature*, vol. 6: "The clear-flowing rice." Annotation: "The clear-flowing river, west of Ye, produces imperial rice." It further states: "The Qi and Huan rivers or the garden." The Qi garden produces bamboo, which was noted early in *The Book of Songs, Wei Feng*, suggesting that bamboo shoots were already well-known in the region.

[27] Zuo Si. Rhapsody on the capital of Wei. In *Selections of refined literature*, vol. 6.

scenes of prosperity were rarely matched elsewhere in the country. Ye County's flourishing economy and strategic location ultimately elevated it far above Ji, Zhuo, and Handan. Although Xu County remained the nominal capital during the final years of the Eastern Han, Ye served in practice as the political and economic center. In comparison, Xu County lagged far behind in wealth and vitality.

The Jiahou Canal, Taolu Canal, and Guangcao Canal

Later, Cao Pi usurped the Han throne and relocated the capital to Luoyang. At the same time, Xuchang (originally Xu County, renamed Xuchang County during the Wei period) and Ye County were designated as auxiliary capitals. The Three Kingdoms era was characterized by a tripartite division among Wei, Shu, and Wu, and aside from their military engagements in regions such as Longshang and the Qinling Mountains, the area near Xiangyang also became a battlefield. As for the rivalry between Wei and Wu, it was particularly intense in the competition for control over the Huainan region (south of the Huai River). Xuchang, situated as it was between these two competing fronts, served as a strategically advantageous location for military command. Although Xuchang lacked significant economic value, it still commanded attention from the influential figures of the time. Historically, the area from Xuchang to the Huai River had been traversed by a tributary of the Hong Canal. After the Hong Canal was blocked, the nearby Ying and Wei Rivers continued to flow. In the late Eastern Han period, when the capital moved to Xuchang, large-scale land reclamation projects were undertaken, and the watercourses were used for irrigation. As a result, the lower reaches of the rivers began to dry up due to a lack of water. Once society stabilized and grain supplies were sufficient, the significance of the land reclamation efforts in Xuchang diminished. The ongoing and intermittent warfare between the Cao Wei and Wu Kingdom necessitated the use of water routes for transportation via the Ying and Wei Rivers. To serve this purpose, Cao Wei successively constructed the Jiahou Canal, Taolu Canal, and Guangcao Canal. The Jiahou Canal was excavated by Jia Kui during his tenure as the governor of Yuzhou, with its construction beginning in the first year of Emperor Xian of Han's Yan Kang era (220 CE).[28] This canal was fed by the Pangguan Reservoir, extending over 200 li in length.[29] The Pangguan Reservoir is located northeast of present-day Xihua County in Henan,[30] and the canal likely used the waters of this reservoir to restore the old course of the Langtang Canal, which converged with the Gu River.

---

[28] Book of Wei: Biography of Jia Kui. In *Records of the three kingdoms*, vol.15; Book of Wei: Annals of Emperor Wen. In *Records of the three kingdoms*, vol. 2: Cao Pi, as King of Wei, led a southern campaign during the first year of Emperor Xian's Yankan reign, and the following year was marked as the first year of Huangchu. In the *Biography of Jia Kui*, it is recorded that the construction of the Jiahou Canal, a 200-li long canal, took place before the Huangchu era and was completed within a single year.

[29] Commentary on canals. In *Water classic*.

[30] Yang Shoujing. *Illustrated commentary on Water Classic*.

The Taolu Canal was excavated by Emperor Wen of Wei in the sixth year of the Huangchu era (225 CE), during his imperial tour to Zhaoling (located east of present-day Yan County, Henan).[31] The canal lay approximately fifteen *li* east of present-day Yan County.[32] Zhaoling is traversed by the Ru River, and the canal was likely constructed by diverting water from this river. Some scholars suggest that the canal may have been identical to the Yin River—a plausible claim, given the canal's name "Taolu," which implies a military function ("defeating the enemy"). This suggests the canal was intended to support grain transportation toward the Huai River, particularly in preparation for military operations against the Kingdom of Wu. By regulating the course of the Yin River and channeling the waters of the Ru River into the Ying River, the canal greatly improved the navigability of the waterway for transport vessels.

The Guangcao Canal was excavated by Deng Ai in the second year of the Zhengshi era (241 CE) during the reign of King Qi of Wei, Wang Fang. Deng Ai had previously proposed: "In the past, when we defeated the Yellow Turbans, we established land reclamation projects and accumulated grain in Xudu to secure the realm. Now that the three strategic fronts have been secured, our focus turns to the Huainan region. Each time we mobilize a major military expedition, the transport of troops consumes more than half our efforts, and the costs are immense—tantamount to undertaking a massive endeavor. The land between Chen and Cai is low-lying and fertile, and can replace the rice fields surrounding Xuchang. Moreover, the waters here flow eastward." The completion of the Guangcao Canal enabled Wei to transport its armies eastward to attack Wu. By navigating the canal, the troops could reach the Jiang-Huai region, where military supplies could be stored, facilitating more efficient campaigns. Furthermore, the construction of the canal helped regulate the local water systems, reducing the risk of flooding in the area.[33] The Guangcao Canal passed through the area between Chen and Cai, making use of the Ying River, and was likely constructed near its banks. It is said that the canal was fed by the Pangguan Reservoir, effectively restoring the old course of the Jiahou Canal,[34] and also utilized the ancient path of the Langtang Canal, extending all the way to the lower reaches of Chen Commandery, where it converged with the Ying River. However, by the time of the Southern and Northern Dynasties, the canal had become dilapidated, with its course overgrown and indistinguishable, described as "a network of intersecting, winding channels, with no way to distinguish them."[35] (See Fig. 4.4).

[31] Book of Wei: Annals of Emperor Wen. In *Records of the Three Kingdoms*, vol. 2.

[32] Xuzhou. In *Essentials of historical geography*, vol. 47.

[33] Book of Wei: Biography of Deng Ai. In *Records of the Three Kingdoms*, vol. 11.

[34] Commentary on canals. In *Water classic*.

[35] Ibid.

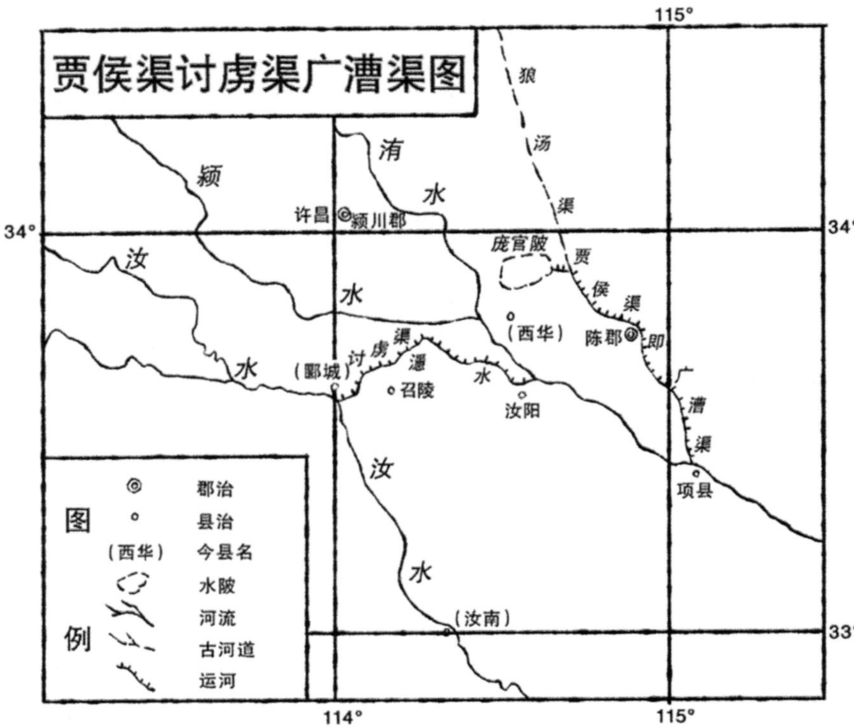

图例 - Legend
郡治 Commandery Seats
县治 County Seats
今县名 Present County Names
水陂 Reservoirs
河流 Rivers
古河道 Ancient Waterways
运河 Canals

Fig. 4.4 The Jiahou, Taolu, and Guangcao Canals

The Restoration of the Bian Canal

While the Guangcao Canal was being excavated, the Bian Canal was also renovated. As mentioned earlier, Cao Cao had previously dredged the Bian Canal, but only a portion of it, from the upstream areas down to the present-day Shangqiu County, after which the project was halted. At this time, due to Deng Ai's suggestion, the Guangcao Canal was constructed, and the Bian Canal was once again revised. This restoration of the Bian Canal was comprehensive, with improvements made from the upstream to the downstream sections, making it fully navigable for boats and ships.[36]

The renovation of the Bian Canal was indeed crucial, but its full potential could only be realized with the smooth operation of the Han Canal to the south of the Huai River. Despite the chaos at the end of the Han Dynasty and the constant warfare between the Wei and Wu states, the Han Canal remained largely intact. In the sixth year of the Huangchu era (225 CE), the conflict between Wei and Wu resumed, with Emperor Wen of Wei personally leading a military campaign to the south. The naval forces sailed from Qiao (present-day Bozhou, Anhui) along the Guo River into the Huai River. Upon reaching the ancient city of Guangling, the emperor observed the troops by the river. His army, consisting of over 100,000 soldiers, stretched across several hundred *li*, with banners fluttering in the wind. Due to an exceptionally cold winter that year, the waterways froze, and the ships were unable to navigate the river, forcing the army to retreat.[37] At the time, Guangling was a district under the jurisdiction of the Wei dynasty, located in present-day Jiangsu's Qingjiang City, situated on the southern bank of the Huai River.[38] The ancient city of Guangling, where Emperor Wen observed his troops, was situated by the river, which likely refers to the present-day Guangling in Yangzhou, Jiangsu. The naval forces led by Emperor Wen were unable to proceed due to the freezing of the waterways, which halted the progress of the ships. This suggests that the naval route they were taking was indeed along the Han Canal. Additionally, it is important to note that the emperor's fleet traveled southward along the Guo River from Qiao. The Guo River was originally a branch of the Hong Canal and was still navigable at the time. Since the fleet traveled down the Guo River, it indicates that the Bian Canal had not yet been fully renovated.

In the early years of the Western Jin Dynasty, Wang Jun, departing from Yizhou (with its capital in Chengdu, present-day Chengdu, Sichuan), sailed downstream along the Yangtze River to launch an eastern campaign against the Wu Kingdom. After the defeat of Wu, Du Yu advised him to lead his fleet along the following route: first, from the Yangtze River into the Huai River, then into the Si River, and finally, from the Bian Canal into the Yellow River, navigating upstream to return to Luoyang, the capital of the Jin Dynasty at the time. If this plan were carried out, it would indeed have been an unprecedented feat.[39] Du Yu's advice was based on the fact that, at that time, the Han Canal had not been blocked, and the Bian Canal, having been dredged by Deng Ai, was once again navigable.

[36] Book of Wei: Biography of Deng Ai. In *Records of the Three Kingdoms*, vol. 11. Annals of Emperor Xuan. In *Book of Jin*, vol. 1.

[37] Book of Wei: Annals of Emperor Wen. In *Records of the Three Kingdoms*, vol. 2.

[38] Treatise on geography II. In *Book of Jin*, vol. 15.

[39] Biography of Du Yu. In *Book of Jin*, vol. 34.

Du Yu Excavates the Yangkou Canal

While stationed at Jingzhou (with its administrative center at Jiangling, in present-day Jiangling County, Hubei Province) Du Yu famously advised Wang Jun to return north via the Bian Canal. Not long after, he initiated canal excavation projects in the area. During the Spring and Autumn and Warring States periods, Jiangling (then referred to as Ying), the seat of Jingzhou, had been the capital of the State of Chu. In earlier times, notable figures such as Sunshu Ao, King Ling of Chu, and Wu Zixu had each undertaken canal construction in the region. However, it appears that, by Du Yu's time, these canals had long since silted up or disappeared. When Du Yu assumed command in Jingzhou, the only remaining natural waterways were the Mian and Han Rivers. Aside from these, there were no navigable routes for hundreds of miles. South of Jiangling lay Baqiu Lake (now known as Dongting Lake). As the critical hub where the Yuan and Xiang Rivers converged, surrounded by mountains and rivers, the area offered both strategic security and logistical advantage. For this reason, establishing transportation links with the region was of vital concern.

Accordingly, Du Yu "opened the Yangkou canal, extending it from the Xia River to Baling over a distance of more than one thousand li. Internally, this project alleviated the navigational dangers posed by the Yangtze River; externally, it established a vital connection for grain transport between Lingling and Guilin."[40] Northwest of modern-day Qianjiang County in Hubei, Yangkou (also known as Zhongxiakou, or simply Xiakou) was the point where water from the Yang River flowed into the Mian River. The Yang River itself originated northwest of Jiangling city and flowed eastward past Jingling County (now the northwest of Qianjiang County) before eventually entering the Mian River at Yangkou. It is said that when Cao Cao pursued Liu Bei to Dangyang, Zhang Fei, at Changban Slope, stood at the rear wielding a spear, covering Liu Bei's retreat. Seizing the opportunity, Liu Bei hastily led a small group of followers along a narrow path, racing toward the Han Ford crossing. After crossing the Han River at Han Ford, he eventually reached Xiakou, which became his place of refuge following the escape.[41] Why this place was also called Xiakou has not been clearly explained in earlier sources. It is possible that the river flowing through the area was then known as the Xia River. Thus, when Du Yu opened Yangkou, he began at the Xia River, transforming it into a navigable canal. (See Fig. 4.5).

The Yang River originated northwest of Jiangling and flowed eastward directly into the Mian River. It appears to have been a naturally navigable waterway. Given this, one may wonder why King Ling of Chu and Wu Zixu undertook canal excavation in this area, and why, several centuries later, Du Yu again found it necessary to carry out further work? This likely relates to the treacherous nature of the Yang watercourse, and possibly to fluctuations in its water volume. While it is no longer possible to verify whether the waterway was obstructed or hazardous at the time, it is reasonable to conclude that the availability of water sources posed a challenge. East of Jiangling stood Lake Lubai, Lake Zhong, and Lake Huguan – three lakes

---

40 Biography of Du Yu. In *Book of Jin, vol. 34.*

41 Commentary on the Mian River. In *Water classic.*

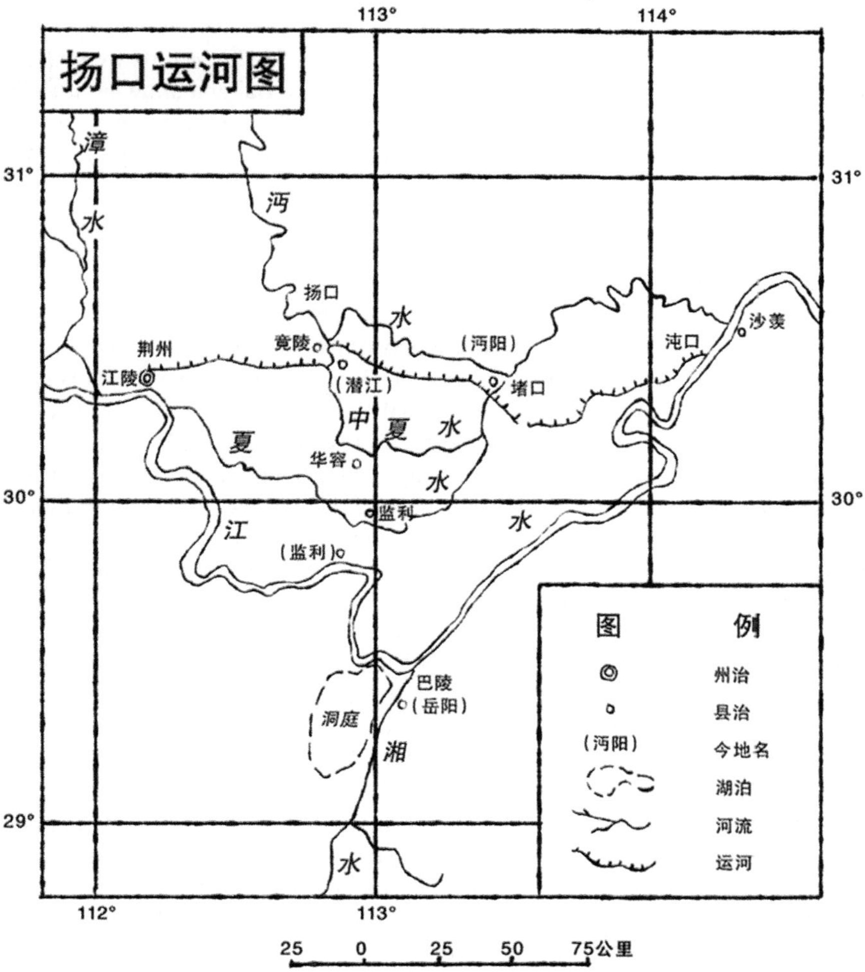

图例 - Legend
州治 Prefecture Seats
县治 County Seats
今地名 Present Names
湖泊 Lakes
河流 Rivers
运河 Canals

Fig. 4.5 The Yangkou Canal

aligned north to south and forming a connected body of water. During the Yuanjia reign of Emperor Wen of the Liu-Song dynasty (424–453 CE), efforts were made to widen and dredge the waterway, connecting it to Lake Bai and allowing it to flow downstream into the Yang River, thereby enhancing the transport capacity of the canal system.[42] This was clearly intended to augment the water supply of the Yang River.

At the time, Du Yu's canal construction was intended"to divert the dangerous waters of the Yangtze into lesser channels and to facilitate external grain transport between Lingling and Guiyang." However, the question of how exactly the dangerous waters of the Yangtze could be made to flow warrants closer examination. The aforementioned efforts to augment the water supply of the Yang River occurred well after Du Yu's time. If the Yang River had a limited water supply, then it is likely that Du Yu encountered this issue during the canal's excavation. How, then, did he solve this problem? It appears that at least some effort was made to address the issue at the time. As previously noted, Lake Lubai, Lake Zhong, and Lake Huguan were interconnected, forming a unified body of water. During the high-water seasons of spring and summer, these lakes would open southward to the Yangtze River. In drier seasons, however, they extended only as far south as the river embankment and water to the north of this divide was therefore retained within the lakes.[43] This area likely served as an artificial watercourse connecting the Yang River to the Yangtze, separated by embankments. It is plausible that Du Yu, in the course of constructing the canal, had already diverted Yangtze water into the Yang River at this location. While creating such a channel would have obviously required opening a new waterway, the primary aim was clearly to supplement the Yang River's water supply. In later years, this channel became obstructed, and the volume of water in this section of the Yang River again became insufficient. Although during the Yuanjia period of the Liu-Song dynasty, a channel was opened from Lake Lubai to feed into the Yang River and replenish its supply, the waterway connecting to the Yangtze was not restored. Diverting water from the Yangtze into smaller rivers was no trivial matter. Without careful planning, such an undertaking risked causing catastrophic flooding. This is what is hinted in the phrase: "to divert the dangerous waters of the Yangtze into lesser channels."

Here, it is also worth briefly addressing the matter of the Xia River. As mentioned earlier, the Yang River may have once been referred to as the Xia River, an inference based on the fact that Yangkou was also known as Xiakou. In fact, there was originally a natural waterway known as the Xia River between the Yangtze and Han Rivers, to which the *Commentary on the Water Classic* devotes an entire section. *The Water Classic* states: "The Xia River originates from a branch of the Yangtze River southeast of Jiangling County." *The Commentary* adds, "At the Yuzhang ferry crossing on the Yangtze, to the east lies Zhongxiakou, the head of the Xia River." This location is likely situated south of present-day Shashi in Hubei Province. From there, the Xia River flows eastward, passing south of Huarong County and Jianli County, before

[42] Commentary on the Mian River. In *Water classic*.

[43] Ibid.

entering the Mian River at Yundu County. The former site of Huarong County lies northwest of today's Jianli; the old town of Jianli lies to its north; and the historical Yundu County corresponds to the southwest of present-day Xiantao, Hubei. The Xia River entered the Mian at Dukuo, located within Yundu County. Dukuo and Zhongxiakou are roughly aligned on an east–west axis. This means that the Xia River, after branching off from the Yangtze, made a broad arc southward before turning northeast to join the Mian River. It was a natural watercourse, not an artificial canal. Not only was it excessively winding—greatly increasing the navigable distance—but it was also situated far from Jiangling and bore no relation to Yangkou. Therefore, it had no connection with the canal excavated by Du Yu.

After branching off from the Yangtze River, the Xia River merged with another distributary of the Yangtze between Huarong County and Jianli County. This distributary diverged from the Yangtze at Zixiakou and flowed northeast to join the Xia River. Zixiakou lay diagonally across the river from the former seat of ancient Huarong County and was situated to the west of present-day Jianli County.[44] Might Du Yu's grain transport route to Lingling and Guiyang have proceeded southward from Jiangling into the upper reaches of the Yangtze, entered the Xia River at Zhongxiakou, and then followed the Xia River to merge with this distributary—emerging from the Yangtze—before eventually returning to the main stream of the Yangtze? A consideration of the local topography suggests this scenario is implausible. This is because the section of the Yangtze River downstream from Jiangling had long been established as a viable shipping route. From Jiangling southward, below Zhongxiakou, the riverbank featured key river ports such as Youkou, Jingkou, and Lunkou. South of Lunkou stood Dacheng, known in historical records as Cangchu City—a site for the storage of grain and supplies, effectively serving as a *di ge* (regional depot).[45] The old city of Baling was likewise the location of Baqiu *di ge* established by the state of Wu.[46] Such *di ge*, used during the Han and Wei periods, were military logistics hubs for storing and transporting provisions.[47] The presence of these facilities along the Yangtze River attests to the frequency and scale of waterborne transport at the time. During the military standoff between Wu and Shu, and later as Wu prepared its defenses against Jin, this stretch of the Yangtze was actively used for troop movements and logistical supply. As Du Yu inherited these well-established routes and infrastructure, there was no need for him to excavate an entirely new waterway. The phrase "to divert the dangerous waters of the Yangtze into lesser channels" should thus not be interpreted as implying the opening of a new navigational route.

However, the issue cannot be considered fully resolved at this point. Not only does it remain unsettled, but it also presents certain ambiguities. In the Southern Song period, Wang Xiangzhi, in his *Record of Scenic Spots Across the Country*, noted that the grain transport canal lay four li north of the county seat of Jiangling. He also cited a memorial from the first year of Emperor Gaozong's Shaoxing reign (1131 CE),

[44] Commentary on the Yangtze River. In *Water classic*.

[45] Ibid.

[46] Book of Shu: Biography of the later sovereign. *History of the Three Kingdoms, vol.33.*

[47] Commentary on the Xiang River. In *Water classic*.

which stated: "Beneath the city of Jiangling Prefecture, there used to be a grain canal. With minor dredging, transport convoys could travel downstream via the Yangtze and Han rivers, reaching Xiangyang and Xiangxian with ease. Field investigations have confirmed that the canal of Jiangling Prefecture was first excavated during the early Jianwu reign of Emperor Yuan of the Jin dynasty (317 CE). It ran from Luoyankou into the Dacao Canal, passing through Lishexue, Zhuankou, and the mouth of the Mian River, directly connecting the three rivers of Xiang, Han, and Yangtze."[48] Wang Xiangzhi further quoted an "*Old Classic*," which recorded: "When Wang Chuzhong served as governor of Jingzhou, he excavated the grain canal, connecting the Yangtze and Han rivers, linking the southern and northern embankments."[49] Wang Chuzhong refers to Wang Dun, who held the post of governor of Jingzhou in the first year of the Daxing reign under Emperor Yuan (318 CE), the year immediately following Jianwu's first year.[50] Thus, both accounts likely refer to the same undertaking. While the exact location of Luoyankou remains uncertain, the section below Luoyankou was known as the Dacao Canal, and the stretch above Luoyankou extending to four *li* north of Jiangling was probably the new channel dug by Wang Dun during the Jin dynasty. As for the Dacao Canal itself and its location, Wang Xiangzhi cited the *Gazetteer of Prefectures and Counties under the Imperial Dynasty*, which states: "The Jianyang River enters Jiangling territory from Huaxi, southwest of Jingzhou, flows through Choutan Market and Wujia Market, then enters the Yangtze at Pojiangkou, connecting with places like Daxiao, Jianli and Zhuankou. Today, it is known as the Dacao River."[51] The Shaoxing-era memorial merely confirms that by the time of Emperor Yuan, the Dacao Canal already existed, but it does not clarify when the canal was originally constructed. The *Imperial Comprehensive Gazetteer of the Great Qing* identifies the Dacao Canal as identical to the Yangkou grain canal once excavated by Du Yu, and locates it just north of Jiangling County.[52] This suggests that Wang Dun's excavation during the Jin dynasty merely restored the western segment of the canal—extending from Luoyankou westward—based on Du Yu's earlier infrastructure.

Here, we may set aside for the moment the section west of Luoyankou and briefly outline the regions through which the Dacao River flowed. The river's upper reaches were fed by the Jianyang River, which flows southeast from present-day Jingmen County. This river, also known historically as the Yang River,[53] is still referred to today as the Jian River. Wang Xiangzhi's claim that "the Jianyang River flows into Jiangling territory from Huaxi in the southwest of Jingzhou" is clearly mistaken. The lower reaches of the Jianyang River must have joined with the Yang River, which was the very waterway utilized by Du Yu when he opened Yangkou. If the matter

[48] Jiangling Prefecture I. In *The record of scenic spots across the country. Vol.64. (The first year of the Shaoxing era corresponds to 1131 CE.)*

[49] Ibid.

[50] Annals of Jin XII. In *Comprehensive mirror for aid in governance, Vol. 90.*

[51] Jiangling Prefecture II. In *The record of scenic spots across the country. Vol.64.*

[52] Jingzhou prefecture I. In *Comprehensive gazetteer of Qing dynasty (Jiaqing edition), vol. 344.*

[53] Jingmen subprefecture. In *Essentials of historical geography, vol. 77.*

ended there, there would be little further cause for debate. However, according to the Imperial Directive of the First Year of the Shaoxing Reign and the Gazetteer of Prefectures and Counties under the Imperial Dynasty, the Daco Canal passed through locations such as Choutan Market, Wujia Market, Pojiang, Daxiao, Jianli, Lishexue, Zhuankou, and the mouth of the Mian River before entering the Han River. Chouan Market lies east of modern Jiangling County; Jianli is north of present-day Jianli County; Zhuankou is southwest of modern Hanyang District in Wuhan; and the mouth of the Mian River refers to the point where the Han River enters the Yangtze. Although the precise locations of several other sites remain uncertain, the inclusion of Choutan, Jianli, Zhuankou, and the mouth of the Mian River provides a general outline of the canal's course. Furthermore, during the Qing dynasty, a canal known as the Cao River was located approximately one *li* southwest of Mianyang Prefecture.[54] Qing-era Mianyang was situated southwest of present-day Xiantao City and bordered Mianyang Lake to the north. This Cao River was likely related to the Dacao River. During the Eastern Jin and Southern and Northern Dynasties, the upper reaches of this canal largely overlapped with the course of the historical Xiayang River, while the lower section, near its confluence with the Yangtze, roughly followed the path of the Zhuan River. However, in those periods, the Xiayang and Zhuan Rivers were not only unconnected but also separated by a significant distance.[55] Aside from the *Comprehensive Gazetteer of Qing Dynasty (Jiaqing edition)*, which identifies the Dacao River as the Yangkou transport canal excavated by Du Yu, no other sources mention the time of its original construction. While the distance along the Mian River between Yangkou (present-day Qianjiang County) and the mouth of the Mian River (in modern-day Wuhan) is relatively long, it is still shorter than the full length of the Dacao River. If the canal were indeed dug by Du Yu, his intended purpose would not have been to establish a northern route to Xiangyang and Mount Xian, but rather a southern route to connect Lingling and Guilin. In that case, a southbound route from Yangkou following the Mian River to its mouth and then entering the Yangtze would have been a more direct and efficient path. From there, one could travel upstream to Baling and farther south to Lingling and Guilin. If so, the construction of the Dacao River might seem redundant. As previously noted, Du Yu's excavation from the Xia River to Baling was intended "to divert the dangerous waters of the Yangtze into lesser channels." Although the Dacao River could, in theory, reach Baling by exiting into the Yangtze at the mouth of the Mian River and proceeding upstream – and then continue southward to Lingling and Guilin – it is difficult to interpret how this would achieve the goal of "diverting the dangerous waters of the Yangtze into lesser channels." Du Yu's excavation at Yangkou directed the Yang River eastward into the Mian River. The Jianyang River, which fed the upper reaches of the Dacao River, originally flowed into the Yang River and continued east. As the primary source of water for the Dacao River, it would have likewise flowed southeast. While the canal may have received additional tributaries further downstream, they too flowed southeastward into the Yangtze. Thus, it is inaccurate to claim that this canal served

[54] Mianyang subprefecture. In *Essentials of historical geography, vol. 77.*

[55] Commentary on the Xia River. In *Water classic.*

to "divert the dangerous waters of the Yangtze into lesser channels." For this reason, identifying the Dacao River as the canal excavated by Du Yu is rather unconvincing. Nonetheless, aside from Du Yu, there is no record of any other canal construction in this region prior to the reign of Emperor Yuan of the Eastern Jin. Therefore, the issue warrants further careful examination.

Thanks to Du Yu's successful canal excavation, the transportation landscape of Jiangling was fundamentally transformed. Even prior to this development, Jiangling had already occupied a position of considerable strategic importance: traveling upriver along the Yangtze River, one could reach the regions of Ba and Shu; heading downstream, one could arrive at Wuchang (present-day E'cheng County, Hubei) and Jianye (modern-day Nanjing). To the south, one could traverse Dongting Lake, ascend the Xiang River, and follow the Li River to reach Panyu. However, access to the north remained hindered by the need to travel a stretch overland before reaching the navigable waters of the Han River. This limitation, a significant flaw in an otherwise well-situated transportation hub, was not remedied until the early Western Jin period, when Du Yu effectively resolved the issue through his canal engineering.

The Completion of the Grand Canal

Today, China's longest north–south transportation artery is undoubtedly the Beijing–Guangzhou Railway. Yet, as early as the time when Du Yu excavated the Yangkou Canal, an ancient north–south inland waterway network had already taken shape. Tracing the development and course of this historical water route from north to south offers a fascinating glimpse into the ingenuity of premodern engineering. Beginning near present-day Tong County, not far from Beijing, the route followed the Lu River and entered the Gu River via the Quanzhou Canal, which had been excavated under the orders of Cao Cao. From the Gu River, it connected to the Hutuo and Zhang Rivers through the Pinglu Canal, another of Cao Cao's hydraulic projects. Continuing from the Zhang River, the route passed through the Licao Canal and the Bai Canal to reach the Yellow River. From there, it entered the upper reaches of the Bian Canal and flowed downward via the Langtang Canal, then into the Ying River and further into the Ru River. From the Ru River, the route proceeded via the Wu River - one of its tributaries - and through a canal connecting the Wu and Zi rivers that dated back to the Spring and Autumn period of the state of Chu. The water then flowed from the Zi River into the Yu River, from the Yu into the Han River, and from the Han, entered the Yangtze River via the Yangkou Canal excavated by Du Yu. From the Yangtze, it passed into Dongting Lake, then into the Xiang River. From the Xiang, it continued via the Ling Canal - dug during the Qin dynasty under Shi Lu - into the Li River, and from there entered the Xijiang (West River), eventually reaching the city of Panyu (modern-day Guangzhou). Though this sequence of waterways may appear complex and circuitous, it is truly astonishing that, long before the invention of railways, the people of ancient China were able to envision and gradually realize such an extensive and interconnected north–south water transport corridor. The earliest components of this system date back to the Spring and Autumn period, but its full integration was not achieved until the early Western Jin. That it took centuries of intermittent construction to complete this monumental network speaks volumes about both the challenges involved and the enduring ambition behind it.

The Excavation of the Pogang Canal and the Jingkou Waterway

During the Three Kingdoms period, Eastern Wu also undertook canal construction, most notably the so-called Pogang Canal (Pogang Du). According to *History of the Three Kingdoms*, in the eighth year of the Chiwu reign (245 CE), Eastern Wu dispatched Colonel Chen Xun to lead 30,000 men —drawn from agricultural garrisons (*tuntian*) and laborers—to excavate a central waterway through Jurong, extending from Xiaoqi to the western city of Yunyang. This project facilitated commercial exchange and included the construction of storage facilities.[56] *The Records of Wu* by Zhang Bo also documents this initiative: "During the Da Huang era of Wu, Chen Xun was ordered to open a waterway in Jurong County and construct twelve weirs to link the various commanderies of Wu and Kuaiji. As a result, ships no longer had to pass through Jingkou."[57] A more detailed account is provided by Xu Song of the Tang dynasty: "(Eastern Wu) assigned Chen Xun to oversee tuntian operations and mobilized 30,000 tuntian troops to excavate a central route through Jurong to the western city of Yunyang, thereby opening a navigable waterway for ships from the Wu and Kuaiji commanderies. This canal, known as Pogang Du, contained fourteen weirs in total. Markets and granaries were constructed along the route. … The canal lay 25 *li* southeast of Jurong County, with the upper seven weirs reaching into Yanling territory and the lower seven extending into Jiangning. From that point on, vessels from the eastern commanderies no longer needed to travel via Jingkou along the Yangtze."[58] Jurong corresponds to present-day Jurong County in Jiangsu Province. The western city of Yunyang was located west of present-day Yanling Town in Danyang County, Jiangsu. Jingkou is now Zhenjiang, Jiangsu; Yanling remains today's Yanling Town in Danyang. The location of Xiaoqi remains unknown. Pogang Canal was situated southwest of modern Zhenjiang, approximately 25 *li* southeast of Jurong County.[59] The area lies near Mount Mao, and the canal most likely traversed the flatter terrain north of Ten-Mount Mao. By cutting through this elevated area, ships coming from the east could proceed westward from Yunyang (now Danyang), then follow the Qinhuai River to reach Jiangning—known during the Three Kingdoms period as Jianye, the capital of Eastern Wu.

[56] Chronicles of the rulers of Wu, II. In *History of the Three Kingdoms, vol. 47.*

[57] Excerpt from the geography section. In *Taiping imperial encyclopedia, vol. 73.*

[58] *Veritable Records of Jiankang, vol. 2.*

[59] Jiangning prefecture, and Zhenjiang prefecture. In *Essentials of historical geography, vol. 21 and vol. 25.*

The excavation of the Pogang Canal was primarily motivated by its shorter, more direct route and the desire to circumvent the treacherous winds and waves of the Yangtze River at Jingkou. *The Records of Wu* by Zhang Bo underscores this point, emphasizing that "ships no longer had to pass through Jingkou." In later periods, Jingkou naturally became a key waterway junction linking Wu Commandery and Kuaiji to the Yangtze River. However, it remains unclear exactly when the waterway connecting to Jingkou was first successfully opened. After the death of King Fuchai of Wu and the fall of his state, historical records concerning transportation in this region became fragmentary and are difficult to verify. It is known that the First Emperor of Qin once journeyed east of the Yangtze on a hunting expedition. It is recorded that the First Emperor of Qin once undertook a hunting expedition east of the Yangtze. His route proceeded by water from Yunmeng down the Yangtze through Danyang to Qiantang, then into present-day Zhejiang, where he ascended Mount Kuaiji. On his return, he passed through the Wu region and crossed the Yangtze at Jiangcheng.[60] Danyang corresponds to present-day Xiaodanyang Town in northeastern Dangtu County, Anhui; Jiangcheng lies north of modern Jurong County, Jiangsu. This journey suggests two key points: first, the emperor did not travel entirely by water near Tai Lake, as there is no evidence — then or later — of any navigable waterway in the Jiangcheng area capable of accommodating boats. Second, Dantu had not yet emerged as a major crossing point. Given his imperial status, the First Emperor would not have risked a crossing at a location prone to dangerous currents and winds. Historical records confirm that the southbound waterway from Jingkou was not excavated until the Eastern Wu period. Near the end of that state, efforts were made to "excavate a canal from Dantu to Yunyang" — in other words, to create a waterway connecting present-day Zhenjiang and Danyang. However, one account notes: "The ridges between Duye and Xiaoxin had to be completely cut through, and the engineering was extremely arduous,"[61] and in the end, the project failed to yield significant results. Since this section of the canal remained incomplete, it raises the question of how vessels traveling westward to Jianye from Wu and Kuaiji commanderies could pass through Jingkou. This issue is addressed in Xu Song's Veritable Records of Jiankang, which states: "Henceforth, ships from the eastern commanderies no longer sailed through the Jingjiang." The Jingjiang was the section of the Yangtze that flowed past Jingkou,[62] and this implies that vessels passing by Jingkou did not necessarily need to enter the Yangtze there. Navigable routes below Jingkou to the Yangtze already existed – such as the ancient Wu River watercourse mentioned earlier – so access to the Yangtze was not limited to the Dantu-Yunyang waterway alone. During the Western Jin, Chen Min once blocked the waters of Malin Stream to create Lake Lian. In the early Eastern Jin, Zhang Kai also established Lake Xinfeng. Both lakes were located north of Danyang and near the aforementioned canal. However, these lakes were initially constructed for irrigation purposes and

---

[60] Annals of Qin Shi Huang. In *Records of the grand historian*, vol. 6.

[61] Prefectures and commanderies section. In *Taiping imperial encyclopedia,* Vol. 170, quoting *Records of Wu.*

[62] Zhenjiang prefecture. In *Essentials of historical geography*, vol. 25.

were not intended to regulate the water volume of the canal.[63] Navigating this route by boat still posed challenges, preventing it from becoming a major thoroughfare. During the Southern Dynasties, when significant military campaigns occurred east of Jiankang (modern-day Nanjing), the Pogang Canal became a critical strategic point for both sides in the conflict, while Jingkou was largely neglected. Near the end of Emperor Wen of Liu Song's reign, Liu Shao assassinated the emperor and usurped the throne. In response, regional forces led by Prince Wang Dan of Kuaiji marched on Jiankang, prompting Liu Shao to breach the embankments at Pogang and Fangshan to impede the advance of the eastern army.[64] Later, during Emperor Ming's reign, when Kong Ji led forces from Kuaiji in support of Prince Xiu of Jinan's rebellion, the emperor immediately dispatched Prince Xiuruo of Baling to garrison Yanling for defense.[65] In both cases, the military forces did not attempt to establish a defensive line at Jingkou to block the eastern armies, nor did the eastern forces attempt to exploit Jiankang's vulnerability by advancing from Yunyang to Jingkou and then sailing up the Yangtze to attack the capital. This indicates that the section of waterway between Jingkou and Yunyang was not regarded as significant by any faction at the time. From the Liu Song dynasty through the Liang and Chen periods, the Pogang Canal continued to serve as the primary transportation artery between Jiankang and the commanderies of Wu and Kuaiji. After the fall of the Southern Dynasties and the loss of Jiankang as the imperial capital, the Pogang Canal gradually silted up and fell into disuse - until it was revitalized with the excavation of the Jiangnan Canal under Emperor Yang of the Sui dynasty.

The Canal Diverting the Yellow River at Shan County into the Luo River

Canal construction during the Western Jin period was limited. Aside from Du Yu's excavation at Yangkou, the only other significant project was the canal opened in the tenth year of Emperor Wu's Taishi reign (274 CE), which diverted water from the Yellow River at Shan County into the Luo River.[66] Shan County, administered jointly with present-day Sanmenxia City, is located in modern-day Henan Province. Following the legacy of Cao Wei, the Western Jin also established its capital at Luoyang.

At the time, grain transported from Guanzhong and Hedong was shipped downstream via the Yellow River and then upstream along the Luo River. However, this route was hindered by the perilous rapids at *Dizhu*, which posed serious challenges for the navigation of grain boats. To the southeast of Shan County runs the Tuo River,

---

[63] Runzhou. In *Yuanhe maps and records of prefectures and counties*, vol. 17.

[64] Records of Song IX. In *Comprehensive mirror for aid in government,* Vol. 127, citing *Book of Song* (Vol. 99).

[65] Biography of Kong Yi. In *Book of Song,* Vol. 84.

[66] Annals of Emperor Wu. In *Book of Jin,* Vol. 3.

which originates in the southern mountains and flows northwest into the Yellow River west of the county seat. Today, this river is known as the Qinglongjian River. On the opposite side of the southern mountain lies the drainage basin of the Luo River, whose largest tributary is the Jian River. The Jian River rises near Guanyintang, slightly southeast of Shan County. The straight-line distance between the headwaters of the Jian and Tuo Rivers is approximately 15 km.

The source of the Tuo River is even closer – only about 8 km – to the headwaters of the Yongchang River, another tributary of the Luo River that joins it near Sanxiang Town, west of Yiyang County. The historical records merely mention that a section of the Yellow River was diverted into the Luo River, without specifying the precise geographic details – whether the diversion was routed through the Jian River or the Yongchang River. This uncertainty requires further inference. Although the headwaters of the Yongchang River lie closer to those of the Tuo River, the Yongchang is considerably smaller in volume than the Jian River and its valley is much narrower, suggesting that the canal likely traversed the mountainous terrain between the Tuo and Jian Rivers. To date, no archaeological evidence or traces of this canal have been discovered, so this remains a hypothesis rather than a definitive conclusion.

The crucial geographical feature here is the mountain range separating the Tuo and Luo River basins. If this ridge could be cut through, the two waterways could be connected, allowing grain boats to bypass the treacherous *Dizhu* rapids and follow a new, safer route. According to historical texts, the canal was successfully excavated. However, how the project was carried out, how the canal was used thereafter, and its actual effectiveness are all absent from historical records, leaving us with little further knowledge. This waterway was undoubtedly of strategic significance, yet later generations failed to recognize its value. During the Sui and Tang dynasties, when the capital was located at Chang'an, immense resources and countless lives were spent trying to navigate the hazardous *Dizhu* stretch for grain transport. It remains unclear why no efforts were made during this time to restore the Western Jin canal. Though the Sui and Tang governments did undertake dredging of the Tuo River, their efforts were intended solely to supply drinking water to the residents of Shan County, rather than to facilitate transportation (See Fig. 4.6).[67]

[67] According to *Essentials of Historical Geography, vol. 48*, "The Liren Canal in Shanzhou (present-day Shan County, Henan) consisted of two branches, one to the north and one to the south. The North Canal was situated to the north of the prefectural city and was excavated under the supervision of Su Cheng in the sixth year of the Kaihuang reign of the Sui dynasty (586 CE). It diverted water from the Tuo River, flowing northwest into the city, and brought considerable benefit to the local population. The southern canal lay to the southeast of the city and drew water from the Xiaoshi area; it was constructed and dredged concurrently with the North Canal. In the eleventh year of the Zhenguan reign of the Tang dynasty (637 CE), the court commissioned Qiu Xinggong to re-excavate the southern canal. There was also another canal known as the Guangji Canal, which was opened in the first year of the Wude reign (618 CE) under the supervision of Zhangsun Cao, the Xingtai (administrative commissioner) of the Eastern Shan Circuit. It brought water into the city to replace the reliance on well water. According to Fu Chang's account in the *History of Jin,* in the fifth year of the Taishi reign of Emperor Wu of Jin (269 CE), a canal was excavated through the southern mountains of Shan County, diverting water from the Yellow River eastward into the Luo River to facilitate grain transport. This marked the origin of hydraulic works such as the Liren Canal."

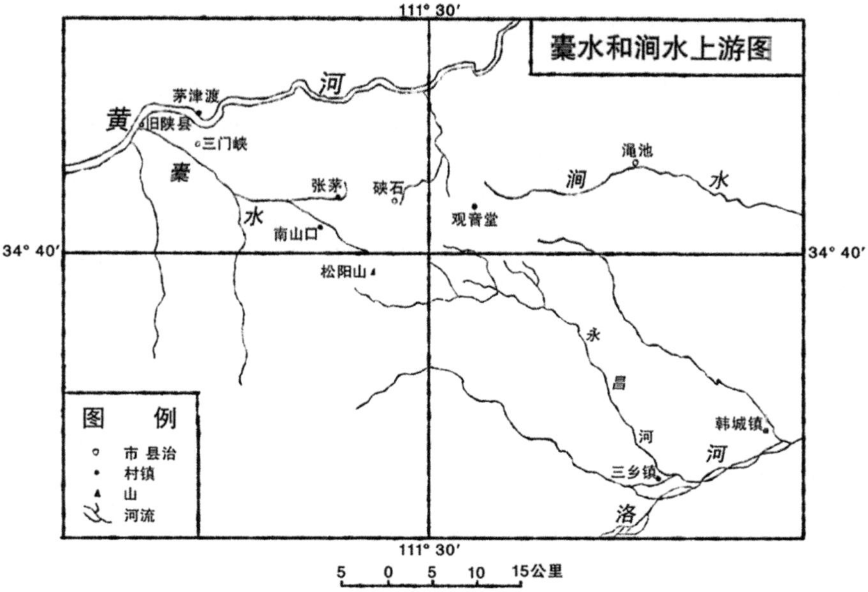

图例 - Legend

市县治 City/County Seats

村镇 Towns and Villages

山 Mountains

河流 Rivers

Fig. 4.6 The upper reaches of the Tuo River and Jian River

The Siltation of the Central Plains Canal

Following Emperor Wu of Jin's diversion of the Yellow River into the Luo River, only three or four decades later, the infamous Yongjia Turmoil broke out (Yongjia being the reign title of Emperor Huai of Jin, 307–313 CE). During this time, various non-Han peoples — including the Xiongnu, Xianbei, Di, Jie, and Qiang — who had long resided throughout northern China, rose up simultaneously, taking advantage of the Jin dynasty's declining power. The entire region north and south of the Yellow River was plunged into war. Continuous and protracted fighting rapidly reduced once-prosperous cities to ruins, while formerly fertile lands turned desolate. The human suffering — mass displacement and death — was overwhelming, reaching a scale scarcely captured in the historical records.

As the remnants of the Jin court retreated south of the Yangtze and Huai Rivers, the Central Plains were effectively abandoned. With no one left to maintain them, the laboriously excavated canals — once crucial arteries for transportation and grain shipment — inevitably silted up and fell into disuse.

Huan Wen Excavates a Waterway East of Juye Marsh

After the Eastern Jin court settled in the south, it continued to harbor ambitions of reclaiming the north. However, constrained by limited national strength, it refrained from acting rashly. It was not until more than fifty years later, under the leadership of Huan Wen, that a serious campaign to recover lost territories was finally undertaken. Huan Wen led three northern expeditions: the first advanced from Jiangling toward Guanzhong; the second from Jiangling northward to Luoyang. The passage in question concerns his third campaign, which targeted the city of Ye. This took place in the fourth year of the Taihe reign of the Deposed Emperor of Jin (369 CE).

For this campaign, Huan Wen divided his forces along two routes. One contingent followed the Guo River northward, aiming to open up the Shimen area near Xingyang and secure a supply line. The other detachment advanced from the Huai River into the Si River, then turned into the Ji River, launching a northern assault on Ye. At the time, the dominant northern state was Former Yan, with its capital at Ye.[68] Huan Wen ordered his subordinate general Yuan Zhen to take the Guo River route, while he himself proceeded via the Huai and Si Rivers.

Why did he not utilize the Bian Canal route? Although the Bian Canal had not yet fully silted up, its water levels were too low to support large-scale transport.[69] To divert water eastward, it was necessary to open the Shimen region near Xingyang. Therefore, Huan Wen assigned Yuan Zhen to advance via the Guo River. Yet even that waterway did not extend continuously to Shimen.

While Huan Wen successfully crossed the Huai River and entered the Si River, the old channel of the He River—which had once connected the Si and Ji Rivers—had completely disappeared due to flooding from the Yellow River and overflow from Juye Marsh.[70] Yuan Zhen dredged the upper reaches of the Guo River, but this proved insufficient, forcing Huan Wen to excavate a new canal east of Juye Marsh. This newly cut watercourse was later known as Huangong Du or Huangong Gou ("Duke Huan's Canal").

The total length of this canal was approximately 300 *li* (about 150 km), and it can be divided into two sections, south and north, separated by a body of water known as Xuexun Marsh, located in the northeastern part of present-day Juye County. The waters of the marsh bifurcate into two branches. The southern branch merges with the Yellow Water flowing from Juye Marsh, then flows southeast past present-day Jinxiang County (to the north of the county seat), eventually joining the Si River. Huan Wen began his excavation at this point in Jinxiang.

---

68 Biography of Huan Wen. In *Book of Jin*, vol. 98.

69 Biography of Xi Jian, with the "Attached Biography of Xi Chao". In *Book of Jin*, vol. 67.

70 The *Book of Song*, in the "Biography of He Chengtian," records: "In the nineteenth year of the Yuanjia reign (442 CE), He Chengtian submitted a memorial entitled *Discussion on Securing the Border*, which stated: 'The Juye Marsh is vast in extent, linking to the Zhu and Si Rivers in the south and connecting with Qing and Qi in the north. An old county seat lies within the marsh itself.'" Juye Marsh is in fact the same as Daye Marsh. It was not originally this large; its vastness resulted from repeated overflows of the Yellow River and the silting up of the He River over time.

The northern branch flows past the walled city of Kanxiang, merges with the Wen River, and continues northward to join the Ji River. This section, approximately 120 *li* in length, was historically known as the Hong River.[71] It is important to note that the "Ji River" referred to here denotes the watercourse formed by the confluence of waters from Juye Marsh and the Wen River.[72] The original section of the Ji River—from its divergence from the Yellow River down to Juye Marsh—had long since silted up and dried. The use of "Ji River" for the eastern section flowing to the sea persisted merely as a geographical convention.

At that time, the Ji River and the Yellow River were closest to each other at the site known as Sidukou, and there existed interconnected waterways between them.[73] Sidukou was located in the eastern part of present-day Linyi County, Shandong Province. As noted by Li Daoyuan, "From the Yellow River one could enter the Ji River, from the Ji River proceed to the Huai River, and from the Huai River reach the Yangtze River directly – the waterways were continuously linked, hence the name Sidukou."[74] Sidukou was likely a key hub of north–south transportation at the time. Traveling upriver from there brought one close to Qiao'ao Fortress, which was located southwest of present-day Chiping County, Shandong. In that period, it stood along the banks of the Yellow River and served as a major military stronghold during the Eastern Jin and Northern and Southern Dynasties. Due to its strategic position overseeing Sidukou – the pivotal junction for north–south water transport – it played a vital role in regional control. It is said that during Huan Wen's northern expedition, "his fleet entered the Yellow River from the Qing River, the ships stretching in an unbroken line for hundreds of *li*."[75] The route they took must have been along this very waterway (See Fig. 4.7).

The Canal Excavated by Diverting Water From the Guang River

Another canal located near the Huangong Canal deserves mention. Well before Huan Wen's northern campaign, Murong Jun of the Former Yan had captured Jizhou and launched an attack against Duan Kan in Qingzhou (present-day southeast Zibo, Shandong). In response, Emperor Mu of the Eastern Jin dispatched Xun Xian to lead a relief expedition. At that time, Murong Jun had stationed Murong Lan with tens of

---

[71] *Water classic: Commentary on* the Ji River; see also Annals of Jin, supplementary notes by Hu. In *Comprehensive mirror to aid in government*, vol. 102.

[72] *The Tongdian (vol. 172), "Administrative Divisions II"* states: "The Qing River is in fact the confluence of the Heze and Wen Rivers." However, Heze lies to the southwest of Juye Marsh, and the waters of the Heze River cannot pass through Juye Marsh to merge with the Wen River; therefore, the text is referring to Juye Marsh itself. The "Qing River" mentioned here actually refers to the Ji River downstream from Juye Marsh.

[73] *Commentary on* the Ji River. In *Water classic.*

[74] *Commentary on* the Yellow River. In *Water classic.*

[75] Chronicle of Jin, part 24. In *Comprehensive mirror for aid in government,* vol. 102.

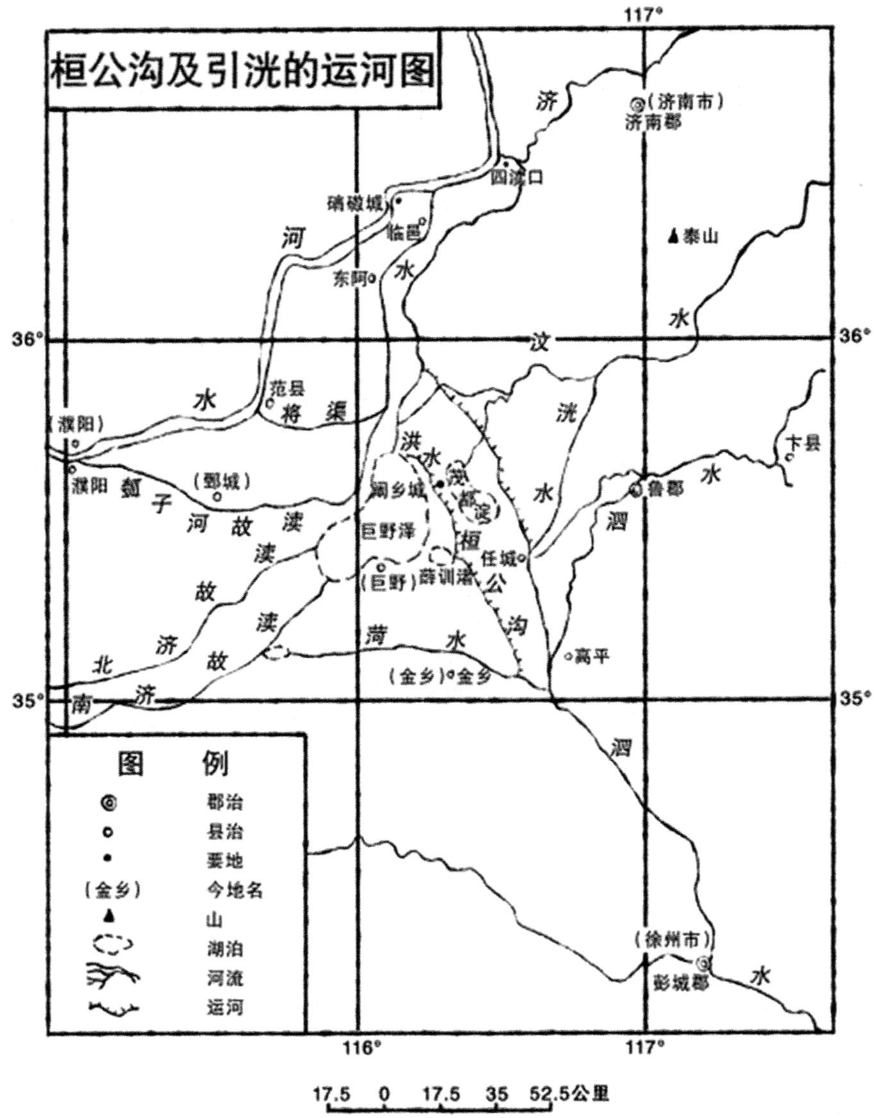

图例 - Legend

郡治 Commandery Seats

县治 County Seats

要地 Strategic Locations

今地名 Present Names

山 Mountains

湖泊 Lakes

河流 Rivers

运河 Canals

Fig. 4.7 Duke Huan's Canal and the Canal diverting the Guang River

thousands of troops in Biancheng, posing a significant threat to the frontier. Xun Xian diverted water from the Guang River to excavate a canal that connected to the Wen River, enabling the transport of troops northward as far as Dong'e for the offensive.[76]

This "Biancheng" does not refer to present-day Yifeng County in Kaifeng, Henan, but rather to Bian County, which belonged to Lu Commandery at the time.[77] The ruins of Bian County are located east of present-day Sishui County, Shandong. The Guang River was a tributary of the Wen River and flowed east of modern Ningyang County, Shandong. Dong'e corresponds to an area southwest of present-day Dong'e County, Shandong. When Xun Xian launched his northern expedition, Murong Lan likely retreated from Bian County to Dong'e, which implies that Xun Xian's troops advanced to that location. The canal thus excavated would have likely drawn from the Wen River, crossed the Ji River, and reached the vicinity of Dong'e.

Nearby Dong'e lay the former course of the Huzi River. The lower reaches of the Huzi River once converged with the Jiang Canal. After the Huzi River dried up, the Jiang Canal remained navigable. The Jiang Canal, located in what was then Fan County (corresponding to present-day eastern Fan County, Henan), was a distributary of the Yellow River. After merging with the Huzi River, it flowed south of Dong'e County and then northeastward, ultimately emptying into the Yellow River to the east of Qiao'ao City.[78] Given that Xun Xian's canal extended to Dong'e, it likely joined with the Jiang Canal.

Although this watercourse is not mentioned by Li Daoyuan, if the account of Xun Xian's canal construction is not apocryphal, the existence of such a canal is highly plausible. It would have served as an additional strategic node along the north–south water transport route, beyond the confluence known as Sidukou. Although Sidukou lies at some distance from Qiao'ao City, this canal was situated much closer and was likely of equal—if not greater—strategic importance. Its existence would have further enhanced Qiao'ao's military significance. For offensives launched from the south, naval forces could enter the Yellow River directly through this waterway; conversely, northern defenders could leverage it to block or control southern fleets ascending the river.

Since Huan Wen's naval forces entered the Yellow River from Qing River, they must have passed through the Sidukou to reach the river, likely bypassing the canal excavated by Xun Xian. Whether they entered the Yellow River via the Sidukou or further downstream near Qiao'ao City, both routes entailed a longer and more circuitous journey. Moreover, due to the limited water flow in the canals, transportation would have been significantly slower. As previously noted, Huan Wen's fleet entered the river from Qing River, with vessels stretching for hundreds of li. Such a laborious and protracted logistical operation inevitably led to delays and hindered

[76] Annals of Emperor Mu; Biographies of Xun Song and Xun Xian. In *Book of Jin*, vol. 8, 75.

[77] Annals of the Jin dynasty, Part 22, with commentary by Hu. In *Comprehensive mirror to aid in government, vol. 100*. According to Hu's note, this refers to Bian County of the State of Lu. This should be understood according to the administrative divisions of the Han period. By the Western Jin, the State of Lu had already been reorganized into Lu Commandery, as recorded in *Book of Jin (vol. 14): Treatise on geography, part I*.

[78] Commentary on the Yellow and Guzi Rivers. In *Water Classic*.

the overall campaign.[79] As Huan Wen continued upstream along the Yellow River and advanced to Fangtou,[80] Yuan Zhen had still not succeeded in opening the Shimen Pass. As a result, the supply of provisions was cut off, and the army faced severe food shortages, forcing a full-scale retreat. On the return journey, they took the overland route, but near Xiangyi (present-day Sui County, Henan), they were overtaken by forces of the Former Yan and suffered a major defeat.

The Qingzhou Gu Constructed by Xie Xuan

The Battle of Fei River marked a pivotal turning point for the Eastern Jin dynasty. After defeating the formidable Former Qin ruler Fu Jian, the Jin forces gained renewed confidence and sought to press their advantage by launching a northern campaign to reclaim long-lost territories. General Xie Xuan led his troops northward, first halting at Pengcheng before advancing to capture the Former Qin's Yanzhou, whose administrative center lay north of present-day Juancheng in Shandong.

During this campaign, faced with treacherous waterways and logistical challenges in grain transport, Xie Xuan adopted the proposal of command officer Wenren Shi to construct water-control infrastructure at Lüliang. Wooden barricades were erected, and seven embankment dams were built, forming a regulated watercourse referred to as *gu* (派), intended to facilitate grain shipment by water. As the next target was Qingzhou (administered from the northwest of present-day Yidu County, Shandong), the watercourse came to be known as the "Qingzhou Gu."[81]

The so-called "Lüliang waters" in this context refer to the Si River as it flows through the Lüliang region, located east of present-day Xuzhou in Jiangsu. This section of the Si River is strewn with massive boulders, creating a hazardous landscape of surging torrents and steep currents—an area historically known for its treacherous navigation. Near Lüliang, a distributary known as the Dingxi branches off from the Si River and eventually rejoins it downstream. During the winter and spring months, this segment of the river often becomes shallow and silted, requiring dredging to remain navigable. Consequently, traffic often diverted to the Dingxi waterway.

Xie Xuan's seven embankment dams were constructed along this difficult stretch to enhance navigation and support grain transport.[82] Some historians believe that Xie excavated a canal at this site—referred to as the Lüliang Weir Canal—and contend

[79] Biography of Xi Jian, with the Attached Biography of Xi Chao. In *Book of Jin*, vol. 67.

[80] Fangtou was located southwest of present-day Xun County, Henan Province, at the confluence where the Qi River flows into the Bai Canal. The name "Fangtou" originated during the time of Cao Cao, who diverted the Qi River into the Bai Canal and placed large wooden barriers there to control the water flow, hence the name.

[81] Biographies of Xie An and Xie Xuan. In *Book of Jin*, vol. 79.

[82] Commentary on the Si River. In *Water classic*.

that with its completion, the grain transport network between the Jianghuai region and the Huai-Yellow River corridor had essentially taken shape.[83]

However, this undertaking was more accurately a project of river channel management rather than true canal excavation. A waterway that has only been regulated cannot be properly called a "canal," particularly as such terminology did not exist at the time. Even if a canal had been dug, it would be an overstatement to claim that the transport system linking the Jianghuai region and the Huai-Yellow corridor had reached full maturity.

Liu Yu Reopened the Bian Canal

After Huan Wen's defeat and retreat, there were several subsequent efforts to reassert control over the Central Plains, though all were relatively small in scale. It was not until Liu Yu rose to power that a major northern expedition was launched once again. Unlike Huan Wen, whose target was the Former Yan entrenched in Ye, Liu Yu aimed to conquer the Later Qin, which then held sway over the Guanzhong region.

Liu Yu's campaign consisted of multiple prongs, two of which were naval forces. One contingent, led by Shen Linzi and Liu Zunkao, set out from Shimen and entered the Yellow River via the Bian River. Their objective was to reopen the Shimen waterway to restore the transport capacity of the Bian Canal. Another contingent, under Wang Zhongde, began dredging work at the Juyue Marsh to connect it with the Yellow River.

Liu Yu himself, following Huan Wen's earlier route, led his own fleet northward, traveling up the Si River into the Qing River, and planned to advance westward along the Yellow River.[84] Because his route mirrored that of Huan Wen, it necessarily passed through the Huangong Canal, located to the east of the Juyue Marsh. As the vanguard, Wang Zhongde led the effort to dredge the Juyue Marsh—essentially to restore the Huangong Canal.[85]

After traversing the Huangong Canal, Liu Yu turned into the Ji River and advanced toward Qiao'ao City via the confluence at Sidukou. As the Ji River approached Sidukou, there stood a city called Luodang on its eastern bank. During the campaign, Liu Yu ordered Yuan Miao, then magistrate of Chang'an, to station troops there, and the city subsequently became known as Yuan Miao City.[86]

Upon reaching Qiao'ao, Liu Yu appointed Xiang Jing as the local commander.[87] Wang Xuanmo, who was the first to breach the city's defenses, was also appointed

[83] Editorial Team on the Grand Canal of the History Department, Yangzhou Normal College. The North–South canal before the Sui dynasty. *Jianghai Academic Journal*, 1961 (1).

[84] Chronicle of Jin, part 39. In *Comprehensive mirror for aid in government,* Vol. 117.

[85] Commentary on the Ji River. In *Water classic*.

[86] Ibid.

[87] Biography of Xiang Jing. In *Book of Song*, vol. 45.

to stay and guard it.[88] These details not only underscore the strategic importance of Qiao'ao but also help trace the route taken by Liu Yu's naval force.

Since Huan Wen's northern expedition, all subsequent campaigns from the Eastern Jin and Southern Dynasties into the north relied on naval operations, which were necessarily tied to navigable waterways. Qiao'ao was a vital node along this strategic water route. Control of Qiao'ao often determined the outcome of the entire campaign. During Xie Xuan's later expedition north,[89] Liu Laozhi also advanced and temporarily held Qiao'ao, likewise moving along this same water route.[90] The repeated use and contestation of this city confirm that the Huangong Canal remained navigable and operational for a significant period of time.

After Liu Yu's defeat of the Former Qin, all territories south of the Yellow River fell under Jin control. Upon returning from Chang'an, Liu Yu, unwilling to take the detour via Qiao'ao City again, ordered the reopening of the Bian Canal.[91] Previously, Huan Wen had been unable to make use of this canal due to his failure to secure Shimen, which controlled the canal's upper reaches. However, having secured victory in his northern campaign, Liu Yu faced no such obstacles, and the restoration of the canal yielded the desired results. Nevertheless, the actual reopening of Shimen should primarily be attributed to Shen Linzi. Among the various armies under Liu Yu's command during the northern expedition, it was Shen Linzi and Liu Zunkao who led the naval forces advancing from Shimen and entering the Yellow River via the Bian Canal. Had Shimen not been opened, their advance would have been impossible. Moreover, Liu Yu's return journey – from Luoyang to Pengcheng (modern-day Xuzhou, Jiangsu) – took barely over a month, a pace unachievable without efficient water transport. It should be noted that the Bian Canal dredged during Liu Yu's campaign had a slightly altered course compared to earlier times. In the distant past, when the Hong Canal system's tributaries flowed eastward, the canal emerged together with the Ji River. During the Eastern Han period, in order to control the flow of water, a sluice gate known as Shimen was constructed at the point of divergence – this was the origin of Shimen in Yingkou. At that time, water diverted from the Yellow River flowed eastward along the southern flank of Mount Guangwu. When Shen Linzi and Liu Zunkao reopened Shimen, torrents once again surged eastward. However, due to landslides that obstructed the flow, the water became blocked. As a result, a new canal was excavated approximately ten *li* (roughly five kilometers) to the north of the original one, restoring the flow to normal.[92] Once the Bian Canal was reopened, Liu Yu traveled downstream from Guanzhong, returning to Pengcheng via the canal. However, following his return, the newly recovered northern territories gradually fell once again into enemy hands, and the benefits of transportation along the Bian Canal ultimately came to be enjoyed by the Northern Wei.

---

[88] Biography of Wang Xuanmo. In *Book of Song*, vol. 73; Commentary on the Yellow River. In *Water classic*.

[89] Biographies of Xie An and Xie Xuan. In *Book of Jin*, vol. 79.

[90] Biography of Liu Laozhi. In *Book of Jin*, vol. 84.

[91] Annals of Emperor Wu II. In *Book of Song*, vol. 2.

[92] Commentary on the Ji River. In *Water classic*.

The Canal Infrastructure of the Northern Wei Dynasty

After the turmoil of the Yongjia period, the construction of canals and the dredging of waterways were primarily undertaken by the Eastern Jin and Southern Dynasties. In contrast, such efforts were rarely seen in the north, likely due in part to the cultural characteristics of the newly established northern ethnic groups. These northern peoples were adept at mounted archery and naturally had little interest in waterways or ships. Even when water transport was occasionally employed, it was typically based on existing natural rivers rather than newly excavated canals. Following the Northern Wei's relocation of its capital to Luoyang, attention began to shift toward these infrastructural concerns.

On one occasion, Emperor Xiaowen traveled to Xuzhou (modern Xuzhou, Jiangsu Province) and, upon returning to Luoyang, chose to travel by boat, heading north along the Si River. His decision to take the water route suggests that he followed the Huan Gong Canal, which had been excavated and dredged earlier by Huan Wen and Liu Yu. This route connected the Si River to the Qing River and from there to the Yellow River, with the imperial entourage encamping at Qiao'ao.

At that time, Cheng Yan, who was accompanying the emperor, submitted a memorial expressing concern over the swift current of the Yellow River and the potential danger of capsizing. Emperor Xiaowen, however, responded that he believed the impoverishment of the old capital at Pingcheng (modern Datong, Shanxi) was due to the lack of navigable transport routes. Now that the capital had been moved to the Yi and Luo region, his goal was to establish transportation links with all regions. He noted that the swift current of the Yellow River had long been viewed as an obstacle to crossing and that, by choosing to travel by boat, he intended to demonstrate that such concerns were unfounded.[93] Later, while in Luoyang, Emperor Xiaowen also proposed excavating a canal alongside the Luo River. He envisioned that, in the event of a future southern campaign, troops could conveniently enter the Luo River nearby, proceed from the Luo into the Yellow River, then from the Yellow River into the Bian River, and from there into the Qing River, eventually reaching the Huai River. In this way, soldiers could disembark directly into battle—as though stepping out the door to engage the enemy – offering exceptional convenience. He estimated that constructing the canal along the Luo River would require fewer than twenty thousand laborers and could be completed within sixty days.[94] The "Qing River" mentioned here refers to the Si River.

The proposed canal project essentially aimed to revive the earlier systems established during the Eastern Han and Wei-Jin periods. It also indicates that after Liu Yu reopened the Shimen (Stone Gate), the Bian Canal did not remain continuously navigable. Had the canal been fully restored at that time, Emperor Xiaowen would not have needed to take a detour via the Si and Yellow Rivers on his return journey from Xuzhou to Luoyang, nor would such elaborate plans have been necessary. In fact, during that period, efforts were made not only to restore the Bian Canal but also

---

[93] Biography of Cheng Yan. In *Book of Wei*, vol. 79.

[94] Biography of Li Chong. In *Book of Wei*, vol. 53.

to rehabilitate the Cai River.[95] The Cai River was formerly known as the Langtang Canal, a part of the ancient Hong Canal system. With the restoration of both the Bian and Cai Rivers, the connecting waterways – namely the Sui, Huan, and Guo Rivers – could once again flow freely. This clearly demonstrates the degree of attention the Northern Wei paid to canal infrastructure following the relocation of their capital. However, these projects were primarily undertaken for military purposes, specifically to facilitate southward campaigns, rather than as part of a long-term developmental strategy. By the end of the Northern and Southern Dynasties period, canal transportation had significantly declined.[96]

[95] Biography of Cui Liang. In *Book of Wei*, vol. 66.

[96] *The History of Song (vol.95): Treatise on canals and waterways* records the Northern Wei project in the second year of the Zhengshi reign (521 CE), when Yuan Qing undertook the dredging of the Yongfeng Canal. For the sake of clarity, this will be discussed in the following chapter.

Chapter 5
The Excavation of Canals During the Sui Dynasty and Its Impact

The Sui Dynasty's Efforts in Canal Excavation

While canal construction certainly played a vital role in facilitating the unification of the country, it became even more essential after unification, serving as the backbone of an integrated national transportation system. Following the unification of China, the Sui Dynasty devoted immense effort to canal excavation, aiming to establish a comprehensive transportation network that would further consolidate the stability and foundations of the state.

Any discussion of the canals of the Sui Dynasty inevitably calls to mind Emperor Yang of Sui—just as mention of the Great Wall evokes the First Emperor of Qin. It is true that Emperor Yang invested the greatest effort in canal construction, to such an extent that these grand projects ultimately contributed to the downfall of the dynasty. Yet the origins of canal excavation during the Sui period trace back not to Emperor Yang, but to his father, Emperor Wen. It was Emperor Yang who inherited and expanded upon Emperor Wen's vision, bringing to completion the ambitious plans that had been set in motion before his reign.

The Guangtong Canal

As previously mentioned, although Chang'an in the Guanzhong region was an ideal location for a capital, the Guanzhong Plain itself was too small to supply sufficient grain to meet the needs of the capital's population. Therefore, establishing the capital at Chang'an necessarily entailed resolving the issue of grain transport via canals. This

N. Shi, *A Historical Geography Research of Canals in China*,
https://doi.org/10.1007/978-981-95-3517-0_5

was the case in the Qin and Han periods and remained so during the Sui and Tang dynasties as well.[1]

In the early Sui period, conditions resembled those of the Western Han: the perilous *Dizhu* Rapids continued to obstruct grain boats, and the winding, silt-laden Wei River made navigation extremely difficult. Upon ascending the throne, Emperor Wen was acutely aware of the seriousness of this issue. In the fourth year of the Kaihuang reign (584 CE), he commissioned Yuwen Kai to excavate the Wei Canal.[2] The imperial edict issued at the time stated: "The Wei River's water levels are unstable, and the channel is shallow and filled with sand, creating frequent blockages. Though the route spans merely a few hundred li, the seasonal fluctuations hinder back-and-forth navigation, and the effort required from boatmen is considerable."[3] The canal under Yuwen Kai's supervision was designed to "divert water from the Wei River, running eastward from Daxing City to Tongguan, over a distance of more than 300 *li*."[4] Daxing City, later known as Wannian County in the Tang dynasty, was located directly east of Chang'an County. Another contributor to the project was Guo Yan, who excavated a canal that "diverted water from the Wei River north of Daxing City, extending eastward to Tongguan, serving as a grain transport route over 400 li in length."[5]

These two accounts differ, leaving the canal's exact starting point open to debate. According to Song dynasty sources, it began at Xingcheng Weir on the Wei River, northwest of Chang'an County.[6] When the grain canal was restored during the mid-Tang period, construction also resumed at this weir. Notably, this site had been the starting point of the grain canal since the Han dynasty and remained so through the Sui.

Topographically, the weir was located eighteen *li* west of the Tang-era Xianyang County.[7] The ancient city of Xianyang County was situated east of modern-day Xianyang City, and Xingcheng Weir is believed to correspond to present-day Diaoyutai in Xianyang. The canal proceeded north of Daxing City, flowed eastward to Tongguan, and emptied into the Yellow River.[8] The point of confluence likely matched that of the Han dynasty grain canal—near the present-day Diaoqiao area in Tongguan County. Some contend that the Sui grain canal entered the Wei River near present-day Huayin County. However, this claim contradicts contemporary Sui sources and overlooks the fact that the elevated terrain northeast of Huayin would

[1] Between the Qin-Han and Sui-Tang periods, other states such as Former Qin, Later Qin, Western Wei, and Northern Zhou also established their capitals at Chang'an. However, given the fragmented nature of these regimes, their territorial scales were quite limited and cannot be compared on equal terms.

[2] Treatise on food and finance; Biography of Yu Wenkai. In *Book of Sui*, vols. 24, 68.

[3] Treatise on food and finance. In *Book of Sui*, vol. 24.

[4] Ibid.

[5] Biography of Guo Yan. In *Book of Sui*, vol. 61.

[6] Song Minqiu. Chang'an. In *Gazetteer of Chang'an*, vol. 12.

[7] Biography of Li Shi. In *Old book of Tang*, vol. 172.

[8] Annals of Gaozu I; Biography of Yu Wenkai. In *Book of Jin*, vol. 1.

have been difficult to traverse. This same challenge applied during the Han dynasty, as discussed earlier.

Construction of this canal began in the fourth year of Emperor Wen's reign during the Kaihuang era, in June. By September of the same year, Emperor Wen visited Ba Water to inspect the canal,[9] indicating that the excavation work had likely been completed. The canal's starting and ending points correspond to those of the Han dynasty canal, and the brevity of the construction period suggests it was carried out in accordance with the established Han regulations. A slight difference between the Sui canal and the Han canal is that the latter, besides drawing from the Wei River, also utilized water from Kunming Pool as a supplementary source, whereas the Sui canal relied solely on the Wei River without diverting additional water sources.

This canal was known as the famous Guangtong Canal, a name likely derived from its passage beneath the Guangtong Granary in Huazhou (with its administrative center in present-day Huaxian, Shaanxi). Established in the third year of the Kaihuang reign (583 CE), the Guangtong Granary served as a key storage hub for waterborne grain transport in the Guanzhong region. It was of equal importance to the Liyang Granary in Weizhou (present-day Qixian, Henan), the Heyang Granary in Luozhou (present-day Luoyang, Henan), and the Changping Granary in Shanzhou (present-day Sanmenxia, Henan).[10]

In the fourth year of the Renshou reign (604 CE), the canal's name was changed following the death of Emperor Wen and the accession of Emperor Yang. To avoid the taboo associated with Emperor Yang's personal name, the canal was renamed the Yongtong Canal. The opening of this canal effectively resolved the grain supply issues facing the capital, Chang'an, and brought tangible benefits to the local populace along its course. As a result, it was also known as the Fumin Canal ("Canal to Enrich the People").[11] (See Fig. 5.1).

---

[9] Ibid.

[10] Treatise on food and finance. In *Book of Sui*, vol. 24.

[11] Regarding the name Guangtong Canal, there are differing accounts in the Book of Sui, specifically in the biography of Guo Yan: "He was appointed as the chief supervisor for canal construction, leading workers to excavate a canal that diverted water from the Wei River, running north of Daxing City eastward to Tongguan, forming a transport waterway over 400 li long. The region inside the pass relied on this canal, which was named the Fumin Canal ('Canal to Enrich the People')." The canal described here was built in the fourth year of the Kaihuang era and lies between the Wei River's southern bank, Daxing City (i.e., Chang'an), and Tongguan—clearly the same canal constructed by Yuwen Kai and from the same period. It was also called the Fumin Canal due to its benefit to the populace. *The History of Northern Dynasty: Guo Yan's biography* refers to it as the "Furen Canal," a name altered in the Tang dynasty to avoid the taboo associated with the character "民" (meaning "people"). *Gazetteer of Chang'an* entry for the Yongtong Canal explicitly identifies it as the Fumin Canal. According to the *Book of Sui, Annals of Gaozu*, in the second year of Kaihuang, a canal was opened diverting Du Yang water to the Sanzhiyuan area. This was the first canal excavation in the Sui dynasty, but it was on a smaller scale, likely intended for irrigation rather than transportation.

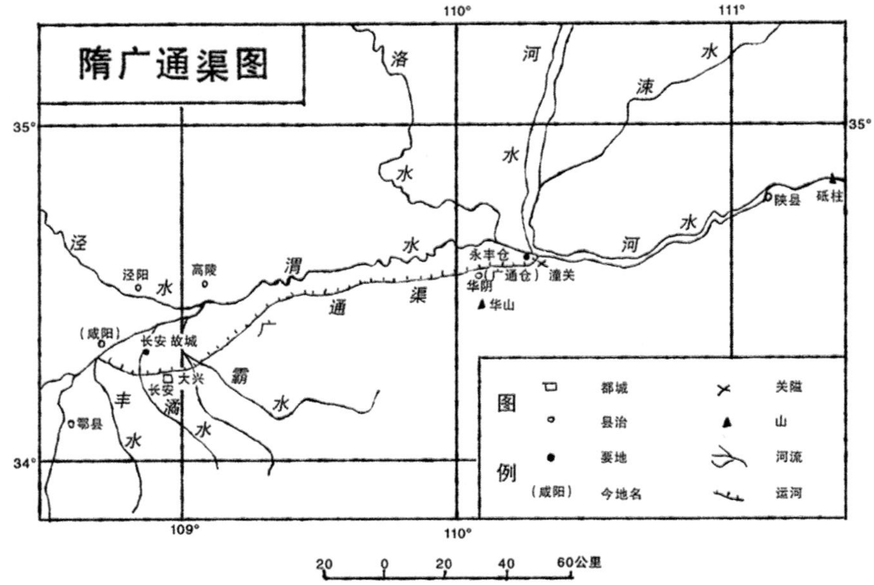

图例 - Legend
都城 Capital Cities
县治 County Seats
要地 Strategic Locations
今地名 Present Names
关隘 Mountain Passes
山 Mountains
河流 Rivers
运河 Canals

Fig. 5.1 The Guangtong Canal during the Sui Dynasty

The Shanyang Canal

In addition to excavating the Guangtong Canal, Emperor Wen of the Sui Dynasty also undertook the construction of the Shanyang Canal. The excavation of Shanyang Canal was part of the preparations for the campaign against the Chen Dynasty; construction began in the seventh year of the Kaihuang era (587 CE), and by the eighth year, a large-scale military expedition was launched.[12] The Shanyang Canal extended southward from the present-day eastern part of Huai'an County in Jiangsu Province, following approximately the former course of the Han Canal.

[12] According to *Annals of Emperor Gaozu, II. Book of Sui, vol. 1*: "In the seventh year of the Kaihuang reign (587 CE), a canal was excavated in Yangzhou – Shanyang Canal– to facilitate grain transport." As noted in the *Book of Sui (vol. 3): Treatise on geography*, Yangzhou at the time referred to Jiangdu Commandery, which governed Jiangdu (modern-day Yangzhou, Jiangsu) and Shanyang (modern-day Huai'an, Jiangsu) counties. The Shanyang Canal was thus located between these two counties.

The Han Canal originally entered the Huai River at the Mokou site, but it appears that by this time Mokou had become silted up or obstructed.[13] As a result, Emperor Wen was compelled to open a new water outlet. The excavation of this ditch was carried out over an extremely brief period, and the actual construction fell short of expectations. Although the ditch was intended to serve the campaign against the Chen Dynasty, it was not effectively utilized in the military operations; part of the southern expeditionary fleet bypassed the canal entirely and traveled via the East China Sea instead of navigating the Han Canal.[14] Eventually, Emperor Yang would undertake a new round of excavation for the same purpose.

The Canals Constructed by Emperor Yang of the Sui Dynasty

Emperor Yang of the Sui Dynasty personally oversaw the construction of four major canals: the Tongji Canal, the Yongji Canal, the Han Canal, and the Jiangnan Canal. Although the Han Canal was based on the earlier work of King Fuchai of Wu, the scale of its reconstruction during the Sui made it nearly equivalent to the excavation of an entirely new waterway. These four canals can be grouped into two distinct systems. The Tongji Canal connects downstream to the Han Canal, which in turn connects to the Jiangnan Canal—together forming a continuous network. The Yongji Canal, on the other hand, functioned as a separate route extending northward to Zhuojun (Zhuojun under the Sui was administered from Ji County, corresponding to present-day Beijing). The former system provided a direct link to the Yangtze River Delta and the Tai Lake region, while the latter ran vertically through the plains east of the Taihang Mountains.

The excavation of these canals served not only economic objectives but also political and military purposes. Although the Eastern Jin and Southern Dynasties controlled only part of China, the economy of the Yangtze River Basin steadily developed, with the Yangtze River Delta and the Taihu Lake region becoming especially affluent. As a historical source observes, "the land is vast and fertile, the people

---

[13] During the Eastern Han period, Chen Min opened a new watercourse at the northern end of the Han Canal. Over time, the original outlet at Moukou gradually silted up and became impassable. Chen Min's undertaking was discussed earlier in the text.

[14] The *Annals of Emperor Gaozu, Part II* in *Book of Sui* (Volume 2) records that during the Sui campaign against the Chen dynasty, the army advanced along eight routes. Among the three eastern routes, Prince Jin (Yang Guang) set out from Liuhe (present-day Liuhe, Jiangsu); Duke of Xiangyi, He Ruobi, departed from Wuzhou (present-day Yangzhou); and Duke of Luocong, Yan Rong, began from Donghai (present-day Lianyungang, Jiangsu). Wuzhou being equivalent to Yangzhou, it is clear that He Ruobi advanced southward via the Han Canal. Prince Jin, departing from Liuhe, likely crossed the Huai River and then proceeded overland. As for Yan Rong, his forces took a detour by sea.

diligent in agriculture; in years of good harvest, several commanderies may be spared from hunger,"[15]—an apt reflection of the region's prosperity.

Yangzhou (with its administrative center at Jiankang, present-day Nanjing, Jiangsu) encompassed the fertile lands of the Wu region. The area was rich in resources—fish, salt, catalpa wood, and Chinese arborvitae—distributed widely throughout the empire. Its abundant production of silk, cotton, and cloth could clothe the entire realm,[16] a level of wealth unmatched by other regions. As early as the Northern Wei period, the imperial court had established markets along its southern frontier specifically to acquire such southern goods.[17] The term "southern goods" largely referred to commodities originating from the Yangtze River Delta and the Taihu Lake region.

The southern terminus of the Han Canal was located in Jiangdu Commandery (modern-day Yangzhou), and the Jiangnan Canal encircled much of the Taihu Basin. The rationale for constructing these waterways, therefore, becomes readily apparent.

Although the Northern Qi regime was geographically confined to the lower reaches of the Yellow River, the region east of the Taihang Mountains remained economically vibrant. Within its territory, the Northern Qi government established official granaries at key river crossings and ferry points across various prefectures to store grain. In years of bumper harvests, the price of one dou of grain fell to as low as nine *qian*—an exceptionally low price. Along the coast, including in Youzhou (centered in present-day Beijing), the salt industry flourished, generating sufficient revenue to support the state and military.[18] Although this region's economy could not rival the Yangtze Delta or the Taihu Basin, the area east of the Taihang Mountains still represented a critical grain-supplying region for the early Sui dynasty.[19] The construction of the Yongji Canal was a direct response to this economic reality.

The eastern Taihang region had once been the heartland of the Northern Qi, while the Yangtze Delta and Taihu Basin had belonged to the Chen dynasty. Although both states had been extinguished, the Sui dynasty remained acutely aware of the residual risks these former territories posed. Suspicion toward Shandong natives (here referring to the region east of the Xiaoshan Mountains, distinct from modern Shandong Province) was particularly evident. During Emperor Yang's reign, a memorial to the throne complained that "the court is now filled with men from Shandong," who "form cliques, promote one another, flatter subordinates and deceive superiors, acting in partisan self-interest."[20] As this region encompassed the former territories of Qi and Chen, such anxieties naturally extended to the local populations and former administrative elites. The construction of canals linking these former centers of power was therefore not coincidental—it was both a strategic and symbolic act to integrate and control them more effectively.

[15] Biography of Kong Jigong and Others. In *Book of Song*, vol. 54.

[16] Ibid.

[17] Treatise on food and finance. In *Book of Wei*, vol. 110.

[18] Treatise on food and finance. In *Book of Sui*, vol. 24.

[19] Ibid.

[20] Biography of Wei Yunqi. In *Old book of Tang*, vol. 37.

The Yongji Canal also had military significance, especially in relation to the planned campaigns against Goguryeo. It provided a logistical backbone for transporting supplies eastward. Nevertheless, among all the major canals, the Tongji Canal and Han Canal received the most attention from Emperor Yang. By this time, the Yangtze Delta and the Taihu region had emerged as prosperous economic zones, and Yangzhou had become one of the most flourishing urban centers in the empire. This wealth and vibrancy compelled Emperor Yang to prioritize canal construction in the region and to personally inspect and enjoy its grandeur.

In the fourth month of the first year of the Daye era (605 CE), construction began on the Tongji Canal. Simultaneously, Emperor Yang dispatched agents to Jiangnan to procure timber and construct tens of thousands of vessels—dragon boats, phoenix barges, yellow dragon ships, red-painted warships, and multi-decked tower boats. By the eighth month of the same year, he had boarded a dragon boat and embarked on a pleasure tour to Jiangdu.[21]

The Tongji Canal

Among the various canals constructed under Emperor Yang of the Sui Dynasty, the Tongji Canal was the earliest. Work on the canal commenced in the first year of his reign (605 CE), and it was made navigable within the same year. Upon his accession, Emperor Yang designated Luoyang as the Eastern Capital,[22] and thus the Tongji Canal began from Luoyang itself. The canal drew water from the imperial gardens west of the city, channeling the flow of the Gu River and the Luo River into the Yellow River. This course likely followed the path of the ancient Yang Canal, originally excavated by Zhang Chun during the Eastern Han dynasty.

After merging with the Yellow River, the canal branched off again at Banzhu (located northwest of present-day Xingyang County, Henan, on the river's southern bank), from where it meandered southeastward, eventually emptying into the Huai River.[23] During the Tang and Song dynasties, this canal became widely known as the Bian Canal or Bian River.[24] However, it should not be confused with the Bian Canal system that developed after the Eastern Han, which was part of the ancient Honggou

[21] Annals of Emperor Yang I. In *Book of Sui*, vol. 3.

[22] Ibid.

[23] Annals of Emperor Yang I; Treatise on food and finance. In *Book of Sui*, vol. 3, 24.

[24] *Yuanhe Maps and Records of Prefectures and Counties (vol. 5): Henan Prefecture* records: "The Bian Canal is located 250 steps south of Heyin County… In the first year of the Daye era under Emperor Yang of Sui, it was re-excavated and dredged, and named the Tongji Canal." *The History of Song (vol. 93): Treatise on Rivers and Canals III* states: "The Bian River was dredged and reopened as the Tongji Canal, diverting Yellow River waters to the Huai River starting from the early Daye period of the Sui. By the Tang dynasty, it was renamed Guangji Canal, but among the people it continued to be commonly called the Bian Canal. In the Song dynasty, the Wuzhang River was renamed Guangji River, and the name Bian Canal became the general designation for the Tongji Canal."

waterway system, comprising the Bian and Huo rivers. That earlier canal network originated near present-day Kaifeng and flowed southeast to join the Si River north of modern Xuzhou. In contrast, the Tongji Canal built in the Sui era flowed directly southward into the Huai River, and thus the two should be clearly distinguished.

By the Tang and Song periods, the Tongji Canal—under its later names Bian Canal or Bian River—had become a critical transportation artery. As a result, it was widely recorded in geographical and administrative texts. The *Yuanhe Maps and Records of Prefectures and Counties* documents the canal's presence in multiple jurisdictions: Heyin County in Henan Prefecture (north of modern Xingyang), Yuanwu in Zhengzhou (west of present-day Yuanyang), Kaifeng in Bian Prefecture, Yongqiu (modern Qixian), Songcheng in Song Prefecture (modern Shangqiu County), Fuli County in Su Prefecture (modern Su County, Anhui), Hong County in Si Prefecture (modern Sixian, Anhui), and Linhuai County (north of present-day Xuyi County, Anhui).

These entries offer detailed descriptions:

- Heyin County: "The Bian Canal lies 250 paces south of the county and is also called the Langdang Canal."
- Yuanwu County: "The Bian Canal, also known as the Langdang Canal, is today called the Tongji Canal."
- Kaifeng County: "In the first year of the Daye era, due to the winding nature of the Bian River making navigation difficult, a canal was excavated southwest of Daliang City to divert its waters. This was called the Tongji Canal."
- Yongqiu: "The city lies just north of the Bian River."
- Songcheng: "The Bian River passes south of the prefectural city."
- Fuli: "South of the county lies the Bian River, with Yongqiao serving as a convergence point for boats."
- Hong County: "The county is adjacent to the Bian River."
- Linhuai County: "The Bian River lies to the west of the county."

The Universal Geographic Record of the Taiping Era also records the presence of the Tongji Canal or Bian River in several counties, including Heyin in Mengzhou, Kaifeng in Bianzhou, Songcheng in Songzhou, Fuli in Suzhou, and Linhuai in Sizhou. It is particularly explicit in its account of Fuli County: "The Bian River lies one hundred steps to the south of the county." In Linhuai County, it states that:

- The eastern wall of Wucheng faces the now-abandoned Tongji Canal.
- Chonggang City lies one li south of the canal.
- Yongtai Lake was formed when the Li River was blocked during the excavation of the Tongji Canal in the third year of the Daye reign (607 CE), causing the surrounding waters to accumulate into a lake.

Of particular historical value is the Kaifeng County entry, which explains that in 605 CE, Emperor Yang ordered the excavation of the Tongji Canal to reroute the difficult and meandering Bian River by cutting a new channel southwest of Daliang.

Later, Wang Cun's Treatise on the *Nine Territories in the Yuanfeng Reign* also documented the presence of the Bian River or Bian Waters in a number of locations, including: Heyin County (Mengzhou); Kaifeng, Yongqiu, Zhongmu, Chenliu, and Yangwu Counties (Kaifeng Prefecture); Xiangyi, Songcheng, Ningling, Gushu, and Xiayi Counties (Yingtian Prefecture); Yongcheng and Zan Counties (Bozhou); Fuli, Linhuan, and Hong Counties (Suzhou); and Linhuai County (Sizhou). He explicitly notes the Tongji Canal under the entry for Kaifeng County.

When arranged according to modern place names, the Tongji Canal passed through present-day Xingyang, Zhongmu, Kaifeng, Qixian, Suixian, Ningling, Shangqiu, Xiayi, Yongcheng, Su County, Lingbi, and Sixian, finally emptying into the Huai River north of Xuyi County.

Certain travel writings from the Tang dynasty also contain similar accounts of the areas through which the Bian River flowed. To illustrate this, we may cite Li Ao's *Record of My Journey to the South* and the Japanese monk Ennin's *Record of a Pilgrimage to Tang China in Search of the Law*. According to Li Ao, after traveling east from Chang'an through Luoyang, he "sailed down the Bian River, which connects to the Huai River." Along the way, he passed through Heyin, Bianzhou, Chenliu, Yongqiu, Songzhou, Yongcheng, Yongkou (present-day Yongqiao, later established as Suzhou), and Sizhou (today west of Xuyi County in Anhui), before entering the Huai River via the Bian Canal.[25] Ennin's account records that he traveled east from Chang'an through Luoyang to Bianzhou, where he boarded a boat in Chenliu County and sailed south along the Bian River. After ten days, he arrived at Sizhou, crossed the Huai River to Xuyi County, and then proceeded to Chuzhou. Although Ennin does not provide detailed descriptions of the places he passed, his arrival at Sizhou implies that he passed through Songzhou and Yongkou. Such travel accounts are unlikely to contain significant errors regarding these routes (See Fig. 5.2).

Although this canal has been silted up for many years, its remnants can still be found. The modern highway between Yongcheng County in Henan Province and Su County in Anhui Province is built upon the ruins of the Bian River, with some sections of the remains rising four to five meters above the ground. During my field investigation in Su County, I wandered along this highway, silently contemplating the historical terrain; I also followed the canal's remnants by vehicle along this road to Yongcheng, observing a landscape much like that of Su County. I have heard that the road from Su County through Lingbi and Sihong to Sihong County was also constructed using the remains of the Bian River, though I regrettably have not had the opportunity to explore it personally. Nevertheless, there remains a stretch of waterway between Sihong County and Hongze Lake that still bears the name Bian River, which is likely the original course of the Tongji Canal.

In this light, the Tongji Canal constructed by Emperor Yang of Sui was distinct from the Bian Canal or Bian River of the Eastern Han through the Six Dynasties period, and naturally also different from the Bian River of the Qin and Han eras. One reason why Emperor Yang did not utilize the original waterway of the Bian River when excavating the canal lies, as previously cited in the *Universal Geographic Record of the Taiping Era*, in the fact that the original "Bian River's course was

---

[25] Li Ao. Record of my journey to the south. In *Collected works of Li Wengong, vol. 18.*

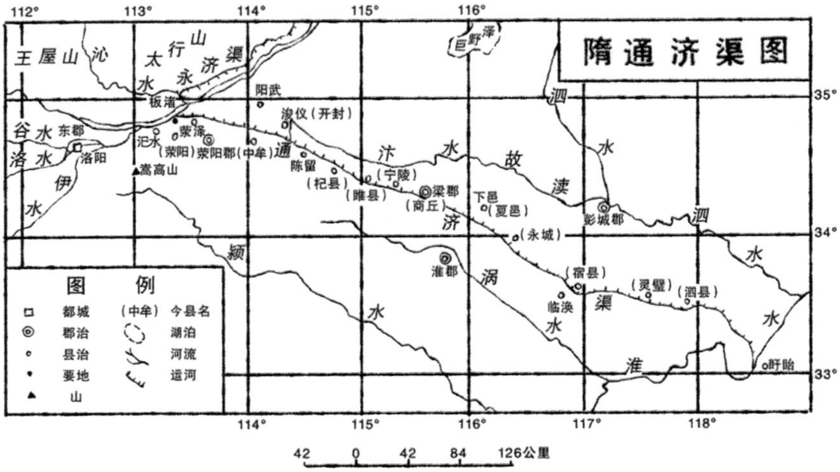

图例 - Legend
都城 Capital Cities
郡治 Commandery Seats
县治 County Seats
要地 Strategic Locations
山 Mountains
今县名 Present County Names
湖泊 Lakes
河流 Rivers
运河 Canals

Fig. 5.2 The Tongji Canal during the Sui Dynasty

winding and tortuous, making navigation difficult for vessels traveling back and forth." Additionally, the perilous terrain of Luliang near Pengcheng, known as the "Hundred Steps," likely contributed to the decision to avoid and not make use of the old Bian River route.

Regarding the Tongji Canal excavated by Emperor Yang of Sui, there have been varying accounts since the Tang dynasty. The *Kunyuan Lu* states: "The Bian Canal, also known as the Langdang Canal, is presently called the Tongji Canal, with its source drawn from the Yellow River." It records that "after Emperor Wu of Liu Song's northern expedition, the Bian Canal became completely silted up; in the first year of the Daye era under Emperor Yang of Sui, it was dredged and reopened, named the Tongji Canal, extending west to the Yellow River and Luoshui, and south to the Yangtze River and Huai River."[26] The *Yuanhe Maps and Records of Prefectures and Counties* similarly notes: "After Emperor Wu of Liu Song's northern campaign, the Bian Canal was entirely silted; in the first year of the Daye era, Emperor Yang of Sui dredged and renamed it the Tongji Canal… It also drew Bian River water eastward

[26] Prefectures and counties VII. In *Comprehensive institutions (Tongdian)*, vol. 177.

from Daliang city into the Si River, connecting to the Huai River."[27] The Comprehensive Mirror in Aid of Governance provides even more detail: "In March of the first year of Daye, one million people from Henan and northern Huai provinces were mobilized to excavate the Tongji Canal... Water from the Yellow River was diverted at Banzhu, flowed through Xingze into the Bian River, and then from Daliang the Bian River water was led into the Si River, reaching the Huai River." Hu Sanxing's annotation explains, "The diversion of Yellow River water into the Bian River, and the Bian River into the Si River, all follow the old river courses."[28] All these records indicate that the Tongji Canal opened by Emperor Yang of Sui was in fact the old watercourse of the Bian Canal flowing east into the Si River from the Eastern Han through the Six Dynasties period, which stands in stark contrast to previously mentioned claims.

How can this discrepancy in accounts be resolved? One approach is to adopt a conciliatory perspective: during the Sui and early Tang periods, navigation commonly followed the route upstream from the Si River into the Bian River, while only from the mid-Tang period onward did the new route become fully navigable. In fact, since the early Tang era, the Bian River between Bianzhou and Sizhou had already been unobstructed, serving as the main route for travelers. During the reign of Emperor Gaozong, the poet Luo Binwang traveled westward along this path. His poem, *Early Departure from Huaikou, Gazing Toward Xuyi*, was composed in relation to this journey.[29] Xuyi, as the point where the Bian River flows into the Huai River, is naturally unrelated to the Si River. During the Kaiyuan era, the poet Meng Haoran also wrote a poem titled Farewell to Magistrate Zhang of Qiao County and Assistant Prefect Shentu of Yue which contains the line: "Setting off by boat along the Bian River at dawn, by evening already moored at the borders of Qiao County."[30] During the Tang dynasty, Qiao County corresponds to present-day Bo County in Anhui Province. At that time, the Bian River did not pass through the city of Qiao County, which is why the poem only mentions the boundary of Qiao County. Regardless of the distance to Qiao County, Meng Haoran indeed traveled by boat along the Bian River to reach the border of Qiao County. His journey was toward Yue, further underscoring that the Bian River had become an important transportation waterway. Although these examples are only a few, they sufficiently demonstrate that the Tongji Canal—or Bian Canal—had no connection with the Si River.

However, identifying the Tongji Canal with the Bian Canal that existed since the Eastern Han Dynasty is not without basis. In the seventh year of the Kaihuang era (587 CE), Emperor Wen of the Sui Dynasty commissioned Liang Rui to reinforce the ancient Han dam located twenty *li* west of Tanghe Yin County, diverting the river into the Bian Canal. This structure became known as the Biankou Dam, also referred to as the Lianggong Dam.[31] This was an effort to restore the Bian Canal dating back

[27] Henan prefecture. *Yuanhe maps and records of prefectures and counties*, vol. 5.

[28] Annals of Sui IV. In *Comprehensive mirror to aid in government*, vol. 180.

[29] *Complete poetry works of Tang dynasty*, vol. 79.

[30] *Complete poetry works of Tang dynasty*, vol. 159.

[31] Henan prefecture. *Yuanhe maps and records of prefectures and counties*, vol. 5; Prefectures and counties VII. In *Comprehensive institutions*, vol. 177.

to the Eastern Han Dynasty. However, the effect at the time was likely limited, which may have prompted Emperor Yang to excavate the Tongji Canal anew.

As previously cited, both the *Yuanhe Maps and Records of Prefectures and Counties* and the *Universal Geographic Record of the Taiping Era* record that Emperor Yang of Sui once excavated a canal southwest of Daliang City to divert the Bian River. This canal—known as the Langdang Canal—was in fact the Tongji Canal. By then, the Bian water diverted by Liang Rui had already flowed into the Tongji Canal, and the original Bian Canal either completely dried up or was reduced to a mere trickle, no longer capable of serving as a vital waterway for boat traffic. The *Kunyuan Lu* mentioned earlier, which was originally cited in the *Tongdian*, appears to have been doubted by Du You despite his reference to it. Du You once noted that near Pengcheng in Xuzhou there was a watercourse called Hou River, also named Bian River, originating from the border of Xiao County.[32] Pengcheng corresponds to present-day Xuzhou, and Xiao County lies to its northwest. It is evident that the Bian River Du You refers to here is the Eastern Han Bian Canal, not the Sui dynasty's Tongji Canal. Du You further states: "In Bian Prefecture, there is the Tongji Canal, which Emperor Yang of Sui excavated to divert Yellow River water to connect the Yangtze and Huai Rivers, also drawing from the Bian water – this is the Langdang Canal."[33] The Bian River flowing from Xiao County to Pengcheng is likely the Bian Canal dating back to the Eastern Han Dynasty. Du You merely mentioned it in passing, indicating that it was no longer the canal capable of handling large-scale grain transport as in earlier times, but rather an ordinary small waterway. As for the Tongji Canal in Bian Prefecture, since it also drew from the Bian water, it was certainly no longer the original Bian Canal. Given that the Tongji Canal connected the Yangtze and Huai Rivers, it could no longer flow eastward into the Si River. Although Li Jifu stated that Emperor Yang of Sui excavated the Tongji Canal drawing Bian River east of Daliang into the Si River and onward to the Huai, he also noted: "Since the Sui dynasty excavated the Bian River, Pengcheng, controlling the strategic Yongqiao pass and guarding the Bian River grain transport route, has held particular military importance."[34] Although Pengcheng lies beside the Si River, the actual Bian route runs through Yongqiao, which later became the seat of Su Prefecture. While Yongqiao was once under the jurisdiction of Xuzhou, it is not very close to Xuzhou proper. The Bian River flowing through Yongqiao could not have turned northward to enter the Si River at Pengcheng. Furthermore, based on the counties and prefectures recorded along the Tongji Canal in the *Yuanhe Maps and Records of Prefectures and Counties*, it is clear that Li Jifu's assertion that the canal flowed into the Si River and then reached the Huai River is mistaken. Sima Guang's *Comprehensive Mirror to Aid in Governance*, which documents the history of the Sui and Tang dynasties using Tang sources, also records this differently. The Tang records may not fully correspond to the actual situation. It is possible that Sima Guang described the canal's original Sui-era construction, while modifications occurred later during the Tang period,

32 Prefectures and counties X. In *Comprehensive institutions*, vol. 180.

33 Prefectures and counties VII. In *Comprehensive institutions*, vol. 177.

34 Xuzhou. *Yuanhe maps and records of prefectures and counties*, vol. 9.

accounting for discrepancies. The *Book of Sui Dynasty*, compiled under Wei Zheng in the early Tang, explicitly states in the "*Annals of Emperor Yang*" that the Tongji Canal was excavated "drawing water from Banzhu to connect to the Huai River," without mention of diverting Bian River into the Si River. Since the *Book of Sui* was compiled within two or three decades after Emperor Yang's canal construction, and given the canal's significant role at that time, it is unlikely that the historians erred so greatly. If the route of the Tongji Canal during the Sui differs from the Bian River's course during the Tang, when and how this change occurred remains unclear. No such alteration is recorded during the early Tang. In the eighteenth year of Emperor Xuanzong's Kaiyuan era (730 CE), Pei Yaoqing proposed improvements to grain transportation, stating that taxes and levies from Jiangnan were sent to Yangzhou's Doumen Gate during February each year, but only crossed the Huai River into the Bian River after April.[35] Pei Yaoqing's account mentions crossing the Huai into the Bian but makes no reference to the Si River, further confirming that the Bian River was unrelated to the Si River.

Some scholars who propose a conciliatory explanation suggest that the section of the Bian Canal between Bian and Si Prefectures did not become fully navigable until after the mid-Tang period. However, by that time, the power of the central dynasty had already declined, and regional military governors had again become unruly and difficult to control. Under such circumstances, it seems implausible that a major canal project could have been initiated—or revived. This argument appears speculative and lacks credibility.

In fact, even after the mid-Tang period, there are still records referencing the Bian River flowing into the Si River. In the early reign of Emperor Daizong, Liu Yan wrote to Yuan Zai regarding grain transport, stating: "The grain ships travel via the Huai and Si Rivers to reach the Bian River, then enter the Yellow River, proceeding westward past *Dizhu*, *Xiashi*, and Shaohua Mountains. Thus, vessels from Chu and Yue can sail directly to the Jianzhang and Changle Palaces—this is a brilliant strategy for securing the state."[36]

Given that by the mid-Tang the Bian Canal between Bian and Si Prefectures had already become fully navigable, Liu Yan's phrase "via the Huai and Si Rivers, reaching the Bian and entering the Yellow River" should be understood as a generalized description, offering little value for precise historical verification.

It is necessary here to briefly trace the origins of the Tongji Canal, as its relationship with the earlier Bian Canal, which had existed since the Eastern Han period, involves both connections and distinctions. The Han-era Bian Canal drew water from the Yellow River at Shimen, whereas the Tongji Canal took its source farther upstream at Banzhu. Shimen was located north of Mount Guangwu, while Banzhu lay even farther to the upper reaches of the river. Today, the Yellow River has shifted southward to the northern foothills of Mount Guangwu, and even the northern slopes of the mountain have been significantly eroded by the river. However, during the Sui and

[35] Treatise on food and finance. In *New book of Tang*, vol. 53.

[36] Biography of Liu Yan. In *Old book of Tang*, vol. 53.

Tang periods, the Yellow River still flowed at a considerable distance from Mount Guangwu, and a county named Heyin was established in the intervening area.

As a result, after diverting from the Yellow River at Banzhu, the Tongji Canal was able to flow eastward along the northern base of Mount Guangwu, remaining separate from the Han-era Bian Canal. Since Mount Guangwu stretches in an east–west direction, the canal had no alternative route but to follow its northern foothills. By the Tang period, the canal's intake had shifted downstream to a point 250 paces south of Heyin County.[37] This adjustment allowed the canal to run directly below the county seat—a proximity enabled by the eastward relocation of its intake. In fact, the construction reused the site of the old Han-era Shimen, though by that time Shimen itself no longer existed.

Although the intake points of the Sui-Tang Tongji Canal and the earlier Bian Canal had changed, they both eventually converged east of Mount Guangwu, following a path broadly consistent with that of the Han-era Bian Canal. While the two canals may have shared a similar route along this stretch, they served different purposes and eventually diverged. Scholars differ on the precise location of this divergence, but the most plausible view places it at Kaifeng County. As noted in both the *Yuanhe Maps and Records of Prefectures and Counties* and the *Universal Geographic Record of the Taiping Era*, during the Daye reign of the Sui dynasty, a canal was excavated southwest of Daliang city to divert water from the Bian River—this became known as the Tongji Canal. Du You's Tongdian offers a similar account. These sources consistently distinguish the Tongji Canal from the "Bian River," which clearly refers to the watercourse in use from the Eastern Han through the Six Dynasties. While the Tongji Canal used the Bian's water, its channel followed a different course.

Some scholars argue that the Tongji Canal split from the old Bian Canal west of Yongqiu County, but this view does not align with the historical evidence. According to the *Yuanhe Maps and Records of Prefectures and Counties*, Yongqiu's current county seat occupies the site of its former city, and "the Bian River flows north of the city." This Bian River is the Tang-era Bian River—that is, the Sui Tongji Canal. By contrast, the Han-era Bian Canal is referred to in the *Commentary on Water Classic* as the Bian River. According to this commentary, the Bian River did indeed pass north of Yongqiu's old city. However, it also notes the presence of another watercourse—the Sui River—north of the same location.[38] This shows that the Han-era Bian Canal did not flow directly beneath the old city of Yongqiu.

Given the substantial distance between the Han Bian Canal and the Sui-Tang Tongji Canal in the Yongqiu area, and the fact that the Sui River lies between them, it is difficult to argue that the Tongji Canal or Tang-era Bian River branched off from the earlier Han Bian Canal. Since the Sui River enters the discussion here, its relationship to the Tongji Canal and the Bian River also warrants clarification.

The Tongji Canal (or Bian River) flowed southeast from Yongqiu County, passing through the counties of Xiangyi, Ningling, Songcheng, and Gushu (modern Gushu Town, southeast of Shangqiu County, Henan). According to the *Treatise on the Nine*

[37] Henan prefecture. *Yuanhe maps and records of prefectures and counties*, vol. 5.

[38] Commentary on the Sui River. In *Water classic*.

Territories in the Yuanfeng Reign, these counties each recorded either "Bian River" or "Bian Water," as previously noted. This same source also records the Sui River in Xiangyi, Ningling, and Songcheng—entries confirmed by the *Yuanhe Maps and Records of Prefectures and Counties*. Since the Han-era Bian Canal lay even farther north than the Sui River, it becomes clear that the Tongji Canal or Bian River was neither identical to the Han Bian Canal nor to the Sui River.

Some have claimed that the Tongji Canal followed the old course of the Sui River east of Bian Prefecture, flowing through Yongqiu, Xiangyi, Ningling, Songcheng, and Gushu, then turning north past the former site of Xiang County and south of Xiao County, before reaching Pengcheng and merging with the Si River. Another branch, they argue, flowed south from Gushu to form the Qi River, which emptied into the Huai.[39] However, according to the *Yuanhe Maps and Records of Prefectures and Counties*, the Sui River was still flowing freely at the time, and the *Treatise on the Nine Territories in the Yuanfeng Reign* confirms that the Bian River of the Tang and Song did not appropriate the Sui's course. Furthermore, both the *Yuanhe Maps and Records of Prefectures and Counties* and the *Universal Geographic Record of the Taiping Era* list the Sui River, Bian River, and the Han-era Bian Canal as distinct entries under Songcheng County, making their separate identities unmistakable.

This evidence shows that the Bian River neither took over the Sui River's course nor was influenced by the Han-era Bian Canal. It could not have followed either waterway eastward to merge with the Si River in Xuzhou. Additional evidence from the Yuanfeng Gazetteer shows that both Cuan County in Bo Prefecture (modern Cuancheng Town, northwest of Yongcheng) and Fuli County in Su Prefecture recorded both the Sui and Bian Rivers—further demonstrating that the two remained distinct waterways.

A final point of clarification is in order: According to the Taiping Gazetteer, the Sui River in Songcheng County flowed five li south of the city. Wang Dan's *Notes of Minister Wang Wenzheng*, a Song-dynasty work, also states that the Bian Canal lay five *li* south of Songcheng. Since both texts are from the Song period, how should we interpret this? Might the Bian River have taken over the Sui River's channel? Based on the detailed and overlapping records in the Yuanhe and Yuanfeng gazetteers—which consistently list both the Sui and Bian Rivers in multiple locations—it seems unlikely that both sources are in error. It is more plausible that the rivers remained distinct, even when they ran in close proximity.

As previously mentioned, some have posited that the Bian River branched at Songcheng, with its southern branch forming the Qi River, a tributary of the Huai River. The *Water Classic: Commentary on the Huai River* states: "The upper reaches of the Qi River originate northeast of Gushu City (modern-day Gushu Town, southeast of Shangqiu, Henan), where it receives water from the Sui River. It then flows eastward past the ruins of Jiancheng County (now southeast of Yongcheng, Henan), turns southeast past Qi County (modern Qixianji, south of Suzhou, Anhui), and continues east into Xiaqiu County (now Sixian, Anhui), flowing north of the old seat

[39] Yan Wenru. A study of the Bian River during the Sui and Tang dynasties. In *Collected essays for the Liaohai anniversary volume*.

of Xiaqiu County. It proceeds southeastward past Tong County (now northeast of Sixian), then east through Xu County (modern south of Sihong, Jiangsu), passing south of the former seat of Daxu County, before finally discharging into the Huai River." This flow route closely resembles that of the Tang and Song Bian River.

However, both the *Yuanhe Maps and Records of Prefectures and Counties* and the *Treatise on the Nine Territories in the Yuanfeng Reign* list the Qi River under Suzhou's Qi County, which, notably, is not a region through which the Bian River passed. This suggests that the Qi River, like the Sui River, was not subsumed by the Bian River. The eastern part of Henan and the northern region of Anhui are characterized by a dense network of rivers, some of which run in relatively close proximity to one another. Thus, it would be inadvisable to draw forced correlations between them, as doing so risks conflating distinct waterways.

The Han Canal

The purpose behind Emperor Yang's excavation of the Tongji Canal was to establish a direct connection between the imperial capital and the prosperous southeastern region. At the time, the most affluent area in the southeast was the Yangtze River Delta. However, the Tongji Canal extended only as far as the Huai River and did not reach this economically vital region. Although the Han Canal had historically served as a link between the Huai River and the Yangtze River, by Emperor Yang's reign, it had experienced multiple course changes and had become so silted and narrow that it could no longer accommodate large vessels. During the Kaihuang reign, Emperor Wen had focused only on improving the waterway entrance from the Huai River into the Han Canal, rather than undertaking a full-scale restoration. As a result, once the Tongji Canal was completed, Emperor Yang promptly moved to rehabilitate the Han Canal's ancient course.

The Han Canal, originally excavated by King Fuchai of Wu, once diverted northeastward, passing through Sheyang Lake. During the Eastern Han period, Chen Min undertook a renovation of the canal, rerouting it away from Sheyang Lake. The new course passed through Fanliang Lake to the north, continued through Jin Lake and Baima Lake, and then rejoined the original northbound route beyond Sheyang Lake. This adjustment allowed the canal to connect with the Huai River at its outlet more directly, significantly shortening the distance compared to the earlier path. Further improvements were made during the Eastern Jin period in two distinct phases. The first took place during the Yonghe era of Emperor Mu (345–356 CE), and the second during the Xingning era of Emperor Ai (363–365 CE). During the Yonghe period, the southern segment of the Han Canal was reconstructed. At the time, the Yangtze River's entry point into the canal had become obstructed, making it difficult to draw water from the river. Consequently, water was diverted from Ouyang Embankment to the southwest. Located within present-day Yizheng County, Jiangsu Province, the intake point at Ouyang Embankment lay sixty *li* (approximately 30 km) from Guangling City. During the Xingning period, renovations focused on the northern section of the Han Canal. As previously mentioned, in the Eastern Han period, Chen Min had rerouted the canal to avoid the lengthy detour through Sheyang Lake by having it pass northward through Jin Lake. However, Jin Lake was prone to violent winds and waves. To address this danger, a new channel was excavated from the southern mouth of the lake along its eastern bank, bypassing the lake entirely and reconnecting at its northern end. This improvement effectively eliminated the navigational hazards associated with traversing the lake.

These developments were recorded in detail by Li Daoyuan in his *Water Classic: Commentary on the Huai River*. However, Li Daoyuan introduced some chronological confusion. He conflated the diversion of Yangtze River water near Ouyang Embankment, a project undertaken during Emperor Mu's Yonghe era in the Eastern Jin, with the canal work overseen by Chen Min during the Yonghe era of Emperor Shun in the Eastern Han. Although both events occurred in similarly named periods, they belonged to different dynasties and were separated by a considerable span of time. In discussing the Han Canal, Li Daoyuan mentioned both undertakings together, but in doing so, he mistakenly merged them, leading to chronological inconsistencies and confusion. Following the Yonghe and Xingning renovations in the Eastern Jin, the Han Canal remained navigable without obstruction. Toward the end of the Eastern Jin period, in the fifth year of the Yixi era (409 CE) under Emperor An, Liu Yu launched his campaign against Southern Yan. His fleet departed from the capital Jiankang (modern-day Nanjing), sailing upstream via the Huai River into the Si River.[40] The route taken was precisely the waterway of the Han Canal.

Later, during the campaign against Later Qin in the twelfth year of the Yixi era (416 CE), although the official histories do not provide a detailed account of the path from Jiankang (modern-day Nanjing) to Xuzhou (present-day Xuzhou, Jiangsu), it is recorded that Wang Zhongde led the vanguard forces and first advanced with a naval fleet into the river system.[41] This strongly suggests that Liu Yu also advanced alongside the fleet, and as overall commander of the naval forces, he could not have bypassed the Han Canal route. In the ninth year of the Taijian reign of Emperor Xuan of the Chen dynasty (577 CE), Wu Mingche took advantage of Northern Zhou's conquest of Qi to launch a northern expedition. On this occasion as well, his forces advanced via the Han Gou route.

When Wu Mingche launched his northern expedition, he was in command of military affairs across the northern and southern regions of Yanzhou and Qingzhou, as well as Qiao Commandery, and served as the governor of Southern Yanzhou. After advancing northward, his son Huijue acted as the interim administrator of the province. This indicates that the campaign originated from Southern Yanzhou, whose administrative center was Guangling—present-day Yangzhou in Jiangsu Province. Since the forces set out from Southern Yanzhou, it was inevitable that they would proceed via the Han Canal. Upon returning from Xuzhou, however, the Zhou general Wang Gui blocked the river route at the mouth of the Huai River, preventing Wu

[40] Annals of Emperor Wu I. In *Book of Song*, vol. 1.

[41] Annals of Emperor Wu II. In *Book of Song*, vol. 2; Annals of Jin 39. In *Comprehensive mirror to aid in government*, vol. 117.

Mingche's forces from sailing southward. Consequently, Wu Mingche was defeated and captured.[42] The blocking of the Huai River's mouth clearly explains why Wu Mingche was unable to re-enter the Han Canal. From the Chen dynasty through to the Sui dynasty, though not many years had passed, there may have been localized blockages in the Han Canal; nevertheless, the main flow likely remained largely unchanged. Especially by the time of Emperor Yang of Sui, the northern and southern sections of the Han Canal had been nearly straightened.

The Han Canal renovated by Emperor Yang of Sui stretched from Shanyang (present-day Huai'an County, Jiangsu) in the north to Yangzi (southeast of present-day Yizheng County, Jiangsu) in the south. Compared to the remnants of the canal constructed by King Fuchai of Wu, it was situated slightly further west. The river course had undergone multiple realignments and was straighter than before, unlike during King Fuchai's time when it detoured through Sheyang Lake. According to Tang dynasty records, this waterway "flows downstream from Huaiyin (southwest of present-day Qingjiang City, Jiangsu) to Shaobo (north of present-day Yangzhou City, Jiangsu) for 350 *li*, and upstream from Shaobo to the river junction for 90 *li*."[43] (See Fig. 5.3).

The Jiangnan Canal

As for the construction of the Jiangnan Canal, it took place five years after the completion of the Tongji Canal and the Han Canal.[44] This Jiangnan Canal extended from Jingkou (present-day Zhenjiang, Jiangsu Province), winding east of Tai Lake, all the way to Yuhang, covering a total distance of 800 *li*. During the Tang dynasty, when Li Ao traveled from Chang'an to Guangzhou by way of Luoyang, he deliberately took a detour along this route.

According to his account, after crossing the Yangtze River from Yangzhou, he first arrived at Runzhou. Runzhou, formerly known as Jingkou, marked the starting point of the Jiangnan Canal. From there, the route passed through Changzhou (modern-day Changzhou), Suzhou (modern Suzhou), crossed the Song River (modern-day Suzhou Creek), and finally reached Hangzhou (present-day Hangzhou).[45]

According to records from the Northern Song period, the Jiangnan Canal passed through several locations including Danyang in Runzhou (present-day Danyang County), Wujin and Jinling in Changzhou (both administrative seats of Changzhou), Wuxi (present-day Wuxi County), Changzhou in Suzhou (administrative seat of Suzhou), as well as Changshu and Wujiang (both present-day counties), Xiuzhou

---

42 Biography of Wu Mingche. In *Book of Chen*, vol. 9.

43 Li Ao. Record of my journey to the south. In *Collected works of Li Wengong, vol. 18.*

44 Annals of Sui V. In *Comprehensive mirror to aid in government*, vol. 181.

45 Li Ao. Record of my journey to the south. In *Collected works of Li Wengong, vol. 18.*

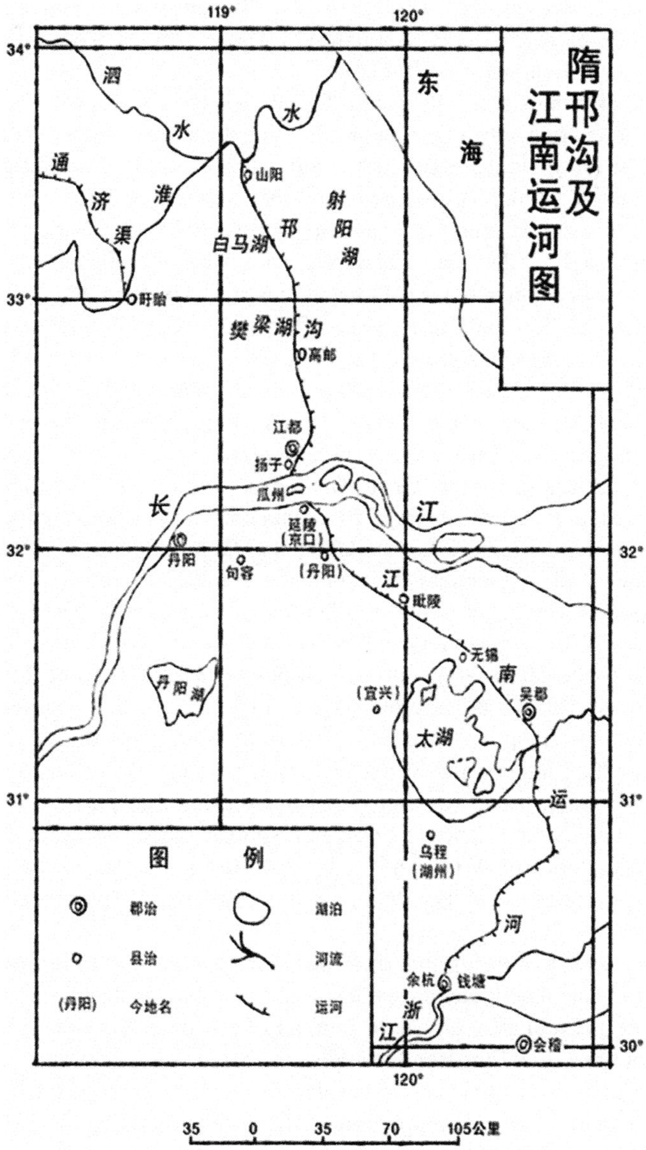

图例 - Legend
郡治 Commandery Seats
县治 County Seats
今地名 Present Names
湖泊 Lakes
河流 Rivers
运河 Canals

Fig. 5.3 The Sui-era Han Canal and the Jiangnan Canals

(whose seat was in present-day Jiaxing County, Zhejiang), and Chongde (present-day southern Tongxiang County, Zhejiang).[46] These regions were precisely the prosperous areas of Jiangnan near Tai Lake.

Like the Guangtong Canal, the Tongji Canal, and the Han Canal, the Jiangnan Canal was also constructed by utilizing or following the traces of older watercourses, rather than being an original creation of the Sui dynasty. South of the Yangtze River, particularly in the Tai Lake region, there has since antiquity existed a landscape of interlacing lakes and crisscrossing waterways—essentially a vast marshland. The effort to connect and regulate these lakes and waterways to facilitate transportation and logistics was therefore a natural and practical undertaking.

In fact, since the Wei and Jin periods, efforts had been continuously made to utilize the marshes and lakes around the Tai Lake region in Jiangnan to open up waterways and facilitate transportation. As previously mentioned, during the Eastern Wu period, Chen Xun excavated a central route through Jurong to connect with commercial centers and construct warehouses. Jurong was home to the Pogang Canal (*Pogang Du*), and it is likely that Chen Xun's excavation took place there. During the Six Dynasties, this waterway served as a canal connecting Jiankang (present-day Nanjing) with the area around Tai Lake. Given that the canal reached Wu Commandery and Kuaiji, it clearly extended beyond the confines of Jurong alone. In the Southern Qi period, Dantu (present-day Zhenjiang, Jiangsu) also had a water route that connected to Wu Commandery.[47] Whether the waterway in Dantu County was directly related to the central route through Jurong excavated by Chen Xun remains uncertain. However, it can be asserted with confidence that this waterway served as a precursor to the Jiangnan Canal later constructed by Emperor Yang of the Sui dynasty. During the Liang period, when Wuxing Commandery suffered agricultural losses due to flooding, there was a proposal to excavate transport canals to divert and discharge the waters of Zhenze Lake. In response, labor was conscripted from the three commanderies of Wu (with its administrative seat in present-day Suzhou, Jiangsu), Wuxing (present-day Huzhou, Zhejiang), and Yixing (present-day Yixing, Jiangsu).[48]

The canals excavated during this time were located near the future route of the Sui-era Jiangnan Canal and may have constituted the water route leading to Wu and Kuaiji or one of its tributaries. Given the number of earlier canals already existing in the area, it is unlikely that Emperor Yang ignored these resources when constructing the Jiangnan Canal. The canal was excavated in the sixth year of the Daye reign (610 CE), five years after the completion of the Tongji Canal and the Han Canal. Considering Emperor Yang's notoriously impatient nature, it is difficult to believe he would have waited five years before completing the full canal system. It is more plausible that, after the Han Canal was completed, the remaining artificial waterways in Jiangnan were still navigable and were temporarily used. However, when the emperor decided to travel to Kuaiji, and it became apparent that these older canals

[46] Liangzhe circuit. In *Treatise on the nine territories in the Yuanfeng reign*, vol. 5.

[47] Treatise on prefectures and commanderies I. In *Book of Southern Qi*, Vol.14.

[48] Biography of Prince Zhaoming. In *Book of Liang*, Vol.8.

could not accommodate large vessels, a new round of excavation became necessary. Even so, it is highly probable that Emperor Yang utilized portions of the existing waterways as the structural basis for the newly constructed canal, even if he did not incorporate them in their entirety.

The Yongji Qu

The construction of the Yongji Canal took place in the fourth year of the Daye reign (608 CE).[49] This canal diverted the waters of the Qin River southward into the Yellow River and extended northward to Zhuo Commandery. The southern section of the canal made use of the remnants of the Bai Canal, originally excavated by Cao Cao,[50] as it had not yet silted up at that time.[51] However, the canals further north—such as the Licao, Pinglu, and Quanzhou canals—have long been left unmentioned.

In the plains east of the Taihang Mountains, waterways frequently overflow and change course, which may have caused the Licao canals to become unnavigable. Originally, the Bai Canal channeled the flow of the Qi River northward from Fangtou, but Emperor Yang of Sui chose not to dredge the old Fangtou route; instead, he redirected the Qin River into the Yellow River. This was likely because Fangtou lay too far to the east, making travel from Luoyang inconvenient. Since the Qin River and Qi River were not naturally connected, Emperor Yang's work involved linking the Qin and Qi rivers, with their point of direct connection near what is now north of Wuzhi County.[52] The canal follows the Qin River into the Qi River. The Qi River flows northward through present-day Neihuang County in Henan Province; Guantao County in Shandong; Linxi County (known as Linqing County during the Sui Dynasty) in Hebei; Qinghe County; and Dezhou in Shandong Province, eventually reaching the sea in

[49] Annals of Emperor Yang I. In *Book of Sui*, vol. 3.

[50] According to volume 17 of the *Yuanhe Maps and Records of Prefectures and Counties*, under "Weizhou," it is recorded that: "The Baigou River in Guantao County was originally called the Bai Canal (Baiqu). It was dredged and transformed into the Yongji Canal by Emperor Yang of the Sui Dynasty and was also known as the Imperial Canal." In the same volume, under "Xiangzhou," it states: "The Yongji Canal in Neihuang County, originally called Baiqu, was dredged by Emperor Yang of the Sui Dynasty and likewise named the Imperial Canal." These records indicate that segments of the Yongji Canal incorporated portions of the old Bai Canal route.

[51] *Book of Sui*, vol. 24: "Treatise on Food and Money" records: "In the 3rd year of the Kaihuang era (583 CE), observing that the capital's granaries remained understocked, the court deliberated on preparations for floods and droughts. An edict thus established bureaus to recruit rice-transport laborers in thirteen riverine prefectures - Pu, Shan, Guo, Xiong, Yi, Luo, Zheng, Huai, Shao, Wei, Bian, Xu, and Ru. Additionally, the Liyang Granary was established in Weizhou." The establishment of the granary in Weizhou was precisely because it was located along the Bai Canal.

[52] According to the *History of the Yuan Dynasty, Vol. 164: Biography of Guo Shoujing*, "Although the Qin River in Huaiqing and Mengzhou was already used for irrigation, there remained residual water behind the weirs and dams, which, after merging with the residual water of the Dan River, flowed eastward to the north of Wuzhi County and discharged into the Imperial River." This stretch of river is very likely a remnant of the Yongji Canal constructed by Emperor Yang of Sui.

what is now Tianjin Municipality. In essence, the Yongji Canal follows the course of the Qi River northward.[53] However, upon reaching Duliu Town in present-day Jinghai County, Hebei, it turns northwest, merging north of present-day Yongqing County with the Sanggan River (today known as the Yongding River, which follows the Sanggan's course). From there, it continues northwest to reach the south of Ji County—the administrative seat of Zhuo Commandery at the time.[54] (See Fig. 5.4).

The Fengyan Canal

It is worth noting a minor but significant hydraulic project known as the Fengyan Canal. This canal served primarily as a water management system, regulating the flow of the Yi and Si Rivers by redirecting them southwestward. The Si River originates to the east of present-day Sishui County in Shandong Province, while the Yi River rises southeast of present-day Qufu County, also in Shandong. In earlier times, the two rivers converged southwest of Qufu. By the Sui dynasty, however, changes in the waterways had shifted their confluence westward, to a location east of Yanzhou City (the administrative seat of what is now Yanzhou County, Shandong). The merged river then flowed southward, frequently inundating the expansive marshlands in the region.

During the reign of Emperor Wen of Sui (581–600 CE), Xue Zhou, then governor of Yanzhou, constructed a dam by piling stones east of the city. This engineering intervention successfully diverted the water flow westward. As a result, the marshlands were transformed into arable farmland, and the waterways became navigable by boats. Given that the lower reaches of the Si River were already favorable for transportation, Yanzhou greatly benefited from this project. The canal thus came to be known in local records as "Xue Gong's Fengyan Canal."[55]

The Impact of the Canal

Within merely six or seven years of ascending the throne, Emperor Yang of the Sui dynasty succeeded in constructing a comprehensive national waterway network. This, in principle, was a highly commendable achievement. However, the emperor pursued it with excessive haste and imposed unbearable burdens on the populace, ultimately turning what could have been a great public good into an instrument of oppression. This harsh governance contributed to his downfall—ending both his life and his dynasty in tragedy. It was only after the fall of the Sui that the true benefits of

---

[53] Hebei circuit, covering the prefectures of Xiang, Wei, Bei, and De. In *Yuanhe maps and records of prefectures and counties,* vol. 16, 17.

[54] Huang Shengzhang. Study on the Yongji Canal. *The inaugural issue of New East.*

[55] Biography of Xue Zhou. In B*ook of Sui*, vol. 56.

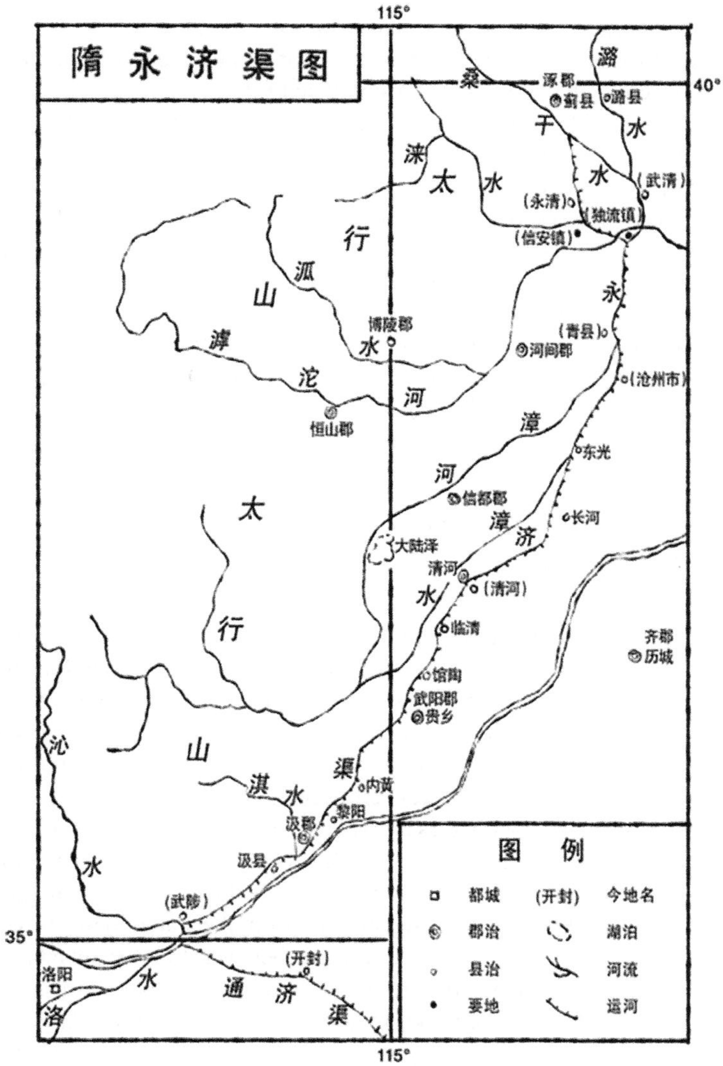

图例 - Legend
都城 Capital Cities
郡治 Commandery Seats
州治 Prefecture Seats
县治 County Seats
要地 Strategic Locations

今地名 Present Names
山 Mountains
运河 Canals
湖泊 Lakes
河流 Rivers
运河 Canals

Fig. 5.4 The Sui-era Yongji Canal

the canal system began to manifest. In the early Tang dynasty, the nation experienced a period of prosperity and strength. While many factors contributed to this flourishing era, the canal system undeniably played a significant role.

Not until the end of the Tianbao era during Emperor Xuanzong's reign—when An Lushan launched his rebellion—did the canals' function face serious disruption. At that time, regions along the Yongji Canal fell under the control of warlords and were effectively severed from the central government. In contrast, the system comprising the Tongji Canal, the Han Canal, and the Jiangnan Canal remained under state control and thus became increasingly vital to the Tang administration. As long as this network remained navigable, the central government's power held firm. But whenever interruptions occurred, the government found itself in disarray, struggling to maintain order. Ultimately, when the southeastern transport routes broke down entirely, it marked the beginning of the end for the Tang dynasty.

Difficulties in Grain Transport Between chang'an and Luoyang

When Emperor Yang of Sui undertook the construction of the Grand Canal, he first established Luoyang as the Eastern Capital, a move that held considerable significance for the canal project. The Tang dynasty inherited and maintained this administrative arrangement. Economically, Chang'an was heavily dependent on the wealth of the regions east of the passes, and the most crucial transport routes were the Yellow River and the Wei River. However, the treacherous section of the Yellow River near the *Dizhu* Rapids had long been regarded as a perilous passage. Li Daoyuan's *Commentary on Water Classic* provides a relatively detailed account of this hazardous stretch. He notes: "From the *Dizhu* downstream to the ferry at Wuhu—a distance of 120 *li*—the river is strewn with jagged rocks rising abruptly from the water, their formations extending all the way to the riverbanks. These are believed to be remnants of the ancient course allegedly channeled by Yu the Great, yet they may still obstruct the flow of the river. Even though the rocky mountainsides were carved through, the river current remains impeded, creating torrents, whirlpools, and surging waves. Altogether, there are nineteen rapids in this section, with the water's velocity comparable to that of the Three Gorges. It has long been known as a dangerous segment that destroys boats and endangers navigation." Li Daoyuan also recounts three historical efforts to improve this section, spanning from the Western Han to the Western Jin: the first in the fourth year of Hongjia during the reign of Emperor Cheng of the Western Han (13 BCE); the second in the second year of Jingchu under Emperor Ming of Wei (238 CE); and the third in the third year of Taishi under Emperor Wu of Jin (267 CE). During the first attempt, the river channel near the *Dizhu* Rapids was widened, but the rocks dislodged from excavation fell into the river and made the currents even more turbulent, exacerbating the danger. The second and third attempts each involved 5,000 men engaged in annual maintenance, but both yielded little substantial result.

Li Daoyuan thus reached the conclusion: "Although successive dynasties attempted to improve this stretch, the waters here have always remained turbulent and fierce, with waves piling up. Merchant ships that passed through would inevitably hesitate, finding the route difficult to traverse. Hence the long-standing reputation of this area as a series of 'multitudinous gorges and rapids.'".[56]

During the Sui and Tang dynasties, successive efforts were made to address this hazardous stretch of river that obstructed grain transport, yet no truly effective solution was ever found. Generally speaking, there were three methods attempted. The first involved carving out towpaths along the riverbanks to enable boats to be hauled upstream. This approach was first implemented during the reign of Emperor Gaozong of Tang by the Chief Architect of the Directorate of Imperial Construction, Yang Wulian. Haulers—men employed to pull the boats—would attach two iron hooks to their chests and use ropes to drag the vessels forward. However, the towing ropes frequently snapped under tension, often resulting in the haulers falling into the river and drowning. Later, during the reign of Emperor Xuanzong, the Prefect of Shan Prefecture, Li Qiwu, constructed a pathway along the mountaintops. Yet this method, too, proved highly laborious and inefficient.[57] Another method involved excavating an entirely new channel across the Yellow River's alluvial flats. This initiative also took place during the reign of Emperor Xuanzong of the Tang dynasty and was known as the "Kaiyuan New River." However, the riverbed in this area was composed largely of rock, which posed significant challenges. To facilitate excavation, workers first burned the rocks and then doused them with vinegar to make them brittle enough to break apart. Nevertheless, the removed stones were discarded into the river itself, which only intensified the turbulence of the water. As a result, boats still had to be hauled upstream manually.[58]

A relatively more practical solution was to shift from river transport to overland transport—an approach already implemented in the early Sui period. At that time, grain and goods were transported overland starting from Xiaoping (in present-day northwest Mengjin County, Henan), and then transferred to river transport at Shan County (modern-day Shanxian, Henan).[59] In the early Tang dynasty, it was stipulated that grain collected as tax from the Jianghuai region be transported to the Hanjia Granary in Luoyang. From there, it would be moved overland to Shan County. This overland route traversed the rugged and lengthy terrain of the Xiao and Han mountain passes, which made transport extremely difficult. During the reign of Emperor Xuanzong, a more practical adjustment was introduced: the establishment of the Jijin Granary east of the Sanmen Gorge and the Yan Granary to its west. Grain transported from the eastern provinces would be unloaded at Jijin, carried overland to Yan, and then transferred back to river transport. This method was briefly revived after the An

[56] Commentary on the Yellow River. In *Water classic*.

[57] Treatise on food and finance III. In *New book of Tang*, vol. 53.

[58] Ibid.

[59] Treatise on food and finance. In *Book of Sui*, vol. 24.

Lushan Rebellion, but it never remained in use for long.[60] Although overland transport avoided the risks of shipwrecks and capsizing, it required far greater labor and incurred higher costs. Ultimately, water transport proved to be the more sustainable and efficient option.

West of the *Dizhu* Rapids, river transport came to an end; prior to the construction of the Guangtong Canal, goods had to be transported upstream via the Wei River. However, the Wei River was shallow and silt-laden, which significantly impeded the movement of grain boats. Although Emperor Wen of the Sui dynasty ordered the excavation of a new canal, it still required frequent dredging; otherwise, it would quickly become obstructed again. When the Guangtong Canal was first completed during the Sui period, it remained relatively navigable without the need for extensive maintenance. By the Tang dynasty, however, dredging became a routine necessity. The most notable effort occurred during the Tianbao reign of Emperor Xuanzong, when Wei Jian oversaw a major dredging project. Historical records note: "Wei Jian restored ancient channels of the Wei River from the Han and Sui dynasties, beginning at Tong Pass and extending all the way to Chang'an, to facilitate the transport of tax grain from the Shandong region. The project diverted the waters of the Ba and Chan rivers, directing them eastward along the Wei River until they converged with the main course at Yongfeng Granary. Additionally, near Changle Slope (now the eastern outskirts of Xi'an, adjacent to the Chan River), an artificial lake was excavated below the Wangchun Pavilion beside the imperial gardens to serve as a harbor for grain ships."[61] After the Guangtong Canal silted up again, transport reverted to using the Wei River, and in some cases, even shifted to overland cart transport.

The New Book of Tang: Treatise on Food and Finance records the following account: "In the early years of the Dahe reign (827–835 CE) during Emperor Wenzong's rule, a severe drought caused the Wei River to dry up. As a result, grain transport vessels had to be dragged through silted channels, leading to significant grain loss and numerous deaths among the boatmen. The situation was deemed so urgent that normal bureaucratic procedures were waived. The magistrate of Xianyang, Han Liao, submitted a proposal to dredge the old transportation infrastructure from the Qin and Han dynasties—namely, the Xingcheng Weir (which extended eastward to the Yongfeng Granary). He proposed that restoring the 300-*li* stretch of river between Xianyang and Tongguan would eliminate the hardships of overland transport. Chancellor Li Guyan opposed the plan, deeming it untimely, but Emperor Wenzong responded, 'If it benefits the people, we need not concern ourselves with superstitious taboos.' The project was carried out, and the weir and canal were successfully restored. As a result, oxen previously used to haul carts were returned to agricultural work, and the Guanzhong region reaped considerable benefits." This dredging effort focused on the Guangtong Canal. The so-called Xingcheng Weir was located approximately eighteen *li* west of Xianyang[62] and marked the intake point where the Guangtong Canal drew water from the Wei River. From the time of its

[60] Treatise on food and finance III. In *New book of Tang*, vol. 53.

[61] Ibid.

[62] Biography of Li Shi. In *Old book of Tang*, vol. 172.

initial construction under Emperor Wen of Sui (r. 581–604) to its dredging under Wei Jian during the Tianbao reign (742–756), roughly 150 years had passed. From the Tianbao reign to the Dahe reign, only 80 to 90 years had elapsed—yet the canal had already become significantly silted. Although the restoration project spanned just 300 li, a relatively moderate undertaking, it nonetheless proved burdensome and vexing for those in power.

Prior to the excavation of the Guangtong Canal by Emperor Wen of the Sui, it was difficult for the emperor to remain in Chang'an should famine strike the Guanzhong region. For instance, in the fourth year of the Kaihuang reign (584 CE), although the construction of the Guangtong Canal had already been completed, the newly built canal likely had not yet enabled a reliable flow of grain transport. At that very moment, Guanzhong was stricken by famine, leaving Emperor Wen with no choice but to travel to Luoyang in search of provisions.[63]

Even after the successful completion and operation of the Guangtong Canal, such difficulties were not uncommon, as the Guanzhong region's climate was generally characterized by frequent droughts and infrequent floods, making crop failures all but inevitable. Emperor Yang of the Sui designated Luoyang as the Eastern Capital precisely to mitigate such risks. During the early Tang period, it became a regular practice for the emperor to relocate to the Eastern Capital to secure provisions whenever Guanzhong suffered from poor harvests. To illustrate the hardship and anxiety this caused, a few examples from the reigns of Emperor Gaozong and Emperor Zhongzong may suffice. In the first year of the Yongchun reign (682 CE), Emperor Gaozong was compelled to move to the Eastern Capital due to famine in Guanzhong, where the price of rice soared to 300 coins per *dou*. The departure was so rushed that some members of the imperial retinue perished from starvation en route. Fearing the threat posed by desperate famine victims along the way and the potential insecurity of the roads, Emperor Gaozong dispatched the Censor Wei Yuanzhong to escort and protect the imperial procession both before and behind.[64] Such arduous and desperate journeys left little room for the projection of imperial dignity. In the third year of the Jinglong reign (709 CE), another famine struck the Guanzhong region, with rice prices reaching 100 coins per dou—slightly better than during the Yongchun era. Nevertheless, in the effort to transport grain from Shandong and the Jianghuai region, eight or nine out of every ten draft oxen perished. At that time, many court officials petitioned for the emperor to once again relocate to the Eastern Capital. Emperor Zhongzong, however, reacted with anger, exclaiming, "What kind of emperor goes begging for food?" Faced with his displeasure, the ministers had no choice but to abandon the proposal.[65] Such long and humiliating journeys to Luoyang in search of sustenance were, in effect, little different from fleeing famine.

At this time, a proposal was put forward to excavate a waterway along the upper reaches of the Dan River, allowing vessels to reach Shangzhou (with its administrative seat in present-day Shang County, Shaanxi). From Shangzhou, the route would

[63] Annals of Gaozu I. In *Book of Sui*, vol. 1.

[64] Annals of Tang 19. In *Comprehensive mirror for aid in government*, vol. 203.

[65] Annals of Tang 25. In *Comprehensive mirror for aid in government*, vol. 207.

ascend the Qinling mountain paths and exit the range at Shimen (located approximately forty li southwest of present-day Lantian County, Shaanxi), continuing north to Lantian, where it could connect with established towpath.[66] If this route had been successfully opened, it would have allowed the transport of tribute grain from south of the Qinling Mountains, thereby avoiding the perilous section of the Yellow River near the *Dizhu* rock formation. This proposal bore similarities to Emperor Wu of Han's earlier efforts to construct the Baoxie Road. Although some work was undertaken at the time, the project ultimately failed to come to fruition. The Dan River was a natural waterway, and its middle and lower reaches were navigable for transport. The section addressed during this period pertained only to the river's upper reaches and did not involve the excavation of an entirely new watercourse in the region. Thus, it cannot be properly classified as a canal. The main road from Chang'an to Shangzhou and regions south of the Qinling Mountains passed through Lantian County. Shimen, however, lay in a remote area approximately forty li southwest of Lantian. Constructing a road through the mountains in that area posed significant challenges. The rationale behind the proponent's choice of such a circuitous route was not clearly explained. Since a stream flowed out of the mountains at Shimen—presumably the Shimen Valley stream,[67] known today as the Tangyu River—it is likely that the suggestion was made with the intention of utilizing this waterway. This same stream had been diverted into the capital[68] in the early Tang period, possibly for urban water supply, but no evidence suggests it was ever used for transport. Naturally, it would not have been capable of supporting vessels, nor could it have been developed into a navigable canal.

Although the transportation of tribute grain to Chang'an was consistently difficult during the Tang dynasty, the city remained the imperial capital—except during the reign of Empress Wu, when the court resided permanently in Luoyang. At the time, several powerful nomadic groups still posed threats in the northwest. While the Tang court was generally able to exert control over them, latent risks persisted. Consequently, even though emperors frequently fled famine-stricken regions, they prioritized the foundational stability of the state and refrained from permanently relocating the capital to the east. However, in the periods following the Tang dynasty, the geopolitical situation in the northwest remained precarious, while the persistent difficulties in transporting goods increasingly influenced decisions regarding the location of the capital. Eventually, the national capital was moved eastward. One might argue, perhaps with some gravity, that this eastward relocation was essentially precipitated by the natural obstacles presented by the *Dizhu* section of the Yellow River and the sediment-laden, shallow waters of the Wei River.

---

66 Biography of Cui Shi. In *Old book of Tang*, Vol.74; *New book of Tang*, Vol.99.

67 Song Minqiu. Lantian. In *Gazetteer of Chang'an*, vol. 16.

68 Canals and waterways II. *Outstanding models from the storehouse of literature*, vol.497.

Yangzhou

The canal extending from the Tongji Canal to the southeast served as the principal artery of domestic transportation at the time. This was not only because it connected the wealthy southeastern regions, but also because it acted as the main thoroughfare linking Yangzhou, Yizhou, southern Hunan, Jiaozhou, Guangzhou, and central Fujian to Chang'an.[69] In essence, this canal system constituted the sole waterway transportation hub connecting regions south of the Yangtze River with the Central Plains. Those traveling far to Lingnan inevitably had to use this route. As mentioned earlier, Li Ao traveled to Lingnan in the fourth year of the Yuanhe era (809 CE) by way of this very route.[70] Those intending to travel to Western Shu (modern Sichuan) did not typically cross the Qinling and Bashan Mountains directly, braving the perils of the plank roads; rather, they often took this indirect route. At the end of the Tang dynasty, Wei Zhuang fled to Shu to escape turmoil by way of these canals. In his poem *Qin Woman's Lament*, he vividly depicts the hardships endured by a woman from Qin traveling to Bashu along this route during times of chaos.[71] Unless this waterway was impassable, travelers would only reluctantly resort to alternative routes. For example, in the second year of the Changqing reign (822 CE) under Emperor Muzong, the poet Bai Juyi was appointed governor of Hangzhou. Due to the inaccessibility of the canal waterways, he took the Xiang and Han rivers route instead. In his *The Memorial of Xie on Assuming the Office of Governor of Hangzhou*, from *The Collected Works of Bai* (Changqing Period)—he notes: "Since the Bian River waterway has yet to be fully navigable, officials must proceed via the Xiangyang and Han River route to reach their posts." This clearly reflects a situation of necessity forcing a detour.

The waterway's vibrant traffic naturally fostered the simultaneous rise of economic centers along its banks. The foremost economic metropolis at that time was Yangzhou (present-day Yangzhou City, Jiangsu Province). Yangzhou's prosperity dates back quite early, originating from the era when King Fuchai of Wu opened the Han Canal; Hancheng (Han City) stood at the waterway's junction where the Yangtze River entered the canal. Hancheng later evolved into Yangzhou. The navigability of the Han Canal often directly influenced Yangzhou's fortunes. By the time Emperor Yang of the Sui Dynasty undertook the construction of the Grand Canal, Yangzhou had already developed into the South's most important economic hub. Emperor Yang's multiple southern tours, ultimately ending with his death in Yangzhou, were largely motivated by the city's wealth and prominence. Following the canal's completion, Yangzhou's prosperity surged even further. Contemporary accounts praised it as "Yangzhou first, Yizhou second,"[72] signifying Yangzhou as the nation's premier economic metropolis, with Yizhou (present-day Chengdu, Sichuan) ranking second. Tang dynasty poets often praised Yangzhou, portraying it almost like

---

[69] Henan prefecture. In *Yuanhe maps and records of prefectures and counties*, vol. 5.

[70] Li Ao. Record of my journey to the south. In *Collected works of Li Wengong, vol. 18.*

[71] Chen Yinke. Reading "The Qin woman's lament". *Tsinghua Journal*, 11 (4).

[72] Annals of Tang 75. In *Comprehensive mirror for aid in government*, vol. 259.

a paradise. Although their descriptions sometimes tended toward exaggeration, they nonetheless reflect the city's remarkable prosperity.

At that time, Yangzhou was not only the foremost economic metropolis within the country but also held a significant position in foreign trade. In the first year of the Shangyuan era under Emperor Suzong (760 CE), Tian Shengong, the military governor of Pinglu Circuit, marched from Ziqing (with its administrative center in present-day Dongping County, Shandong Province) down the Ji and Huai Rivers into Yangzhou. There, he plundered the residents' wealth, ransacked houses, and massacred several thousand merchants of Persian and Central Asian origin.[73] The significant number of Persian and Central Asian merchants residing in Yangzhou clearly demonstrates the city's flourishing foreign trade. When compared to the Warring States and Qin-Han periods, Yangzhou occupied a position similar to that of Tao (modern Dingtao County, Shandong). However, while Tao was solely a domestic economic center, Yangzhou was not only the most prosperous city within the country but also served as a vital port for international commerce. In contemporary terms, Yangzhou at that time could be likened to Shanghai today.

Bianzhou

Situated in the opposite direction from Yangzhou was Bianzhou, which corresponds to present-day Kaifeng in Henan Province. While Yangzhou was located at the southern entrance of the Han Canal, Bianzhou lay at the northern entrance of the Tongji Canal. These two cities, positioned at opposite ends of the canal, effectively controlled its two termini. Although Bianzhou's prosperity could not rival that of Yangzhou, it nevertheless stood as a significant economic hub within the Central Plains. However, when the Tongji Canal was initially opened, Bianzhou's economic potential had yet to be fully realized. Although the administrative unit of Bianzhou was first established during the reign of Emperor Xuan of the Northern Zhou, it was abolished in the second year of the Daye reign of Emperor Yang of the Sui (606 CE), and the counties of Kaifeng and Xunyi were incorporated under the jurisdiction of Zhengzhou.[74] In the second year of the Daye reign (606 CE), the Tongji Canal had already been opened to navigation. Despite this development, Bianzhou was still abolished as an administrative unit, indicating that it had not yet attained significant strategic importance at the time. Zhengzhou, also known as Yingyang Commandery during the Sui dynasty, lay to the east of Liang Commandery, whose administrative seat was at Songcheng (present-day Shangqiu County, Henan). According to the *Geographical Treatise of the Book of Sui*, the local customs of these two commanderies were described as "highly valuing agriculture and placing great importance on ritual propriety and moral education," suggesting that they had not yet been notably influenced by the Tongji Canal. By the Tang dynasty, however, the situation changed

[73] Biography of Tian Shengong. In *Old book of Tang*, vol. 124.

[74] Bianzhou. In *Yuanhe maps and records of prefectures and counties*, vol. 7.

dramatically, particularly from the mid-Tang period onward. At that time, Daliang, the location of Bianzhou, had become "a strategic hub in the empire, commanding both land and water routes, serving as a vital chokepoint for the Hebei region, and functioning as a critical artery for grain transport between the Jianghuai region and the two Lakes."[75] This strategic importance became even more pronounced during the period when the regional military governors (fanzhen) held significant power.

When the troops in Bianzhou expelled their military governor Liu Shining, Lu Zhi, in a court discussion of the matter, remarked that "the Liang-Song region (present-day eastern Henan) occupies a critical strategic location: it can exert pressure on the Qi and Lu regions (modern Shandong) while also controlling the vital arteries of transportation in the Jianghuai area (the Yangtze and Huai River basins)."[76] Such a region, he argued, could not be overlooked. Later, when the troops of Bianzhou recalled the beneficence of Liu Shining's father, Liu Xuanzu, they requested that his former subordinate—and nephew—Han Hong be appointed as military governor. The Tang court acceded to this request, issuing an imperial edict that declared: "The Liang-Song region, as a hub of land and water transportation, serves both as the lifeline of the grain transport system and as a crucial bulwark for the protection of the imperial house."[77] This clearly set it apart from ordinary prefectures. Following the fall of the Tang, the significance of Bianzhou was further underscored by the fact that four of the Five Dynasties chose it as their capital. The implications of this are self-evident.

Runzhou

Another major economic hub was Runzhou, located directly across the Yangtze River from Yangzhou. Known as Jingkou during the Southern and Northern Dynasties, Runzhou corresponds to present-day Zhenjiang in Jiangsu Province. As early as that period, Jingkou had already become a significant stronghold in the southeast, primarily because of a waterway that connected it to the heartland south of the Yangtze River. After Emperor Yang of the Sui dynasty excavated the Jiangnan Canal, this location continued to serve as a crucial river-mouth junction linking to the canal network, making it easy for the area to regain its former prosperity.

Runzhou was officially established in the fifteenth year of the Kaihuang reign under Emperor Wen of the Sui (595 CE), but at the beginning of the Daye era, it was abolished as a prefecture. Its administrative seat was reorganized as Yanling

---

[75] Liu Kuanfu. Inscription on the wall of the judicial office in Bianzhou. In *Complete literature works of Tang dynasty*, vol.740.

[76] Lu Zhi. Memorial on the expulsion of Liu Shining from Bianzhou. In *Collected works of Minister Lu Xuangong*, vol. 20.

[77] Bai Juyi. Imperial edict to Han Hong. In *Collected works of Bai Juyi from the Changqing era*, vol. 40.

County under the jurisdiction of Yangzhou.[78] This reclassification might suggest that Runzhou held limited importance at the time. However, upon closer examination, this was not actually the case. As noted in the *Geographical Treatise of the Book of Sui*, "Jingkou (modern-day Zhenjiang, Jiangsu) connects eastward to Wu and Kuaiji commanderies, southward to the Yangtze River and the Taihu Basin, and westward directly to the capital Jiankang, making it a strategic and prominent regional metropolis." The references to Wu and Kuaiji here clearly indicate the Wu and Kuaiji commanderies. The rivers and lakes to the south of Jingkou undoubtedly refer to Tai Lake and the Zhejiang water system—there is little ambiguity in this. As for the phrase "westward directly to the capital Jiankang," it is somewhat general in scope, but it likely includes Jiangning, which, during the Eastern Jin and Southern Dynasties, was part of the old capital Jiankang. In the Sui period, Jiangning was designated as Danyang Commandery. The Book of Sui highly praised Danyang, stating: "Danyang (present-day Nanjing, Jiangsu), as the former imperial capital, had commercial districts and marketplaces whose prosperity could rival those of the twin capitals, Chang'an and Luoyang." In other words, Danyang was comparable to the empire's foremost cities.

The continued flourishing of Danyang was undoubtedly due in part to the lingering cultural and political influence of the Eastern Jin and Southern Dynasties. However, its ability to rely on and connect through Jingkou was also a key contributing factor. In fact, positioned directly across the river from Yangzhou, both cities were vital termini of the Grand Canal system. Given Yangzhou's prosperity, it is implausible that Runzhou would have languished in obscurity. That said, since Yangzhou was located north of the Yangtze River, it served as the principal departure point for goods transported from both the upper and lower reaches of the river, as well as from regions south of the Yangtze, destined for Chang'an and Luoyang. In this respect, Runzhou could not quite match Yangzhou's logistical significance. Nevertheless, Runzhou still held considerable strategic and economic weight. For a prolonged period during the Tang dynasty, the regional governor of Zhexi Circuit (Western Zhejiang Circuit) administered the circuit from Runzhou[79]—testament to the city's economic vitality and its critical geopolitical location. Since the Six Dynasties period, Jiangning—corresponding to present-day Nanjing—had long been a major city in the southeast and had even served as a national capital. However, during the Tang dynasty, the administrative seat of the Zhejiang Western Circuit was not established in Jiangning, but rather in Runzhou. This choice undoubtedly reflects the fact that, at the time, Runzhou's condition surpassed that of Jiangning. Historically, Jiangning was known as Jinling, but by this period, some people had begun to refer to Runzhou by the name Jinling as well.[80] Such prosperity did not diminish even in the Song dynasty. A Song

[78] Geographical treatise. In *Book of Sui, vol. 31.*

[79] According to the Old Book of Tang (vol. 41): *Geographical Treatise V*, "Since the reign of Emperor Daizong of Tang during the Yongtai era (765–766), Runzhou (present-day Zhenjiang, Jiangsu Province) has long served as the administrative seat of the Inspectorate of Zhejiang West Circuit."

[80] Li Deyu states in the *Preface to the Fu of Drums and Pipes. Collected Works of Duke Li Weigong, Huichang First Rank Collection: Separate Collection, Vol. 1*: "In former years, I held the tally

poet once wrote, “The waterways opened by Yu the Great and later dredged by the Sui dynasty continue to bring benefits across the ages, allowing boats from the lands of Wu and Chu to pass unhindered.” Another poem states, “The sails of the boats still reflect the green hues of the mountains of Min,”[81] subtly revealing the enduring vitality of this maritime network. Particularly in the latter verse, one can discern that seagoing vessels from the coastal regions of Fujian (Min) still frequented the harbors of Runzhou.

Weizhou and Youzhou

Along the banks of the Yongji Canal, two economic centers also emerged: one located in the southern section of the canal, corresponding to present-day Daming County in Hebei Province, known as Weizhou; the other situated at the northern terminus of the canal, which was Zhuojun during the Sui Dynasty and Youzhou during the Tang Dynasty—today’s Beijing. Youzhou’s position at the canal’s endpoint naturally evokes the example of Linzi during the Warring States and Han periods. Linzi was also located at the terminus of a waterway transportation system, which contributed to its sustained prosperity as an economic hub. Although the local products surrounding Youzhou were far less abundant compared to those of Linzi, Youzhou possessed an advantage that Linzi did not: its role in foreign trade. Youzhou was a crucial gateway for trade with the northern frontier, much as Yangzhou was a major port for overseas commerce. Just as Yangzhou was home to numerous Persian merchants and settlers, Youzhou was inhabited by many non-Han peoples from beyond the northern borders. Among them was An Lushan, one of the principal instigators of the An Lushan Rebellion in the mid-Tang period. It was precisely Youzhou’s wealth and prosperity that enabled An Lushan to gather sufficient strength to rebel. After the An Lushan Rebellion, the Lulong Jiedushi—whose seat was Youzhou—although nominally under Tang control, operated almost as an independent state. The affluence of these local regions was undoubtedly a significant factor contributing to the rise of such rebels and separatist regimes.

Weijun and Ye were located close to each other. The stark contrast between the rise of one city and the decline of the other was largely due to the shifting influence of the canal in these two locations. As previously mentioned, Ye’s prosperity resulted from the excavation of the Baigou Canal; naturally, when the Baigou ceased to flow, Ye began to decline. After Ye was destroyed, the Northern Zhou deliberately sought to replace it with Anyang (present-day Anyang City, Henan Province). During the reign of Emperor Xiaowen of the Later Wei, Xiangzhou was established

---

and took office in Jinling (present-day Nanjing, Jiangsu).” The Jinling mentioned here refers to Runzhou. Li Deyu once served as Inspector of Zhejiang West, whose administrative seat was in Runzhou, thus naturally not referring to Jiangning.

[81] Wang Xiangzhi. Zhenjiang Prefecture. *Topographical records of the imperial realm*, Vol. 7, quoting poems by Xiong Shumao and Cai Tianqi.

at Ye. Later, the Northern Zhou moved the Xiangzhou administrative seat to Anyang and renamed Anyang as Ye.[82] These were merely human interventions. Given that Anyang was both distant from the Zhang River and lacked a canal to replace the Baigou, it was difficult for it to continue Ye's former prosperity. The construction of the Yongji Canal, which supplanted the old Baigou route, enabled Weizhou to assume Ye's former status. Because the Yongji Canal facilitated the transport of goods to and from the Jianghuai region, Weijun prospered even more.[83] The fact that Tang Dynasty officials responsible for the Hebei Circuit managed Weizhou instead of Ye reflects this shift.[84] However, it should be noted here that Qinghe, located north of Weizhou, also lay adjacent to the Yongji Canal. During the Tang dynasty, to supply the northern garrisons, all the taxed cloth transported from the Jianghuai region was accumulated at Qinghe, which at the time was known as the "Northern Treasury of the Realm."[85] Clearly, Qinghe was also a fairly important city within the Yongji Canal basin. However, Qinghe was still less significant than Weizhou. Qinghe's status as the "Northern Treasury of the Realm" was due to the accumulation of taxed cloth transported from the Jianghuai region. The concentration of Jianghuai's taxed cloth in Qinghe was because of its relative proximity to Youzhou in the north, allowing military supplies to be conveniently accessed. In terms of the Yongji Canal, Qinghe was located roughly in the middle reaches of the waterway, which meant it could not compare to Weizhou or Youzhou in importance. Unlike Youzhou, which served as a major distribution center for goods arriving from the northern frontier, or Weizhou, where supplies from Jianghuai first arrived, Qinghe did not hold such pivotal roles. Weizhou's advantageous geographical position made it an important urban center at the southern end of the Hebei Plain. After the An Lushan Rebellion, the three Hebei garrisons were established precisely within the Yongji Canal basin. The Weibo military governor, based in Weizhou, resisted imperial orders, much like the Lulong military governor. Although Weizhou, located near the Central Plains, was under long-term control by Hebei military governors alongside places like Youzhou, it still benefited from the Yongji Canal's resources, thus maintaining a certain degree of prosperity. This prosperity continued through the late Tang and Five Dynasties periods. Since the Later Tang established this place as a secondary capital1, through the Later Jin, Later Han, Later Zhou, and even into the Song dynasty, although the name of Weizhou underwent successive changes, the system of maintaining it as a secondary capital was never abolished2. During the Southern Song period, Liu Yu, under the protection of the Jurchen Jin dynasty, established the so-called puppet state of Qi, with its capital also located here. The fate of Weizhou mirrored that of Yecheng; as the transportation capacity of the Yongji Canal declined, Weizhou likewise fell into decline.

---

[82] Xiangzhou. In *Yuanhe maps and records of prefectures and counties*, vol. 16.

[83] Geographical treatise III. In *New book of Tang*, vol. 39.

[84] Ibid.

[85] Biography of Yan Zhenqing. In *New book of Tang*, vol. 153.

The Extension of the Guangtong Canal

Emperor Wen and Emperor Yang of the Sui dynasty constructed a nationwide network of waterways, to the immense benefit of future generations. During the subsequent Tang dynasty, this transportation network saw no major expansions. Instead, efforts were directed toward dredging and minor repairs. Dredging ensured that the existing infrastructure remained functional, while repairs served to reinforce the network and increase its density. The western terminus of the Guangtong Canal during the Sui dynasty reached only as far as Chang'an. In the Tang dynasty, however, a new canal—called the Shengyuan Canal—was excavated to the north of the Wei River (渭水), west of Chang'an, effectively extending the Guangtong Canal further westward. The project was undertaken during the reign of Empress Wu Zetian. The Shengyuan Canal drew water from the Qian River northwest of Guo County (present-day Baoji County) and flowed eastward, ultimately emptying into the Wei River at Xianyang. The canal's primary purpose was the transportation of timber from the regions of Qi and Long Prefectures. Qi Prefecture was centered in present-day Fengxiang County, Shaanxi, and Long Prefecture in present-day Long County, Shaanxi.[86]

Given its function as a timber transport route, this canal differed from typical irrigation canals used for agriculture. Since the canal was named "Shengyuan," some mistakenly interpreted its name as indicating that the Qian River was diverted to flow across a plateau. This, however, is a superficial reading of the name. The Qian River flows southward along the western foothills of the steep Wuzhi Plateau. Since the river valley lies considerably lower in elevation, it would not have been possible to draw canal water uphill onto the plateau. Such an interpretation therefore reflects a misunderstanding of the local topography. The seat of Guo County was situated south of the Wuzhi Plateau, within the Wei River valley. The Shengyuan Canal likely drew water from a slightly lower southwestern point of the plateau, flowing eastward along its base. The canal ran along the edge of the plateau and at an elevation higher than that of the Wei River valley—hence the name "Shengyuan," which connotes "rising above the plain."

Later, in the third year of the Xiantong reign of Emperor Yizong of Tang (862 CE), an additional diversion was constructed at the eastern end of Baoji County, drawing water from the Wei River into the canal.[87] The intake point was likely located downstream from the confluence of the Qian and Wei Rivers. This enhancement was presumably made necessary when the Qian River's volume proved insufficient. The Shengyuan Canal continued eastward to Wugong County, where it joined the Liumen Weir and merged with the Chengguo Canal. East of the Liumen Weir, the Shengyuan

[86] *The New book of Tang, vol. 37: geographical records* states, "Northwest of Guo County in Fengxiang Prefecture lies the Shengyuan Canal, which diverts water from the Qian River to Xianyang. In the early years of the Chui Gong era, water from Qi and Long was transported into the capital." How could water from Qi and Long have been transported to the capital? Even if it were possible, what practical use would it have served? It is likely that the text mistakenly refers to Qi and Long water when it should have said Qi and Long timber.

[87] Song Minqiu. Xingping. In *Gazetteer of Chang'an*, vol. 14.

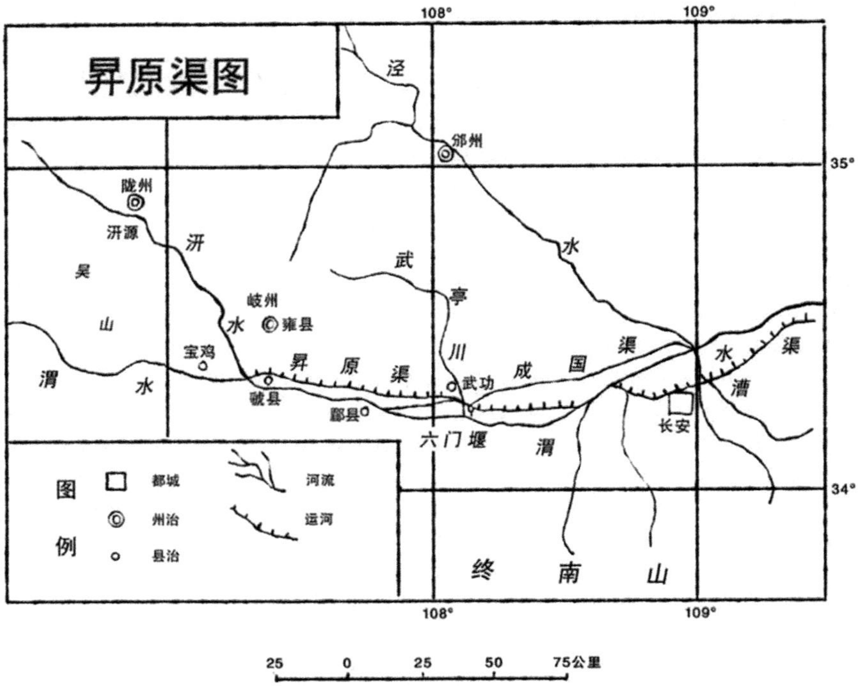

图例 - Legend
都城 Capital Cities
州治 Prefecture Seats
县治 County Seats
河流 Rivers
运河 Canals

Fig. 5.5 The Shengyuan Canal

and Chengguo Canals once again separated into independent channels. By the time it reached Xingping County, the Chengguo Canal ran eleven *li* (approx. 5.5 km) north of the county seat, while the Shengyuan Canal lay fifteen *li* (approx. 7.5 km) to the south, signifying a notable divergence in their courses.[88] Some have suggested that the Shengyuan Canal comprised two branches: one drawing water from the Wei River at Mei County and flowing eastward into Xianyang County before rejoining the Wei; the other drawing water from the Qian River northwest of Guo County and, with its middle and lower reaches known as the Chengguo Canal. This view, however, confuses the relationship between the two canals, overlooking the fact that the Chengguo Canal originally lay south of the Shengyuan Canal and only after passing the Liumen Weir, did it shift to the north (See Fig. 5.5).

During the Tang dynasty, several minor canals that existed in the Guanzhong region and outside the city of Chang'an also deserve brief mention. In the second

[88] Ibid.

year of the Tianbao reign of Emperor Xuanzong (743 CE), Han Chaozong diverted water from the Wei River into the city of Chang'an and constructed a pool in the Western Market to store timber.[89] In the first year of the Dali reign of Emperor Daizong (766 CE), Li Gan excavated a transport canal at the mouth of a valley in the hills to the south of Chang'an, which entered the imperial park through the Jingfeng and Yanxi Gates and was used to convey firewood and charcoal.[90] However, these canals were all relatively short in length. Also worth noting is that in the eighth year of the Wude reign of Emperor Gaozu (625 CE), Jiang Xingben constructed the Wujie Weir in Qianyuan County, Longzhou, to divert water from the Longchuan River for transport purposes.[91] Qianyuan County, a subordinate county of Longzhou, corresponds to present-day Long County. Apart from the Qian River, no other major rivers are found in this area, so the so-called Longchuan River most likely refers to the Qian River itself. Even if the Qian River was navigable, it could only support a short waterway. Later, in the second year of the Kaiyuan reign of Emperor Xuanzong (714 CE), Jiang Shidu opened the Fushui Canal in Huayin,[92] but this was merely intended to supplement the volume of the Wei Canal and cannot be regarded as a transport canal in the proper sense.

The Expansion and Improvement of the Bian River System

The Bian River flows southeast from Bianzhou. East of Bianzhou, the former region of Dingtao had become disconnected from this transportation system. To address this deficiency, Empress Wu Zetian ordered the construction of the Zhan Canal in the first year of the Zaichu era (689 AD), linking Bianzhou with the regions of Cao (whose administrative center was in present-day Dingtao County, Shandong) and Yan (in present-day Yanzhou County, Shandong) to facilitate the transport of tax

[89] Annals of Emperor Xuanzong. In *Old Book of Tang*, vol. 9; Geographical Treatise I. In *New Book of Tang*, vol. 37; Biography of Han Chaozong. In *New Book of Tang*, vol. 118; Song Minqiu. Chang'an. In *Chang'an Zhi*, vol. 12; and In *Examination of City Wards in the Two Tang Capitals (vol. 4)*: *Transport Canals*, Xu Song argues that because Longshou Plateau blocks the northern side of Tang-era Chang'an, it would have been impossible to divert the Wei River into the city, and thus he suggests that the water diverted by Han Chaozong came from the Yu River instead. I previously referenced Han Chaozong's account when discussing forests in the middle reaches of the Yellow River. However, during the Kaiyuan and Tianbao periods, the transport canals remained navigable, and these canals did in fact draw water from the Wei River, which flowed north of Chang'an. It is therefore not implausible that Han Chaozong diverted water from the canal into the Western Market. Xu Song seems not to have considered this possibility; his argument, based solely on the obstructive presence of Longshou Plateau north of Chang'an, that the Wei River could not have been diverted into the city, appears insufficient to fully resolve the issue.

[90] Annals of Emperor Daizong. In *Old book of Tang*, vol. 11; Geographical treatise I. In *New book of Tang*, vol. 37; Biography of Li Gan. In *New book of Tang*, vol. 145; Song Minqiu. Chang'an. In *Chang'an zhi*, vol. 12.

[91] Geographical treatise I. In *New book of Tang*, vol. 37.

[92] Ibid.

grain.[93] The Zhan Canal was constructed by diverting water from the Bian River into the Bai Canal. This Bai Canal began north of present-day Kaifeng and should not be confused with the Bai Canal into which Cao Cao diverted the Qi River. The Zhan Canal was a precursor to the Guangji River developed during the Song dynasty. It ran diagonally northeast from Bianzhou, passed through Caozhou, and emptied into the Juye Marsh. Since Yanzhou lay to the east of the marsh, this enabled the transport of tax grain from both Cao and Yan.

After the opening of the Zhan Canal under Empress Wu, further improvements continued into the Tang dynasty. In the early Changqing reign (821–824 AD) under Emperor Muzong, Cui Hongli oversaw the construction of the Mangshan Canal in Yanzhou, extending from Huangdui to Qingqiu.[94] The location of Qingqiu remains uncertain, but Huangdui, a known administrative settlement, corresponds to present-day Yutai County, Shandong.[95] To the northeast of present-day Jiaxiang County lies Mengshan, and to its east was the Huangong Canal.[96] This canal had originally been excavated by Huan Wen during his northern campaign against the Former Yan dynasty, east of the Juye Marsh. The Huangong Canal was still extant during the Northern Song period.[97]

At the time, Cui Hongli served as the military governor (Jiedushi) of the Tianping Army, whose headquarters was in Yunzhou—corresponding to modern Dongping County, Shandong. When Wang Zhixing, the military commissioner of the Xusi Circuit (headquartered in present-day Xuzhou, Jiangsu), ordered the transportation of grain for military purposes, he demanded the mobilization of 5,000 carts. Recognizing the difficulty of the long journey, Cui Hongli undertook the excavation of the Mangshan Canal to facilitate this supply route. The governor of Xusi Circuit was based in Xuzhou, which corresponds to present-day Xuzhou City, Jiangsu. From Xuzhou, the Si River could be navigated upstream, reaching Yanzhou. The Mangshan Canal constructed by Cui Hongli was essentially the dredging of the Huangong Canal. Since the Huangong Canal passed through Mangshan, it came to be known as the Mangshan Canal. The Mangshan mentioned here refers to present-day Mengshan in Jiaxiang County, and the confusion between "Mang" and "Meng" likely resulted from phonetic similarity. Because Yanzhou was already connected to the Bian River via the Zhan Canal, the opening of the Mangshan Canal further extended the Bian River's transportation network to the southwestern regions around Mount Tai (See Fig. 5.6).

Wu Zetian not only expanded the transportation network at the northern end of the Bian River but also undertook similar improvements at the northern terminus of the Han Canal. In the fourth year of the Chuigong era (688 CE), she ordered the excavation of a new grain transport canal near Lianshui County. This canal connected southward to the Huai River and northward to the regions of Haizhou, Yizhou, and

[93] Geographical treatise II. In *New book of Tang*, vol. 38.

[94] Biography of Cui Hongli. In *New book of Tang*, vol. 164.

[95] Jingdong Circuit. In *Treatise on the nine territories in the Yuanfeng reign*, vol. 1.

[96] Jining prefecture. In *Essentials of historical geography*, vol. 33.

[97] Yunzhou. In *Universal geographic record of the Taiping era*, vol. 13.

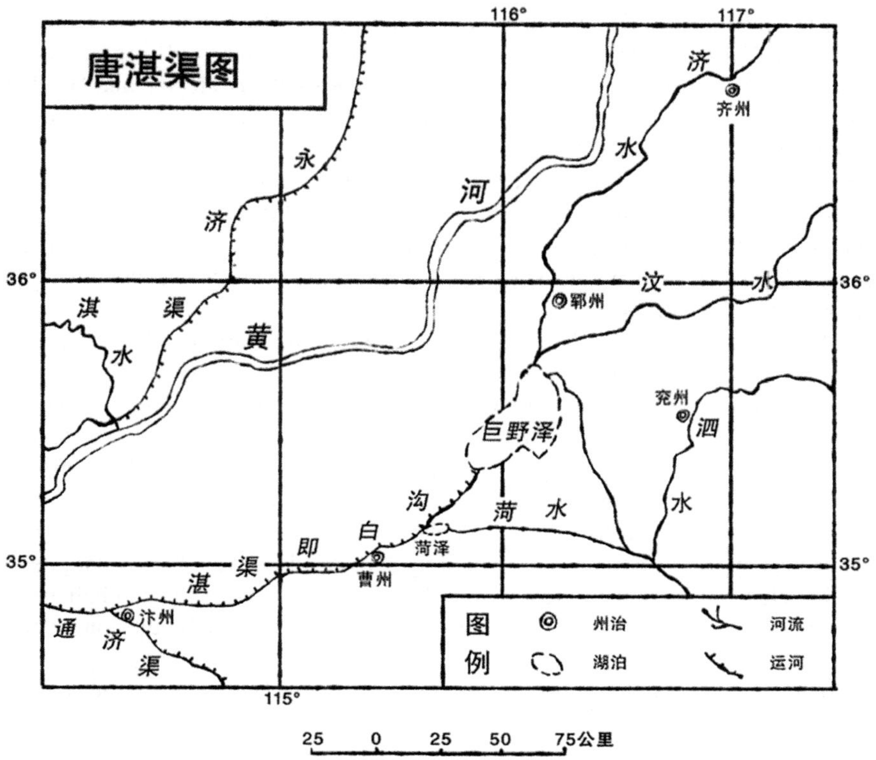

图例 - Legend
州治 Prefecture Seats
湖泊 Lakes
河流 Rivers
运河 Canals

Fig. 5.6 The Tang-era Zhan Canal

Mizhou (corresponding to present-day Lianyungang in Jiangsu, Linyi in Shandong, and Zhucheng in Shandong, respectively).[98]

The Shu River, which originates in the northern part of present-day Yishui County, Shandong, flows through present-day Ju County and Linshu. The modern Xingshu River flows southeast from Linshu County and enters the sea north of Lianyungang. During the Tang period, the river flowed south from Linshu County, passed through present-day Xinyi County and Shuyang County in Jiangsu, and eventually entered the sea. Present-day Shuyang County corresponds to the Shuyang County of the Tang dynasty. To the south of Shuyang County lies Lianshui County, which borders the Huai River. Therefore, this new canal most likely passed through the areas of Shuyang

[98] Geographical treatise II. In *New book of Tang*, vol. 38.

and Lianshui counties. Today, Lianshui County is still home to the Zhonglian River, Donglian River, and Xilinan River, which are likely remnants of this new canal. Since the Shu River flows through the territories of Mizhou, Yizhou, and Haizhou, the new canal served to connect the Huai River in the south with Haizhou, Yizhou, and Mizhou in the north. To the west of the Shu River lies the Yi River, which originates in present-day Yiyuan County, Shandong, and flows south through Linyi County. During the Tang dynasty, the Yi River reached Xiapi County (present-day southern Pi County, Jiangsu), where it entered the Si River. By connecting the Si River with the Huai River, the new canal shortened the travel route, avoiding a long detour along the Yi River.

More than two decades later, in the first year of the Taiji era (712 CE), under Emperor Ruizong, another canal was dug in Xuyi County (present-day Xuyi in Anhui). Known as the "Straight Canal" (Zhihe), it ran from Xuyi directly to Yangzhou.[99] The very name of the canal highlights its purpose: if navigable, it would allow grain transport vessels to bypass the Han Canal entirely, crossing the Huai River directly into the Bian River system. This was likely intended to avoid the treacherous floodwaters of the Huai River and to replace the older Han Canal route. However, the project appears not to have been entirely successful, as grain boats continued to use the Han Canal. Navigating the Huai River remained a significant challenge at the time, prompting various efforts at improvement—among them, the construction of the Straight Canal.

In the early years of the Kaiyuan era (713–741AD), Qi Han served as the governor of Bianzhou. He constructed a new waterway near the mouth of the Bian River where it enters the Huai River. This waterway extended from Xucheng to the Huai River, a distance of only eighteen li. Although short, this canal avoided the perilous currents of the Huai River, thereby facilitating the grain transport operations.[100] Xucheng was a county under Sizhou during the Tang dynasty, located to the northwest of present-day Xuyi County, Anhui. Today, Xuyi County, situated on the southern bank of the Huai River, directly faces Linhuai County of Sizhou, which marks the point where the Bian River meets the Huai River. The Huai River flows northeastward from the south of Linhuai County, creating a large bay where the Bian River and the Huai River meet. The construction of this new canal not only bypassed dangerous sections of the Huai River but also shortened the distance for shipping. However, it proved to be too short, and many issues remained unresolved.

Qi Han later returned to his post as governor of Bianzhou and sought to address these problems comprehensively. At the time, the shipping route between the Huai and Bian rivers extended 150 *li* from Hong County (present-day Si County, Anhui) to Linhuai. The traditional method of navigating this stretch involved using bamboo ropes to tether boats, which were then pulled by oxen. Even with this method, navigating the swift currents was difficult. To improve this situation, Qi Han initiated the construction of the Guangji New Canal, which extended thirty *li* east from Hong County, entering the Qing River, and then followed the Qing River for over a hundred

[99] Ibid.

[100] Biography of Qi Han. In *New book of Tang*, vol. 128.

li. After exiting Qing Shui, the canal was to be extended another eighteen li to enter the Huai River at Huaiyin (present-day north of Qingjiang, Jiangsu). The goal of this new canal was to bypass the dangerous rapids of the Huai River. However, upon its completion, the new canal proved no less turbulent, with the water flow equally fast and the riverbed filled with jagged stones, making navigation just as difficult as before. Consequently, the canal was abandoned.[101] The new canal was closely connected to the Qing River (also known as Qing Shui). At the time, the Qing River was navigable and could even serve as a substitute route for grain transport in place of the Bian River, suggesting that it was not a minor stream but likely a reference to the Si River. However, many scholars have associated the ancient course of the Si River in this region with the present-day Yuhuang River. If this identification is correct, the distance from Hong County to the Yuhuang River would greatly exceed thirty li, beyond the reach of the Guangji New Canal. This casts doubt on the identification, and it is possible that the ancient course of the Si River lay to the west of the present-day Yuhuang River. A conclusive resolution awaits future field research and on-site investigation.

The Waterway Where the Han Canal Enters the Yangtze River

The water outlet where the Han Canal flows into the Yangtze near Yangzhou remained largely unchanged for a long time following its relocation during the Yonghe era (345–356 CE) of Emperor Mu of the Eastern Jin dynasty. At that time, the canal's water was diverted via the Ouyang Dike, located in what is now Yizheng County. This intake point was sixty *li* from Guangling, resulting in a rather circuitous route. It was not until the 25th year of the Kaiyuan reign (737 CE), when Qi Han served as governor of Runzhou, that attention was drawn to the issue. Runzhou corresponds to present-day Zhenjiang in Jiangsu Province, located across the Yangtze River from Yangzhou. Grain shipped north from Runzhou had to travel along the Yangtze for sixty *li*, detouring past Guabuwei, where boats were often capsized or damaged by wind and waves, leading to heavy losses. In response, Qi Han rerouted the grain transport path to Jingkoutang, allowing vessels to cross the river directly—reducing the journey to just twenty li, a significant improvement in efficiency.[102] Jingkou was the historical name for Dantu County, the administrative center of Runzhou. The ferry crossing was located about three li northwest of present-day Zhenjiang. In the Tang dynasty, it was known as Suanshan Ferry, also referred to as Jingkou Port.[103] The "Jingkoutang" mentioned likely refers to this very location.

---

[101] Chronicle of Emperor Xuanzong II. In *Old book of Tang*, vol. 9; Literary works: biography of Qi Han. In *Old Book of Tang II*, vol. 190.

[102] Literary works: Biography of Qi Han. In *Old book of Tang II*, vol. 190.

[103] Zhenjiang Prefecture. In *Essentials of historical geography*, vol. 25.

After the grain had been ferried across the Yangtze, what became of it? Qi Han addressed this as well by excavating a new watercourse known as the Yilou River, which extended for 25 *li* and reached Yangzi County, thereby linking up with the original grain transport route. This ensured that grain boats could pass smoothly without obstruction.[104] One important detail warrants attention: during the Tang dynasty, the governor of Runzhou administered Dantu County, which lay on the southern bank of the Yangtze River. Yet, the Yilou River dug by Qi Han was located on the northern bank—territory originally belonging to Yangzhou. How, then, could the governor of Runzhou undertake a canal project in land under Yangzhou's jurisdiction? In fact, the Yangtze River between Yangzhou and Runzhou was especially wide, with the northern bank reaching up to the outskirts of Yangzhou city. Guazhou, located midstream in the Yangtze, was under Runzhou's jurisdiction. Over time, the river's current shifted southward, causing the waterway north of Guazhou to gradually silt up and form new land.[105] Although this newly formed land lay to the north, it remained part of Runzhou's administrative domain. As a result, Qi Han was able to excavate the Yilou River in this area, thereby ensuring that the Han Canal could still connect to the Yangtze River south of Yangzhou.

Following the efforts of Qi Han, the canals near Yangzhou continued to receive modifications and enhancements. In the fourth year of the Zhenyuan reign (788 AD) under Emperor Dezong, Du Ya constructed a canal to the west of Yangzhou, tracing the eastern foothills of Shugang (Shugang Hill). He diverted water from the Pei River to the outskirts of the city, where it entered the Grand Canal, thus aiding the transport of tribute grain.[106] Later, in the second year of the Baoli reign (826 AD) under Emperor Jingzong, Wang Bo observed that the section of the Grand Canal running through the city of Yangzhou was shallow. During periods of drought, grain transport boats struggled to navigate it. To remedy this, he excavated a new canal segment starting from Sifang Port, approximately seven li west of the southern Changle Gate. The waterway curved eastward and joined the old canal at the Chanzhi Temple Bridge. This new stretch, measuring 19 li in length, was somewhat deeper, which greatly facilitated navigation.[107] These projects were intended not only to supplement water sources and dredge sediment, but also to address problems caused by the dynamic flow and sedimentation patterns of the Yangtze River.

---

[104] Literary works: Biography of Qi Han. In *Old book of Tang II*, vol. 190.

[105] Shi Nianhai. *On the economic region of Yangzhou and the lower Yangtze River during the Tang dynasty*.

[106] Geographical treatise V. In *New book of Tang*, vol. 41.

[107] Biography of Du Ya. In *Old book of Tang*, vol. 146; Geographical treatise V. In *New book of Tang*, vol. 41.

Improvements to the Southern Grand Canal

During the Tang dynasty, the Southern Grand Canal underwent significant expansion. The key figures behind these developments were Meng Jian, who created the Mengdu Canal, and Li Su, who developed the Yuanhetang Canal. The Mengdu Canal, located 40 *li* west of Wujin County (present-day Changzhou, Jiangsu, which was the capital of Changzhou at the time), was constructed by Meng Jian in the eighth year of the Yuanhe era (813 AD) during the reign of Emperor Xianzong. The Mengdu Canal, stretching 41 *li*, connected the northern part of the canal to the Yangtze River, enabling the flow of water southward for grain transport. It is said that Meng Jian initiated the canal construction due to specific circumstances,[108] though the exact reasons for its creation are unclear. The canal was likely named after Meng Jian, indicating his involvement in its development. The Yuanhe Canal, located in Changshu County (present-day Changshu, Jiangsu), was built by Li Su in the fourth year of the Yuanhe era (809 AD). It extended from the southern part of Changshu County all the way to the northern part of Changzhou County (modern Suzhou, Jiangsu), linking with the Southern Grand Canal.[109] This canal was later renamed Yunhe Canal.

In the tenth year of the Yuanhe era (815 AD), Meng Jian also opened a new river in Kuaiji County, Yuezhou (present-day Shaoxing County, Zhejiang, which was the capital of Yuezhou at the time). This new river was located five *li* north of the city of Kuaiji and marked the beginning of the Eastern Zhejiang Canal during the Song dynasty. Ten *li* northwest of Kuaiji County, Meng Jian also constructed a navigation canal, which was named the "Yundao" (transportation canal), indicating that it was navigable.[110] Additionally, in the 18th year of the Zhenyuan era (802 AD) during Emperor Dezong's reign, a navigable canal was constructed in Yuqian County (present-day Yuqian, Zhejiang).[111] This canal, which was 30 *li* long, allowed for boat traffic, though the precise starting and ending points are unknown. While it was small in scale, it effectively served as a minor canal for local water transport.

The new river in Kuaiji and the waterway in Yuqian mentioned here do not fall within the scope of the Southern Grand Canal. However, for the sake of convenience, they are included here.

---

[108] Biography of Meng Jian. In *Old book of Tang*, vol. 163; Geographical treatise V. In *New book of Tang*, vol. 41.

[109] Geographical treatise V. In *New book of Tang*, vol. 41.

[110] Biography of Meng Jian. In *Old book of Tang*, vol. 163; Geographical treatise V. In *New book of Tang*, vol. 41.

[111] Geographical treatise V. In *New book of Tang*, vol. 41.

The Various Canals of Hebei

The Yongji Canal in Hebei also saw the construction of many branches during the Tang Dynasty. The waterway of the Yongji Canal, starting below Weizhou (now Hebei's Daming County), essentially corresponds to the present-day Wei River and the southern section of the Grand Canal within Hebei Province. As the watercourse lay predominantly on the eastern side, the rivers flowing down from the Taihang Mountains all had a necessary connection with the Yongji Canal. People in the Tang Dynasty devoted considerable effort to this project. During the Yonghui reign (650–655 AD) of Emperor Gaozong, Li Linggui oversaw the excavation of a branch of the Yongji Canal in Weizhou, directing its flow into Xinshi to facilitate commercial travel. His work earned the admiration of the local population.[112] Later, in the twenty-eighth year of the Kaiyuan era (740 AD) under Emperor Xuanzong, Lu Hui undertook the relocation of part of the canal in Weizhou, diverting water from the Shihui depression to the Wei Bridge west of the city, in order to enable the transportation of goods from the Yangtze and Huai river regions.[113] The fact that Weizhou could access and transport products from the Jianghuai region via the Yongji Canal speaks to the city's flourishing prosperity.

In addition, there were several notable canals in various regions: in Beizhou Jingcheng (now Weixian, Hebei), there was the Zhangjia River; in Jizhou Nangong (now Nangong County), there was the Tongli Canal; in Tangyang (now Xinhe County, Hebei), there was the Tangyang Canal; in Zhaozhou Zhaoqing (east of Longyao County, Hebei), there was the Fengshui Canal; in Cangzhou Qingchi (now Cang County, Hebei), there were the Qingchi Canal, the Wudi River, and the Yangtong River; Wudi (now Wudi County, Shandong) also had the Wudi Ditch; in Dezhou Pingchang (now in Depeixian, Shandong), there was the Xinha River; in Yingzhou Hejian (now Hejian County), there was the Changfeng Canal; and in Jizhou Yuyang (now Jixian, Tianjin), there was the Pinglu Canal.[114] These canals, both directly and indirectly, were all connected to the Yongji Canal.

The Zhangjia River in Jingcheng County, Beizhou, was originally an ancient channel from the Western Han period.[115] Near the former city site of Guangzong County, the river split into two branches; the left branch flowed north through Cheng County and into the Jiang River northwest of Nangong County.[116] The Zhangjia River excavated during the Tang dynasty followed the course of this historical waterway. As for the Tongli Canal in Nangong County, some sources claim it diverted water from the Zhang River to irrigate farmland,[117] suggesting that it served no transportation function. However, *the New Book of Tang* does not include such a description. The

[112] Biography of the Sons of Gaozu: Prince Chu Zhiyun. In *Old book of Tang*, vol. 64.

[113] Geographical treatise III. In *New book of Tang*, vol. 39.

[114] Ibid.

[115] Geographical treatise. In *Book of Han*, vol. 28.

[116] Commentary on the Yellow River. In *Water classic*.

[117] Jizhou. In *Grand unified gazetteer of the Qing dynasty*, vol. 49.

Tangyang Canal in Tangyang County, located thirty *li* southwest of Tangyang, originates from the Julu County and flows into Nangong County.[118] Nangong County, being traversed by the old Jiang River channel, was likely the river into which the Zhangjia River flowed, and the Tangyang Canal most likely also fed into this ancient channel of the Jiang River. The Fengshui Canal in Zhaoqing County, Zhao Prefecture, was likely the result of efforts to improve the Fengshui River. The Fengshui River is the downstream portion of the Sha River,[119] with the old channel in Longping County lying about ten li east and flowing into the current Ningjin County region,[120] ultimately joining the Zhang River. During the Tang Dynasty, Cangchi County had three main transport canals: the Qingchi Canal, the Yangtong River, and the Wudi River. According to the *New Book of Tang*, the Qingchi Canal was located twenty *li* southeast of Cangchi County and merged with the Maoshi River, while another canal, seventy *li* southeast, connected to the Zhang River, diverting floating waters from the region. This is likely the so-called Qingchi Canal. The Fushui River originated from a diversion of the Yongji Canal south of Dongguang County. It flowed northeastward, passing ten li south of Cangzhou, before ultimately emptying into the sea to the northeast.[121] *The New Book of Tang* also mentions the Yangtong River, which lay fifteen li southeast of Qingchi County. This river was also an artificial canal, though its exact origin and endpoint are unclear. In Wudi County, there was also the Wudi Ditch, which provided an outlet to the sea. Although the Wudi River and the Wudi Ditch are referred to by distinct names, they are, in fact, historically a single watercourse. The separate nomenclature arises from their traversal of two different counties and the disparate periods in which records or gazetteers were compiled. This particular river originates from a branch of the Qinghe River in Dongguang County, flowing south through Nanpi County.[122] To the northeast of Nanpi County lies Qingchi County. The Wudi River mentioned in the *New Book of Tang* located fifty-seven *li* southwest of Qingchi County, which corresponds accurately to the described location. It also flows twenty *li* south of Rao'an County,[123] eventually reaching Wudi County, before continuing eastward into the sea. As for the new river in Pingchang County, it is identified as the Majia River.[124] In Hejian County, there are two Changfeng Canals: one is located one hundred li to the northwest of the county, opened during the Zhenguan period, and the other lies five *li* southwest of the county, opened during the Kaiyuan period. The route of the earlier canal remains unclear, but the later one drew water from the Hongtuo River and flowed through Shucheng and Pingshu counties before emptying into the Qi River.[125] Shucheng is

[118] Geographical treatise III. In *New book of Tang*, vol. 39.

[119] Shunde prefecture. In *Comprehensive gazetteer of Qing dynasty (Jiaqing edition), vol. 30.*

[120] Zhaozhou prefecture. In *Comprehensive gazetteer of Qing dynasty (Jiaqing edition), vol. 51.*

[121] Cangzhou. In *Universal geographic record of the Taiping era*, vol. 65.

[122] Commentary on the Qi River. In *Water classic.*

[123] Cangzhou. In *Yuanhe maps and records of prefectures and counties*, vol. 18.

[124] Dezhou. In *Yuanhe maps and records of prefectures and counties*, vol. 17.

[125] Geographical treatise III. In *New book of Tang*, vol. 39.

located to the northeast of present-day Hejian County, and Pingshu is in the present-day Dacheng County. The Qi River is equivalent to the Yongji Canal. The rivers and canals mentioned here—Zhangjia River, Tongli Canal, Tangyang Canal, Fengshui Canal, Qingchi Canal, Yangtong River, Wudi River, Wudi Ditch, and Changfeng Canal—are all related to the Hongtuo River and the Zhangshui River. Both the Hongtuo River and the Zhangshui River are large rivers situated to the east of the Taihang Mountains. Since both of these rivers feed into the Yongji Canal, the canal's influence extended across a broad swath of territory east of the Taihang range. Thus, the success of the Yongji Canal and the related canal infrastructure led to the creation of an extensive and interconnected water transport network in the region east of the Taihang Mountains (See Fig. 5.7).

As for the Pinglu Canal in Yuyang, Jizhou, it runs alongside the sea and serves as a waterway for transporting goods, designed to avoid the dangers of the sea.[126] This represents a different type of canal, one that is not part of the Yongji Canal system.

The Yaoxian Canal and Other Smaller Canals

In addition to the grand canal system constructed by Emperor Yang of the Sui Dynasty, there were several smaller-scale canals worth mentioning. Near the salt pools in Hedong, there was a small canal flowing southwest, known as the Yaoxian Canal, which was specifically used for transporting salt produced in the salt pools. The canal was named after Yao Xian, who dredged it during the Sui Dynasty's Daye period (605–618 CE). In fact, its origins can be traced back to the Yongfeng Canal, which was dug by Yuan Qing during the reign of the Later Wei Dynasty (504–508 CE). However, the Yongfeng Canal fell into disrepair shortly after its completion, rendering it largely ineffective. The Yaoxian Canal remained functional until the early Song Dynasty, at which point it began to silt up. During the reign of Emperor Renzong of Song, the canal was dredged once more, but the name was changed back to Yongfeng.[127] The Yaoxian Canal is located to the south of the Su River and is not far from it. The lower reaches of the canal merge with the Su River, which was already a well-known waterway in ancient times. In the early Tang Dynasty, Xue Wanche also dug the Su River Canal along the side of the Su River, extending from Wenxi to Linjin, running parallel to the Yaoxian Canal.[128] Although the Yaoxian Canal and the Su River Canal were not part of the canal system constructed by Emperor Yang of the Sui Dynasty, they were still connected via the course of the Yellow River and could be integrated into the broader waterway network.

---

[126] Ibid.

[127] Treatise on canals and waterways V: Rivers of Hebei, In *History of Song*, vol. 95.

[128] The dredging of the Su River Canal by Xue Wanche occurred in the 17th year of the Zhenguan era (643 CE). See Geographical treatise III. In *New book of Tang*, vol. 39.

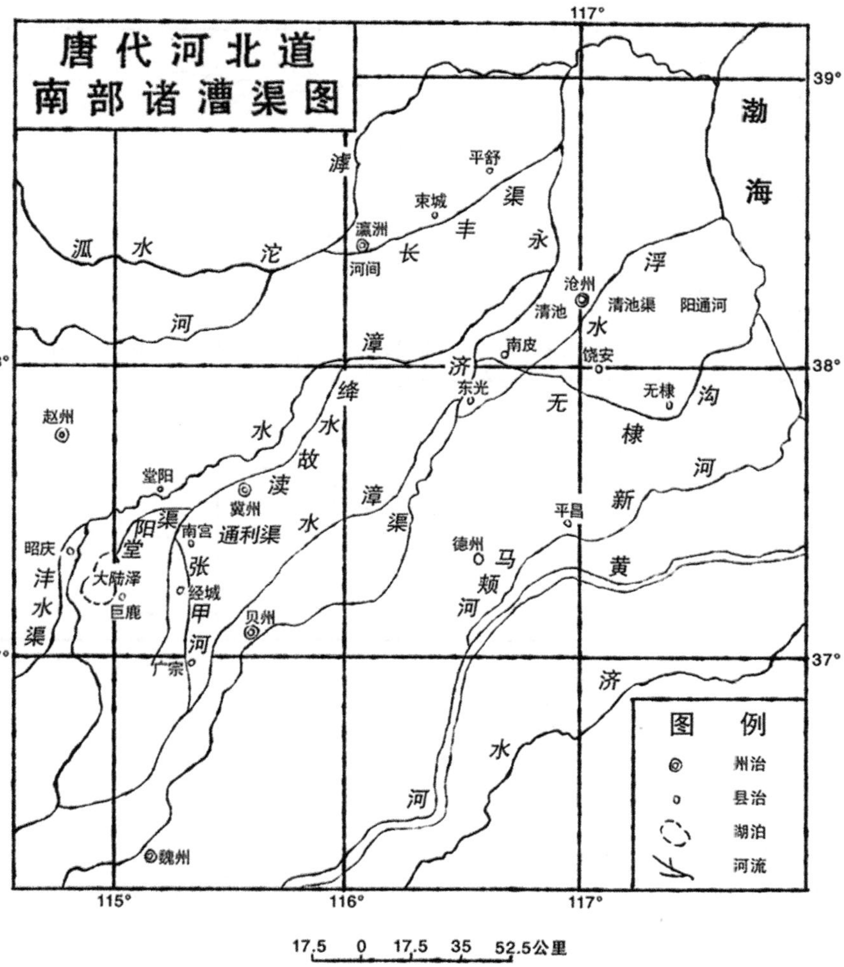

图例 - Legend
州治 Prefecture Seats
县治 County Seats
湖泊 Lakes
河流 Rivers

Fig. 5.7 Major Canals in Southern Hebei Circuit during the Tang Dynasty

Additionally, it is worth mentioning the Yongtai Canal in Wuling (present-day Changde County, Hunan)[129] and the Xinyuantou Canal in Wenjiang (present-day

129 Geographical treatise VI. In *New book of Tang*, vol. 40.

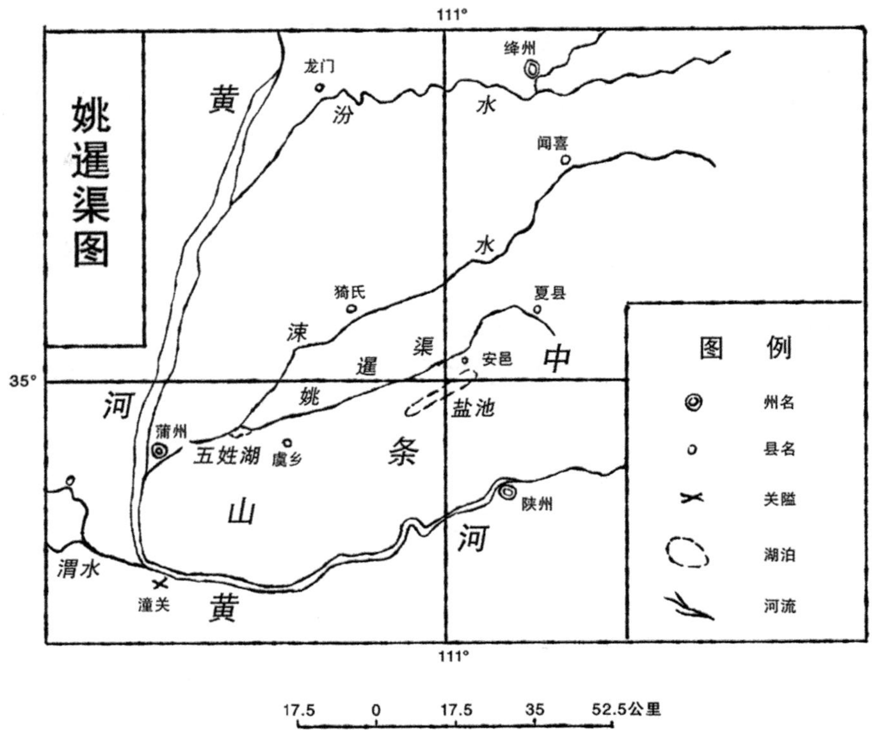

图例 - Legend
州名 Prefecture Names
县名 County Names
关隘 Mountain Passes
湖泊 Lakes
河流 Rivers

Fig. 5.8 The Yaoxian Canal

Wenjiang County, Sichuan).[130] While their scale was small, they still provided significant convenience within their respective localized regions. These smaller canals, though less extensive, played an important role in facilitating transportation and improving local infrastructure (See Fig. 5.8).

[130] Geographical treatise VI. In *New book of Tang*, vol. 42.

Grain Transport System in the Early Tang Dynasty

The exact amount of grain transported to Chang'an via the canals during the Sui Dynasty is not documented in detail, and thus a definitive figure cannot be provided. However, in the early Tang period, much like in the early Han, the empire was newly established, and expenditures were relatively modest. As a result, the amount of grain required for transport was not very large. During the reigns of Emperor Gaozu and Emperor Taizong, the annual grain transport was only about 200,000 *dan*, which was sufficient to meet the needs. By the later years of the Kaiyuan era, under the management of Pei Yaoqing, who oversaw the grain transport system, a total of 7 million *dan* of grain was transported over three years, averaging more than 2 million *dan* per year—ten times the amount of the early years. In the early years of the Tianbao era, when Wei Jian was appointed as the transport official, the annual grain transport reached 4 million *dan*, the highest recorded during the Tang Dynasty.[131] This excellent state of transportation continued until the end of the Tianbao period, when the An Lushan Rebellion broke out, marking the disruption of this system.

During the Kaiyuan and Tianbao eras, the volume of grain transported to Chang'an was extraordinarily high—indeed, a rare phenomenon throughout the Tang dynasty. However, the bulk of this transported grain did not come from the southeastern coastal regions, but rather from various localities in the lower reaches of the Yellow River. During the Kaiyuan era, Pei Yaoqing, who was in charge of managing the grain transport system, had suggested increasing the transport of rice and millet from the Jiangnan region to supplement Chang'an's supply. In reality, however, his policy focused on augmenting the transport of grain taxes from Jin, Jiang, Wei, Pu, Xing, Bei, Bo, and Ji prefectures—all of which were located along the lower Yellow River, not in the southeastern coastal provinces.[132] Jin and Jiang prefectures were administered from present-day Linfen and Xinjiang in southwestern Shanxi; Wei, Xing, and Bei from present-day Daming, Xingtai, and Qinghe in southern Hebei; Pu from Juancheng in Shandong; and Bo and Ji from Liaocheng and Chiping, also in Shandong. Taken together, these areas were all situated in the lower Yellow River region. Later, when Wei Jian undertook the rehabilitation of the Han and Sui grain transport canals, his efforts likewise focused solely on facilitating the transport of tax grain from Shandong.[133] At that time, the term "Shandong" could, in a broader sense, encompass parts of Jiangdong as well, but in practice it still primarily referred to the lower Yellow River basin. This was because, since the early Sui dynasty, agriculture in the Yellow River region had continued to develop, yielding substantial local supplies of rice and millet—making long-distance reliance on grain from the Jiangnan or Taihu basin largely unnecessary. In the twenty-fifth year of the Kaiyuan era (737 CE), the implementation of the he-di (purchase-storage) system in the Guanzhong region even led to the suspension of grain transport from east of the pass.[134] The grain

[131] Treatise on food and finance III. In *New book of Tang*, vol. 53.

[132] Ibid.

[133] Ibid.

[134] Annals of Tang (30). In *Comprehensive mirror for aid in government*, vol. 214.

taxes from the southeastern provinces were instead converted into cloth production. This shows that the grain from the Yellow River basin was sufficient to meet local needs, and supplying Chang'an with grain through the transport system was not particularly challenging. Under the Tang dynasty's institutional system, a Grand Granary (*tai cang*) was maintained in the capital, while each prefecture and county had its own Official Granary (*zheng cang*). In addition, Public Granaries (*changping cang*) were established to stabilize market prices, and Emergency Granaries (*yi cang*) were maintained to guard against scarcity.[135] These storage systems began to be established from the early Tang period onward; while some were later repurposed or relocated, the overall system remained largely intact.[136] It was this institutional framework that allowed Wei Jian, during the Tianbao era, to divert grain from prefectural *yi cang* in order to purchase lighter goods as a form of tribute to curry favor.[137] Likewise, Yang Guozhong was able to draw upon *yi cang* stores from across the empire to fill the emperor's private reserves in an effort to win imperial favor.[138] These examples indicate that the granaries were still well-stocked during this period. Due to regional differences in agricultural productivity, the quantity of grain stored in granaries varied. In the eighth year of Tianbao (749 CE), only four circuits—Guanzhong, Hebei, Hedong, and Henan—had more than one million *dan* of grain stored in their official granaries. Among these, Henan held the largest quantity, with more than 5.8 million *dan*. In terms of *yi cang*, only Hebei and Henan stored over ten million *dan*, followed by Hedong with over seven million, and Guanzhong with more than five million. Jiangnan stored just over six million *dan*. Likewise, the largest amounts held in *changping cang* were in Hebei and Henan.[139] This period was marked by relative political stability and continuous development of agricultural infrastructure—particularly irrigation—in the Yellow River basin, which clearly led to agricultural expansion and productivity.[140] It was precisely this robust agricultural base in the north that minimized the need to draw on grain from the southern regions, particularly the lower Yangtze and Taihu basin. In other words, during this period, the southern sections of the Grand Canal and the broader grain transport network south of the Yangtze had yet to assume a significant role.

---

[135] Treatise on official ranks. In *Old book of Tang*, vol. 43.

[136] Treatise on food and finance. In *Old book of Tang*, vol. 49.

[137] Treatise on food and finance. In *Old book of Tang*, vol. 49.

[138] Biography of the imperial in-law Yang Guozhong. In *New book of Tang*, vol. 247.

[139] Food and finance. In *Tongdian*, vol. 12.

[140] Shi Nianhai. *The development of agriculture in the yellow river basin and its adjacent regions between the Kaihuang and Tianbao periods.*

The Importance of Jiangnan's Grain Transport System After the Mid-Tang Dynasty

After the outbreak of the An Lushan Rebellion, most parts of the Yellow River basin were subjected to severe turmoil and devastation. The population declined sharply, farmland was largely abandoned, and agriculture fell into a prolonged state of stagnation. Even after the rebellion was quelled, the rise of autonomous regional warlords meant that tribute and tax revenues no longer flowed to the central government. With the imperial treasury depleted, the court increasingly relied on the Lower Yangtze region—particularly the Taihu Basin—to supply grain for both military and civilian needs. The Lower Yangtze and Taihu Basin areas had seen steady agricultural development since the Southern Dynasties, and this progress continued and even accelerated during the Sui and early Tang periods.

Following the An Lushan Rebellion, the population of the Yellow River basin, especially in Henan Prefecture (home to Luoyang, the Eastern Capital of the Tang dynasty), declined dramatically. Once a flourishing region, Henan Prefecture had as many as 120,000 households during the Kaiyuan era and ranked among the empire's most densely populated regions. Yet by the Yuanhe era of Emperor Xianzong (806–820 CE), only 18,000 households remained—a loss of six-sevenths.[141] By contrast, the Yangtze River basin—sheltered from warfare—witnessed a substantial increase in population due to large-scale southward migration. Suzhou (modern Suzhou, Jiangsu Province), for example, grew from over 60,000 households during the Kaiyuan era to more than 100,000 by the Yuanhe era.[142] It is said that one-third of Suzhou's population at that time consisted of gentry migrants from the Central Plains.[143] These elite households likely brought with them large numbers of dependents and retainers, thereby expanding the local labor force and further driving agricultural and economic development. Even regions in the middle reaches of the Yangtze—such as Tan (modern Changsha, Hunan), Gui (modern Guilin, Guangxi), and Hengyang (Hunan)—experienced new waves of growth and became increasingly significant as grain-producing areas.[144]

With these shifts, the Tang government became increasingly dependent on the wealthy lower Yangtze River basin—particularly the Taihu region—as its primary source of tribute grain and economic revenue. This reliance was not only a matter of strategic planning for high-ranking officials within the Three Bureaus of Transport; it was also frequently discussed among members of the scholarly elite. "The expenditures of the military and the state rely on resources from the Jianghuai region"—that is, the area between the Yangtze and Huai Rivers.[145] In reality, the regions

141 Henan prefecture. In *Yuanhe maps and records of prefectures and counties*, vol. 5.

142 Suzhou. In *Yuanhe maps and records of prefectures and counties*, vol. 25.

143 Inscription on the wall of the Wu County assembly hall. In *Complete prose of the Tang*, vol. 519.

144 Biography of Liu Yan. In *Old book of Tang*, vol. 123.

145 The Emperor Xianzong's edict of July 23rd, 814 CE. In *Complete prose of the Tang*, vol. 63.

levied for such contributions extended far beyond the Jianghuai area and encompassed what were known as the Eight Southeastern Circuits.[146] These included: Zhexi (Western Zhejiang), with its administrative seat in modern Zhenjiang, Jiangsu; Zhedong (Eastern Zhejiang), based in modern Shaoxing, Zhejiang; Xuanshe, based in modern Xuancheng, Anhui; Huainan, based in modern Yangzhou, Jiangsu; Jiangxi, based in modern Nanchang, Jiangxi; E'yue, based in modern Wuhan, Hubei; Fujian, based in modern Fuzhou, Fujian; and Hunan, based in modern Changsha, Hunan. These regions together covered the present-day provinces of Zhejiang, Jiangxi, and Fujian; most of Jiangsu and Anhui south of the Huai River; the greater part of Hunan; and eastern Hubei—comprising virtually the entire middle and lower Yangtze basin. The tribute grain and tax revenues extracted from these areas were continuously transported to Chang'an via the Bian River, the Grain Transport Canal, and other major sections of the Grand Canal system.

The Contest for Control Over Transport Routes

At the outbreak of An Lushan's rebellion, the Yongji Canal in Hebei became cut off from the rest of the canal network. Although the Yongji Canal itself was not silted up or destroyed during the chaos, it could no longer be utilized by the central government. After An Lushan's rebel forces crossed the Yellow River, the Bian River also ceased to function as a viable transport route. At that time, the only option for transporting grain to the Guanzhong region was to follow the Yangtze River upstream, then continue along the Han River to Xiangyang, from where the grain would be conveyed to Shangyu and finally delivered to the capital.[147] Shangyu refers to the area of Shang Prefecture, which corresponds to present-day Shangxian and Danfeng counties in Shaanxi. The route from Xiangyang to Shangxian would naturally have followed the Dan River. Today, river vessels on the Dan can reach as far as Danfeng County, and it is likely that conditions were similar during the Tang period. This suggests that north of the Dan River, transportation required overland haulage to cross the Qinling Mountains.

Later, as An Lushan's rebel forces grew increasingly powerful, river transport along the Yangtze was rendered nearly impossible. This disruption dealt a severe blow to the Tang central government. Even after the An Lushan Rebellion was eventually suppressed, the empire was left fractured, with powerful regional warlords defying imperial authority. The central government found itself effectively competing with these autonomous military governors for control over the empire's vital transport arteries. During this period, several key locations became strategically indispensable: Bianzhou in the upper reaches of the Bian River; Yongqiao (in present-day Su County, Anhui) in the lower reaches; Guokou on the Huai River; and Xiangzhou (modern Xiangfan, Hubei) and Dengzhou (modern Deng County, Henan) along the Han River.

[146] Annals of Tang (53). In *Comprehensive mirror for aid in government*, vol. 237.

[147] Treatise on food and finance. In *New book of Tang*, vol. 53.

Each of these places was capable of controlling access to the transport routes. The stability—and even survival—of the central government depended on whether it could maintain control over these critical nodes.

The underlying tensions eventually came to the surface in the early years of Emperor Dezong's reign. In the second year of the Jianzhong reign period (781 CE), Li Zhengji, who held power in the regions of Ziqing and surrounding prefectures, joined forces with Tian Yue of Weibo and Liang Chongyi, who controlled Xiang and Deng prefectures. Together, they openly rebelled against the Tang court.[148] The territory occupied by Li Zhengji corresponded to much of present-day Shandong Province. Tian Yue controlled the southern part of modern Hebei and the western region of Shandong—situated to the northeast of the Bian River—while Liang Chongyi held southwestern Henan and central Hubei, lying to the river's southwest. The coordination among these three forces posed a comprehensive threat to the Bian River's transportation network. Li Zhengji even deployed troops to seize control of key transport chokepoints—Xuzhou, Yongqiao, and Guokou—thereby formally severing the transport artery.[149]

In this critical situation, Zhang Wanfu, the governor (cishi) of Haozhou (headquartered in present-day Fengyang, Anhui), stationed himself at Guokou. There, he dispatched grain-laden vessels to the capital. The forces from Ziqing, though massed along the shore, merely glared and dared not attack, allowing the grain boats to continue their journey upstream.[150] Yet this was a temporary expedient and could not be relied upon as a long-term strategy. Fortunately, Li Zhengji died not long afterward. His son concealed the news of his death in an attempted usurpation, but Li Wei surrendered Xuzhou to the Tang, and Li Na (Li Zhengji's son) was ultimately defeated by imperial forces. As a result, the grain transport route was reopened.[151]

However, unexpectedly, in the following year—the third year of the Jianzhong reign (782 CE)—Li Xilie, who had been transferred to Xuzhou (modern Xuchang, Henan), allied himself with Li Na and conspired to launch an assault on Bianzhou (modern Kaifeng, Henan). Bianzhou, as the strategic hub of the Bian River system, was no less important than Yongqiao or Guokou. Moreover, it was a major stronghold in the Central Plains; its fall or retention carried significant implications. By the end of the fourth year of the Jianzhong reign (783 CE), Bianzhou was finally captured by Li Xilie, resulting in the complete severance of the grain transport route from the Yangtze and Huai regions.[152]

With the transport route cut off, the granaries of Chang'an fell into scarcity. The resulting instability affected not only the general populace, but also the imperial guards, who reportedly cried out in the streets: "You ask us to serve as soldiers, yet provide neither grain nor pay—are we to be treated as criminals?" Eventually, Han Huang, the military governor of the Zhenhai Circuit (headquartered in present-day

---

[148] Annals of Emperor Dezong. In *Old book of Tang*, vol. 12.

[149] Biography of Li Zhengji. In *New book of Tang*, vol. 123.

[150] Biography of Zhang Wanfu. In *Old book of Tang*, vol. 152.

[151] Biography of Liu Xuanzuo. In *Old book of Tang*, vol. 145.

[152] Biography of Li Xilie. In *New book of Tang*, vol. 225.

Zhenjiang, Jiangsu), dispatched a shipment of rice to Shan (modern Sanmenxia, Henan). When news of the shipment reached Chang'an, Emperor Dezong was overwhelmed with joy. So great was his relief that he momentarily cast aside imperial decorum and hurried to the Eastern Palace to inform the Crown Prince himself, exclaiming: "The rice has arrived at Shan! We, father and son, shall live!".[153]

Although this grain transport crisis was temporarily alleviated with the recapture of Bianzhou[154] and the tightening of control over Xuzhou,[155] smaller-scale conflicts over the transport routes continued to arise intermittently. The Tang court remained persistently anxious about the security and accessibility of these vital arteries of grain conveyance.

Du You's Proposal for the Establishment of a New Transportation Route

At a time when regional military governors such as Li Zhengji, Tian Yue, and Liang Chongyi openly defied imperial authority and the grain transport system faced severe disruption, a new canal project was proposed by Du You.[156] Observing that both the Bian River and Han River routes had become obstructed and impassable, Du advocated for the revival of the ancient Langtang Canal—an integral part of the historic Honggou waterway system.

His proposed route began south of Bianzhou, departing through Xunyi County after ten li, then entering the Pipa Ditch. From there, it would cut through the Cai River to reach Chenzhou (with its administrative seat in present-day Huaiyang County, Henan), before merging with the Ying River, which in turn flowed into the Huai River. The route would then continue via Shouzhou (modern Shou County, Anhui) into the Fei River.

Du also proposed excavating through Jiming Ridge, the upper source of the Fei River, to establish a connection with Chao Lake. From Chao Lake, vessels could then proceed onward to the Yangtze River, completing a continuous inland waterway for grain transport and military logistics.[157]

153 Annals of Tang 48. In *Comprehensive mirror for aid in government*, vol. 232.

154 Annals of Emperor Dezong. In *Old book of Tang*, vol. 12.

155 Biography of Zhang Jianfeng. In *New book of Tang*, vol. 158.

156 Treatise on food and finance III. In *New book of Tang*, vol. 53.

157 Treatise on food and finance III. In *New book of Tang*, vol. 53; *Taiping huanyu ji (vol 1): Kaifeng prefecture* states: "The Pipa Canal lies eleven li south of Kaifeng County. It flows from the western boundary of Zhongmou County into the Tongji Canal. When Emperor Yang of the Sui intended to travel to Jiangdu, he ordered a canal to be dug southwest of the city of Daliang to divert water from the Bian River—this was the Lang Canal. According to *Old maps*, 'Its shape resembled a pipa (a Chinese lute), hence the name.'" In *Comprehensive mirror for aid in government (vol. 227): Annals of Tang (43)*, Hu's commentary notes, "The Cai River was anciently known as Pipa Canal and was located in Xunyi County." Furthermore, Yue Shi identified Pipa Canal as Lang Canal, which subsequently became the Cai River. This waterway originated as an old transport route from the Qin

Much of this proposed route traced ancient waterways that had silted up over time due to years of neglect. The segment from Bianzhou to Chenzhou had only fallen out of use after the Sui dynasty opened the Tongji Canal, and it passed through flat plains, making it relatively easy to dredge. The most labor-intensive section was that of Jiming Ridge. Although this ridge was mentioned in ancient records, in reality it had never been fully developed as a canal route. Nevertheless, the ridge spanned only a little over forty li, so even if a proper canal could not be dug, goods could still be transported overland using carts without too much difficulty. If this ancient waterway could be successfully restored, not only could grain boats that once entered the Han Canal sail directly from Baisha (in present-day Yizheng County, Jiangsu) to Dongguan (in present-day Chaoxian, Anhui) and into Chao Lake, but grain shipments from regions such as Jiangnan, Qianzhong (modern-day Guizhou), Lingnan, Shu, and Han would no longer need to detour through the Han Canal into Bian. In that case, even though rebels like Li Na and Tian Yue held control of the Bian River, and Liang Chongyi occupied the Xiang and Deng regions, they would no longer pose a threat to the transportation network. However, not long after this proposal was put forward, Li Na's forces collapsed from within, and the Bian River route was reopened. As a result, Du You's canal project was ultimately abandoned.

The Dredging of the Cai River

The necessity of restoring the old course of the Langtang Canal was readily apparent to any discerning observer of the time. The Bian River was the sole transport artery linking Guanzhong with the affluent southeastern regions. While this posed no issue in times of peace and stability, it became a major concern in periods of turmoil, when the central government lost control over regional military governors. Under such conditions, how could a single transport line be reliably maintained? Even if the Han River could serve as a partial substitute, the transport of grain over the Zhongnan Mountains presented immense logistical difficulties.

Before Du You submitted his proposal for a new transport route, Li Peng had already undertaken dredging operations along the waterway between Chen and Ying prefectures. At the time, the upper reaches of the Bian River had been rendered impassable due to the rebellion of Li Lingyao, the military commander of Bianzhou, compelling Li Peng to urgently restore this alternative route to mitigate the crisis.[158]

Roughly a year after Du You submitted his proposal, yet another rebellion broke out—this time led by Li Xilie, who, in alliance with Li Na, once again severed the Bian River route. It was only then that Li Mian successfully carried out the dredging of the Cai River. The Cai River corresponded broadly to the ancient Langtang Canal, a part of the historic Hong Canal system. Li Mian did not address the section at Jiming

---

and Han dynasties and appears to have no direct connection to Emperor Yang of Sui's excavation of the Tongji Canal.

[158] Biography of Li Peng. In *Old book of Tang*, vol. 132.

Ridge, likely because at that time the central government still retained control over the Huai River's waterway, making it unnecessary to open an additional passage through the ridge.[159]

Li Mian's successful dredging of the Cai River substantially alleviated the transportation difficulties confronting the imperial government. In particular, it greatly facilitated the Tang court's subsequent military campaign against Wu Yuanji, the overreaching military governor in Cai Prefecture (present-day Runan County, Henan), and contributed significantly to the eventual suppression of his rebellion.[160]

Since the An Lushan Rebellion, the Tang court had devoted its full efforts to maintaining this crucial transportation waterway. The hardship involved in doing so far exceeded that experienced during Emperor Yang of the Sui's initial excavation of the canal. The challenges faced were manifold and severe—by comparison, the original construction seemed far less arduous. In the end, when the court's strength was utterly exhausted, the Tang dynasty itself fell.

[159] Biography of rebels–Li Xilie. In *New book of Tang*, vol. 145.

[160] The *Old book of Tang* (vol. 15): Annals of Emperor Xianzong (2) states: "In the eleventh year of the Yuanhe reign (816 CE), the post of 'Transport Commissioner of the Huai and Ying Rivers' was first established. Grain from the Yangzi River depot was transported upstream from Huaiyin to Shouzhou—a distance of forty *li*—then entered the mouth of the Ying River, and continued upstream to the Shenqiu region in Yingzhou. From there, it traveled five hundred *li* to reach Xiangcheng, then a further five hundred *li* upstream into the Yun River, and finally three hundred *li* more to be delivered to Yancheng. This route delivered 500,000 *shi* of grain and 15.5 million bundles of straw, saving 76,000 *guan* (strings of cash) in transportation costs via the Bian River."

Chapter 6
The Eastward Shift of the Political Center and the Obstruction of the Canal System

The Decline of Guanzhong and the Eastward Relocation of the Capital

After the fall of the Tang dynasty, subsequent regimes paid little heed to Guanzhong when choosing their political centers or capitals. From that time on, no dynasty ever again established its capital in the region. Once hailed as a land of strategic and material abundance—"a terrestrial and celestial granary"—Guanzhong gradually declined and slipped from the national consciousness. This decline had already begun to manifest in the waning years of the Tang. The silting and obstruction of canal systems made it increasingly difficult to transport grain from the prosperous southeast to the northwest. At the same time, the region suffered repeated invasions from foreign powers and disruptions from unruly regional warlords, turning Guanzhong into a frequent battlefield. Its former prosperity naturally faded into memory. As the Tang dynasty drew to a close, even an episode of capital relocation took place: Zhu Wen, under the pretext of protecting Emperor Zhaozong, coerced the court into moving from Chang'an to Luoyang. This move laid bare the reality that the once-glorious Guanzhong—once praised as both "land and sea" and "heaven's granary"—could no longer command the political loyalty or emotional attachment of the empire's rulers.

Zhu Wen rose to prominence through the power he amassed in Bianzhou, a city located along the Bian River. His rise was intimately linked to the transportation advantages afforded by the river. After usurping the Tang throne, he established his capital in his hometown of Bianzhou. Economically, Bianzhou was a prosperous metropolis; yet it lay on the vast open plains, entirely lacking in natural defenses—hardly an ideal site for an imperial capital. Zhu Wen was certainly aware of this. So, after residing there for a year or two, he moved the capital to Luoyang. When his son ascended the throne, the court returned once again to Bianzhou. But in such a declining age, a feeble monarch could hardly be expected to pursue plans with lasting vision.

N. Shi, *A Historical Geography Research of Canals in China*,
https://doi.org/10.1007/978-981-95-3517-0_6

Among the Five Dynasties, the Later Jin was the weakest, preoccupied with short-term survival and devoid of any long-term strategic vision. Zhu Wen had established his capital in Bianzhou because it was the original base of his rise to power. Later Jin likewise made Bianzhou its capital—but this decision was motivated purely by economic considerations. One might say it was seduced by the convenience of the canal; though in reality, by that time, the canal was no longer fully navigable.

The so-called Five Dynasties were, in essence, a continuation of the regional military governors who had carved out territories during the waning years of the Tang. Because they held sway over the Central Plains, later historians classed them within the orthodox dynastic lineage. By this point, although the lower reaches of the Bian River remained under Later Jin control, the regions south of the Huai River had already formed a separate political sphere. The strategic value of the Bian River lay in its ability to connect with the lands south of the Yangtze and Huai Rivers; without this southern linkage, it amounted to no more than half a canal. Yet even this "half-canal" was enough to entice the Later Jin ruler and his ministers into relocating the capital from Luoyang to Bianzhou.

The official edict announcing the relocation emphasized the advantages of transportation, declaring: "In the aftermath of successive wars and the widespread suffering of the populace, laborers are numerous but the granaries stand empty. For years, the conscription of carts and the hauling of grain and fodder have strained the people daily. The state has long relied on canal transport, yet it cannot meet our needs. Now, Bianzhou stands at a vital juncture of land and water routes, blessed with natural features of rivers and mountains. It is a land of abundant storage, a nexus that connects in all directions. Having inspected it personally, I find it well suited to our purposes. Let it be raised to the status of a capital, to benefit both military affairs and civilian livelihoods. Bianzhou shall henceforth be designated the Eastern Capital, with the establishment of Kaifeng Prefecture therein."[1] This clearly reveals that Later Jin's decision to relocate the capital was based solely on the convenience of grain transport via canal. Beyond that, no other factors appear to have been considered.

After the fall of Later Jin, the Later Han succeeded, but its reign lasted only four years and naturally left little mark. It was not until the rise of Later Zhou that a truly remarkable emperor appeared—Emperor Shizong. He led vigorous military campaigns both to the north and south, demonstrating exceptional strategic prowess. Among the rulers of the Five Dynasties, he was undeniably one of the most outstanding. Yet fate proved cruel: his reign lasted a mere six years before his untimely death. From the establishment of the capital at Bianzhou by Later Jin until the death of Emperor Shizong of Later Zhou, nearly two decades passed amid turmoil and upheaval. Throughout this entire period, the question of relocating the imperial capital was never seriously broached. There was neither the opportunity nor the determination to pursue such a change. Consequently, the Later Zhou court, like its predecessors, simply adhered to the precedent set by Later Jin and remained entrenched in Bianzhou.

---

[1] Book of Jin III: Annals of Gaozu III. In *Old history of the Five Dynasties*, vol. 77.

The Relationship Between the Song Dynasty's Capital Location and the Canal System

By the early Song period, the previously fractured realm was gradually reunified, and the foundations of the new dynasty steadily stabilized. As a result, the issue of where to establish the capital again came to the fore. Emperor Taizu of Song, as the founding monarch, possessed both ambition and foresight, aspiring to emulate the grandeur of the Han and Tang dynasties. Seeking to break with the entrenched practices of the Five Dynasties era, he considered relocating the capital to Chang'an.

In the ninth year of the Kaibao reign (976 CE), he traveled west to Luoyang, intending first to establish it as the capital before eventually moving further west. However, his younger brother Guangyi (the future Emperor Taizong) and a number of ministers objected, arguing that Luoyang was not suitable. In response, Taizu remarked, "The recovery of Henan is not yet complete, but in the end we shall reside in Chang'an." When Guangyi asked for his reasoning, Taizu explained: "I wish to move westward, to take advantage of the strategic natural defenses afforded by mountains and rivers, to reduce the number of idle troops, and to follow the precedent of the Zhou and Han dynasties in order to bring peace to the realm."

Guangyi countered that "it is virtue, not geographical defensibility, that truly matters," and repeatedly urged him to return to Bianzhou. After persistent remonstrations from his brother and other ministers, Emperor Taizu ultimately abandoned the relocation plan and retained Bianzhou as the capital.[2] Bianzhou, which had undergone several name changes during the Five Dynasties period, was by then officially called Kaifeng Prefecture, though it was commonly referred to as Daliang.

To speak plainly, Emperor Taizu of Song faced compelling practical difficulties that ultimately prevented him from relocating the capital to Luoyang or Chang'an. Since the An Lushan Rebellion in the Tang dynasty, military authority had often been concentrated in the hands of regional governors, creating a political structure in which provincial power outweighed that of the central government. By the time the Song dynasty was founded, this imbalance had persisted for nearly two centuries, and the instability it caused was well understood by Emperor Taizu. Determined to reverse the decline, he methodically dismantled the independent power bases of the regional warlords and consolidated military control in the central government, thereby completely inverting the earlier structure: the center was now strong, while the periphery was weakened.

While this solution addressed a long-standing problem, it also created new challenges. With hundreds of thousands of imperial guards now stationed in the capital, the demand for provisions—along with the salaries of court officials and the daily sustenance required by Kaifeng's large population—imposed a staggering and continual burden on the state's resources. Although Emperor Taizu remained inclined to move the capital, he could not overlook the logistical difficulties of provisioning such a massive operation once relocated. Kaifeng's strategic advantage lay

[2] Wang Cheng. Biography of Li Huaizhong. In *Eastern capital historical records*, vol. 28.

in its integration with the Grand Canal system, which ensured the steady transport of grain and supplies from the Jianghuai region. In contrast, Luoyang and Chang'an, at that time, lacked similarly reliable waterborne transport routes.

During Emperor Taizu's visit to Luoyang in the ninth year of the *Kaibao* reign (976 CE), when he revealed his intentions to relocate the capital, one minister reportedly admonished him: "Each year, Bian receives four to five million *hu* of grain from the Jianghuai region by canal, enough to sustain an army of several hundred thousand. The treasury and the imperial guards are all concentrated there. If Your Majesty truly intends to move the capital to Luoyang, I fail to see any advantage."[3]

In the end, such concrete logistical realities outweighed Emperor Taizu's strategic aspirations, and the plan to relocate the capital was abandoned. He did not fully appreciate how, in the Han, Sui, and Tang dynasties, grain had been successfully transported to Luoyang and further on to Chang'an—nor did he adequately account for the sophisticated canal networks, such as the Weishui Canal, that had once made such long-distance provisioning possible.

Later, during the reign of Emperor Renzong of Song, the question of relocating the capital once again entered political debate. Fan Zhongyan addressed the matter directly, observing: "Luoyang is strategically secure, whereas Bian lies exposed on all sides in an open plain. In times of peace, it is suitable to reside in Bian, but should conflict arise, the court must be based in Luoyang. We ought to gradually accumulate reserves and restore the palaces there." At that time, Lü Yijian held power as chief minister. When Emperor Renzong sought his opinion on Fan's suggestion, Lü—who had long been at odds with Fan Zhongyan—dismissed it outright, remarking, "This is yet another of Fan Zhongyan's impractical and overreaching ideas."[4]

Among the various proposals for capital relocation advanced during the Song period, Fan Zhongyan's was arguably the most pragmatic and strategically sound. Yet it was rejected by those in power and never moved beyond the realm of discussion. The chronic weakness of the Song state and the narrowness of its strategic horizons can, in no small measure, be traced back to the decision to anchor the dynasty's capital in Kaifeng. In essence, the court's attachment to the logistical conveniences afforded by the Grand Canal came at the cost of a broader, long-term strategic vision (See Fig. 6.1).

Main Transport Canals and Their Dredging in the Song Dynasty

Because the Song dynasty had established its capital at Kaifeng, it could not afford to neglect the canal system. The principal transport arteries of the period were the Bian River, the Yellow River, the Huimin River, and the Guangji River. Of these, the Yellow River was a natural waterway and needs no elaboration here. The Bian River

[3] Ibid.

[4] Biography of Fan Zhongyan. In *History of Song*, vol. 314.

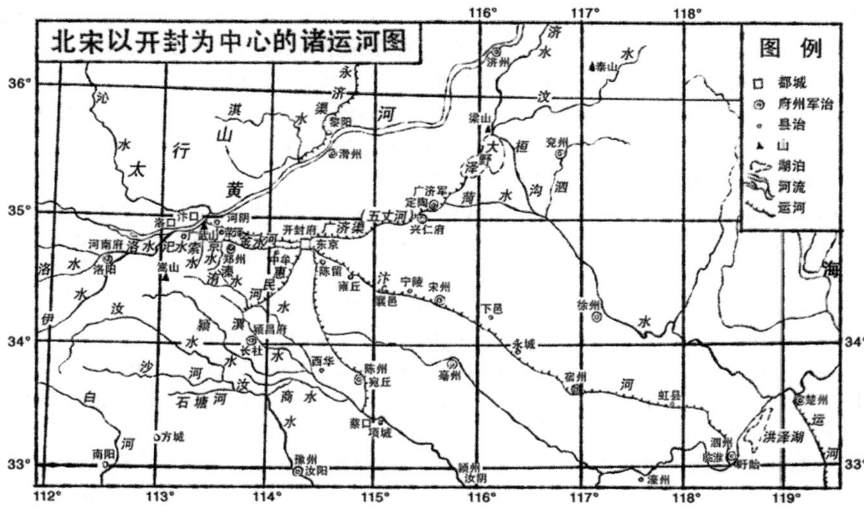

图例 - Legend
都城 Capital Cities
府州军治 Military Administration under Prefectures and Provinces
县治 County Seats
山 Mountains
湖泊 Lakes
河流 Rivers
运河 Canals

Fig. 6.1 Transport canals centered on Kaifeng in the northern song

was inherited from the Sui and Tang dynasties; the Huimin River was the former Cai River of the Tang; and the Guangji River originated from the Zhan Canal excavated under Empress Wu, later known as the Wuzhang Canal or Wuzhang River.

Although these canals were all legacies of the Tang dynasty, by the late Tang and the turbulent Five Dynasties period, social and political disorder had caused many waterways to silt up and become obstructed. Nevertheless, even during this chaotic era, some efforts were made to maintain or restore canal traffic. A few examples illustrate this broader trend.

One notable case concerns the Yongji Canal. After the An Lushan Rebellion, the Hebei region (known as *He-shuo*) was effectively controlled by autonomous military governors, making it difficult for the Tang court to exercise authority there. Whether the Yongji Canal remained navigable during this period is uncertain. During the Later Liang, Luo Shaowei, military commissioner of Tianxiong Commandery (headquartered in Weizhou, modern Daming County, Hebei), governed what had once been the powerful Weibo military base. By this time, Weizhou's military power had diminished, and it was subordinate to Zhu Quanzhong, founder of Later Liang.

With no imminent military conflicts, Luo accumulated large reserves of grain, "piled like mountains." Meanwhile, the capital was facing severe food shortages.

To alleviate this, Luo proposed felling timber in the Taihang Mountains, transporting it down to Qimen in Anyang (modern Qi County, Henan), and constructing 300 boats. These vessels would carry one million *dan* of grain annually, traveling by water to Luokou, where the Luo River meets the Yellow River.[5] Although this route was described as entering Luokou from the Yellow River, the grain originated in Weizhou, which relied heavily on the Yongji Canal, suggesting that the canal was still used to transport supplies southward.

Another instance dates to the third year of the Changxing reign (932 CE) under Later Tang's Emperor Mingzong, when Zhao Dejun, military commissioner of Youzhou (modern Beijing), excavated a new canal known as the "Southeast River." This waterway extended 165 *li* (about 82 km), from Wangmakou to Yukou, and was 65 paces wide and 12 feet deep.[6] The exact locations of Wangmakou and Yukou remain uncertain, but the name "Southeast River" suggests it lay southeast of Youzhou, near the confluence of the Yongji Canal and the Sanggan River (modern Yongding River). Since the Sanggan River was a natural watercourse that required no excavation, this new canal likely formed the northernmost extension or feeder of the Yongji Canal.

Zhao Dejun memorialized the project to the imperial court, emphasizing that the canal was designed to facilitate grain transport, with boats capable of carrying 1000 *dan*. Beyond its practical value as a transport artery linking the northern frontier with the Central Plains, the canal also allowed Zhao to demonstrate loyalty and secure favor from the Later Tang court.[7] This record also indicates that, despite the eastern Taihang plains being under military governors' control, most sections of the Yongji Canal remained navigable well into the late Tang and Five Dynasties periods.

During this same period, however, the water supply for the Bian River became a pressing issue. In the second year of the Tongguang reign (924 CE), Later Tang's Emperor Zhuangzong ordered Zhu Qing, prefect of Caizhou, to dredge the Suo River to supplement the canal.[8] Caizhou was headquartered at Ruyang (modern Runan County, Henan), while the Suo River lay in what is now Xingyang, Henan.[9] When the Ji River had flowed freely in earlier centuries, the Suo River joined it downstream. After the Ji River dried up and the Bian River was constructed, the Suo River became a tributary feeding into the Bian system.

Although Zhu Qing was prefect of Caizhou, his mandate to dredge the Suo River indicates that the river did not necessarily lie within his jurisdiction. The Suo River, a minor river, had not previously been considered significant for canal engineering. Its sudden prioritization at this time suggests that the volume of water entering the Bian system via Biankou had become insufficient, necessitating remedial action.

---

[5] Book of Liang: Biography of Luo Shaowei. In *Old history of the Five dynasties*, vol. 14.

[6] Book of Tang: Annals of Mingzong IX. In *Old history of the Five dynasties*, vol. 43.

[7] Book of Jin: Biography of Zhao Dejun. In *Old history of the Five dynasties*, vol. 98.

[8] Annals of Later Tang II. In *Comprehensive mirror for aid in government*, vol. 273.

[9] Commentary on the Ji River. In *Water classic*.

A particularly noteworthy figure in this context is Emperor Shizong of the Later Zhou. He began overseeing canal restoration in the second year of his reign (955 CE), as part of preparations for a military campaign against the Southern Tang. Because the campaign required a strong naval force, Shizong first ordered the dredging of the Bian Canal[10] downstream from Yongqiao, which had by that time become unnavigable.[11]

Although vessels could theoretically travel from the Huai River into the Han Canal, practical difficulties remained: the water level of the Han Canal was higher than that of the Huai River. To maintain the Han Canal's elevation, a dam known as the Beishen Weir had been built at the junction where the canal met the Huai. This arrangement kept the canal's water level stable but made it impossible for large Huai River vessels to enter. Conveniently, a nearby waterway—the Guan River—provided an alternative. Shizong ordered the Guan River widened and dredged, transforming it into a navigable link between the Han Canal and the Huai River. As a result, even large warships could now travel in single file along the Han Canal and ultimately reach the Yangtze River.[12]

At the same time, Emperor Shizong oversaw the restoration of the Bian River, thus reactivating the principal water transport arteries that had been in use since the Sui and Tang dynasties.[13] In the fourth year of the Xiande reign (957 CE), he also ordered the dredging of the Wuzhang River, diverting water from the Bian River northward

---

[10] Suzhou. In *Universal geographic record of the Taiping era*, vol. 17. It states: "In the seventh year of the Yuanhe reign (812 CE), Suzhou was reestablished, with its administrative center relocated to Yongqiao, which was originally the site of Hong County."Hong County corresponds to present-day Si County in Anhui Province.

[11] Annals of the Later Zhou III. In *Comprehensive mirror for aid in government*, vol. 292. It states:"In the second year of the Xiande reign (955 CE), eleventh month, on the day yichou, the Bian River, which had burst its banks in the late Tang, had left the southeastern region of Yongqiao a marshland. The emperor, planning an attack on the Southern Tang, first ordered Wu Xingde, military commissioner of Wuning, to mobilize laborers and dredge along the former embankment, channeling the water eastward to Sishang. Many officials believed the project could not succeed, but the emperor said: 'Within a few years, its benefits will surely be realized.'"The term Sishang here likely refers to the area near Sizhou, as the Bian Canal during the Sui and Tang dynasties did not directly flow into the Si River.

[12] Annals of the Later Zhou V. In *Comprehensive mirror for aid in government*, vol. 294. It states: "In the fifth year of the Xiande reign of the Later Zhou (958 CE), first month of spring, on the day jichou, the emperor wished to send warships from the Huai River into the Yangtze, but was blocked at the Beishen Weir and could not proceed. He planned to cut a channel through the Guan River, northwest of Chuzhou, to open a passage. An envoy was dispatched to survey the site and reported back that the terrain was unfavorable and the project would require great effort. The emperor personally inspected the area, laid out a plan, and ordered laborers from Chuzhou to dredge the channel. Within ten days it was completed, with minimal labor. Hundreds of large warships reached the Yangtze, astonishing the Southern Tang forces, who regarded it as a divine feat."Chuzhou here refers to present-day Huai'an County, Jiangsu Province.

[13] Annals of the Later Zhou V. In *Comprehensive mirror for aid in government*, vol. 294. It states: "In the third month of the fifth year of the Xiande reign (958 CE), the Bian River mouth was dredged, and its waters were directed into the Huai River. From then on, waterborne transport between the Yangtze and Huai regions became navigable."Hu Sanxing's commentary adds: "This was the same transport route used in the Tang dynasty. Since the Jianghuai region had been under separate control, grain transport had ceased and the waterways became silted. Now, they were dredged and restored."

into it.[14] This project increased the Wuzhang River's water volume, enabling boats from the Qi and Lu regions (modern Shandong) to reach Daliang (Kaifeng). This first dredging focused on the upper reaches of the river.

Two years later, a second dredging effort extended the Wuzhang River as far as Dingtao (modern Dingtao County, Shandong), where it connected to the Ji River.[15] This phase concentrated on the lower reaches of the Wuzhang River and involved key areas such as Caozhou (modern northwest Caoxian, Shandong), Jizhou (modern southern Juye County, Shandong), and Liangshan Marsh (formerly Juyue Marsh).[16] Both Caozhou and Jizhou were situated along Liangshan Marsh. At that time, Dingtao was under the jurisdiction of Caozhou, and it was likely in this area that the Wuzhang River entered Liangshan Marsh. The region had once formed part of the ancient Ji River's course, but by the Five Dynasties and Song periods, that waterway had largely silted up. Fortunately, the Ji River still flowed unimpeded northeast of Liangshan Marsh, allowing boats on the Wuzhang River to pass through the marsh and enter the Ji River. From there, goods could be transported to Yunzhou (modern Dongping County, Shandong) and, with a short overland segment, further east to Qingzhou (modern Yidu County, Shandong).

In the sixth year of the Xiande reign (959 CE), the Cai River was also dredged. This project redirected water from the Bian River into the Cai River to facilitate grain transport to Chen (modern Huaiyang County, Henan) and Ying (modern Fuyang County, Anhui).[17] Together, these projects laid the foundational structure for several of the major transport canals used by the Song dynasty. Emperor Shizong also undertook improvements to the Yongji Canal in Hebei, which served as a critical transport route during his northern campaign against the Khitans.

The re-dredging of the Yongji Canal took place in the sixth year of the Xiande reign (959 CE). At that time, the lower reaches of the canal—north of Qianningjun (modern Qing County, Hebei)—had already been ceded to the Khitan by Shi Jingtang. The restoration effort thus focused on the stretch from Cangzhou (whose administrative seat was southeast of present-day Cangzhou, Hebei) to the southern boundary of Qianningjun. Emperor Shizong's northern campaign proceeded along this route: from Cangzhou, the army passed through Qianningjun and advanced directly to Duliukou (present-day Duliu Town, located north of Jinghai County, Tianjin). From there, they turned into the Baigou River (also known as the Baima River), continuing

[14] Annals of Later Zhou IV. In *Comprehensive mirror for aid in government*, vol. 293.

[15] Book of Zhou: Annals of Shizong VI. In *Old history of the Five dynasties*, vol. 119.

[16] Annals of Later Zhou V. In *Comprehensive mirror for aid in government*, vol. 294.

[17] Book of Zhou X: Annals of Shizong. In *Old history of the Five Dynasties*, vol. 119; According to the Biography of Wang Pu. In *History of Song*, vol. 249, "At the beginning of the Xiande reign of the Later Zhou, the military governorship of Huazhou was established, and Zuo (father of Pu) was appointed as its prefect. Not long after, he was transferred to take charge of Yingzhou. … Within the territory of the prefecture, there existed an old commercial canal, three hundred li from the Huai River, which had long since silted up. Zuo dredged and cleared the canal, thereby restoring its navigability." This effort likely coincided with Emperor Shizong's dredging of the Cai River. After these successive rounds of restoration and dredging, the canal system from the Sui and Tang dynasties had, for the most part, been restored to its original form.

westward to Yijin Pass and Waqiao Pass, recovering territory south of these strategic points. Yijin Pass was established as the administrative center of Bazhou, while Waqiao Pass became the seat of Xiongzhou.[18]

The official responsible for restoring the Yongji Canal downstream from Cangzhou at the time was Han Tong. Historical sources record that, south of Qianningjun, he "repaired damaged embankments and opened thirty-six auxiliary outlets, thereby reaching Yingzhou and Mozhou."[19] Yingzhou was headquartered in present-day Hejian County, Hebei, while Mozhou was based in modern Renqiu County, Hebei. These "auxiliary outlets" were constructed in areas where water did not naturally reach, serving as spillways to release excess floodwater[20] rather than as new navigable channels leading directly to Yingzhou or Mozhou.

In reality, neither Yingzhou nor Mozhou had any direct waterway connection with Cangzhou or Qianningjun; both were located close to the Hutuo River, and at that time they were still under Khitan control. It would have been impossible for Han Tong to excavate canals directly into these territories, and the military situation left no opportunity for such deliberate hydraulic undertakings.

As part of Emperor Shizong's northern campaign, Bazhou was established at Yijin Pass, with the counties of Wen'an and Dacheng (both formerly of Yingzhou) incorporated into it. Likewise, Xiongzhou was set up at Waqiao Pass, with the counties of Guiyi (east of modern Xiong County) and Rongcheng (modern Xiong County), formerly belonging to Mozhou, annexed. Yingzhou and Mozhou were thus key initial targets of the campaign.

Han Tong's work on the Yongji Canal north of Cangzhou restored navigability for military transport, enabling troops and supplies to move downstream and connect via a branch of the canal into the Baigou River, which ultimately provided access toward Yingzhou and Mozhou. Therefore, the phrase "reaching Ying and Mo" does not mean that new canals were excavated directly from Cangzhou or Qianningjun into those prefectures; rather, it signifies the use and enhancement of existing waterways for strategic military logistics.

The Integration of Economic and Political Centers

After the death of Emperor Shizong of the Later Zhou, the dynasty quickly came to an end. The Song dynasty rose in its place, inheriting the legacy left behind by Shizong. Emperor Shizong had established his capital at Kaifeng, from which several canals radiated outward, with Kaifeng as their central hub. Though Kaifeng was not an ideal location for a capital in traditional terms, the convergence of these canals transformed it into a major transportation nexus. Thus, selecting Kaifeng as the capital was a deliberate attempt to unify the economic and political centers in a

[18] Annals of Later Zhou V. In *Comprehensive mirror for aid in government*, vol. 294.

[19] Ibid.

[20] Ibid.

single location—an approach that set the Song apart from earlier dynasties such as the Qin, Han, Sui, and Tang.

The Importance of the Bian River in the Song Dynasty

Although the Song dynasty relied on four major transport routes—the Bian River, the Yellow River, the Huimin River, and the Guangji River—the volume of grain transported via the Bian River far surpassed that of the other three. Surviving records clearly demonstrate its critical importance.

In the early years of Emperor Taizong's reign—specifically in the sixth year of the *Taiping Xingguo* era (981 CE)—the Bian River annually transported 3 million *dan* of rice and 1 million *dan* of beans from the Jianghuai region. By contrast, the Yellow River carried only 500,000 *dan* of millet and 300,000 *dan* of beans; the Huimin River brought 400,000 *dan* of millet and 200,000 *dan* of beans from the Chen and Cai regions; and the Guangji River transported the least, just 120,000 *dan* of millet from the Qi and Lu regions.[21]

As national expenditures grew, grain shipments increased annually, most notably along the Bian River. The Jianghuai region, as the empire's principal granary, naturally produced more grain than other regions, but the Bian River's dominance was nonetheless overwhelming. By the end of Taizong's reign (995–997 CE, *Zhidao* era), the river carried 5.8 million *dan* of grain annually. By the beginning of Emperor Zhenzong's reign (1008 CE, *Dazhong Xiangfu* era), the figure had risen to 7 million *dan*. During Emperor Shenzong's reign, a formal quota was established: 6 million *dan* of grain were to be shipped annually along the Bian River, compared with 620,000 *dan* on the Guangji River and 600,000 *dan* on the Huimin River.

Grain transported via the Guangji and Huimin Rivers was primarily used to provision troops stationed near the capital, so the majority of grain delivered directly to the imperial court came from the Bian and Yellow Rivers.[22] Later, during the campaigns against Western Xia, the northwest's grain supply was requisitioned locally for the armies, leaving the capital almost entirely dependent on the Bian River. The Guangji Canal even became impassable on several occasions,[23] further increasing the Bian River's burden.

The Song dynasty valued the Bian River above all other transport routes.[24] This is vividly illustrated by an event in the final years of Emperor Taizong's reign, when the river burst its banks near Kaifeng. The emperor personally supervised the repairs,

[21] Treatise on food and finance, part 1 (III): grain transport. In *History of Song*, vol. 173.

[22] Treatise on canals and waterways III: the Bian River. In *History of Song*, vol. 93.

[23] Treatise on food and finance, part 1 (III): grain transport. In *History of Song*, vol. 173.

[24] Treatise on canals and waterways III: the Bian River. In *History of Song*, vol. 93.

and members of the imperial entourage—including princes and senior officials—waded into the mud to assist.[25] This episode reveals the immense significance the Bian River held for the survival and stability of the dynasty.

Kaifeng in the Song Dynasty

After succeeding the Five Dynasties, the Northern Song established its capital in Kaifeng, where the city soon reached the zenith of its prosperity. This flourishing urban life was inseparably linked to the development of the canal system. Among the major waterways of the Song dynasty, the most important were the Bian River, the Cai River, the Wuzhang River, and the Jinshui River—all of which flowed through Kaifeng. In other words, these canals converged at Kaifeng and radiated outward in all directions, connecting the capital with regions across the empire.

Among these canals, the Bian River occupied a uniquely prominent position. Streets and marketplaces lined its banks in dense succession, forming a bustling urban corridor. The celebrated painting *Along the River During the Qingming Festival* by Zhang Zeduan—an enduring masterpiece—vividly captures the thriving scenes along both sides of the Bian River in and around Kaifeng. The section depicting the Rainbow Bridge (*Hongqiao*) is especially famous. In the river below, large barges and merchant vessels, laden with goods, sail steadily past. On the bridge above, carts and wagons rumble by, alongside pack animals and laborers bearing heavy loads, all moving ceaselessly in both directions.

Remarkably, the Rainbow Bridge had no supporting piers beneath it; instead, it was built using massive timber beams cantilevered across the span, giving the appearance of a rainbow suspended in midair—true to its name. The construction of this bridge demonstrates the considerable advancement of local architectural and engineering techniques. Many such bridges appear to have existed, suggesting that this form of design was fairly common at the time.

It was precisely because goods from all directions were transported along these canals that the streets and alleys of Kaifeng bustled with prosperous trades and industries. Meng Yuanlao's *Records of the Eastern Capital: A Dream of Splendor* offers a vivid, almost eyewitness account of this flourishing urban life. He writes: "From the East Corner Tower streets, heading east lies Panlou Street. On the south side of this street is the Eagle Shop, which exclusively sells hawks and falcons. Elsewhere, shops deal in pearls, fine silks, aromatic medicines, and bedding. Southward is a lane called Jie Shen, famed as a market for gold, silver, and brightly colored silks. The buildings are grand and imposing, with broad storefronts standing in dense rows, awe-inspiring to behold. Each transaction often involves sums in the millions, a sight that astonishes onlookers."

He continues: "At Jiuqiao Gate Street, taverns and markets face one another, with colorful pavilions on either side and embroidered banners fluttering, obscuring the

[25] Annals of Taizong II. In *History of Song*, vol. 5.

sky." He even notes that the night market at Zhouqiao remains bustling until the third watch of the night. Though these examples are but a few, they already paint a clear picture of the vibrant prosperity that characterized Kaifeng during the Song dynasty.

Issues Concerning the Bian River Waterway

The Bian River was undeniably crucial to the security and stability of the state, yet its waterway was frequently beset by problems. Since the river drew its water from a diversion of the Yellow River—a river notorious for its heavy sediment load—silting and sediment accumulation in the Bian River were inevitable.

To ensure a consistent water flow, regular dredging was essential. In the early Song period, it was mandated that dredging be carried out annually. Yet even then, breaches in the embankments occurred frequently, often alarmingly close to Kaifeng itself.[26] During Emperor Zhenzong's reign, in the Dazhong Xiangfu era (1008–1016 CE), the dredging cycle was relaxed to once every three to five years. Later still, there were intervals as long as twenty years without a single dredging operation.

Over time, the unchecked buildup of sediment caused riverbeds in some stretches to rise higher than the surrounding urban areas. By the latter half of the Northern Song, from Kaifeng's East Water Gate downstream to Yongqiu (modern Qixian County, Henan) and Xiangyi (modern Suixian County, Henan), the riverbed stood more than one *zhang* and two *chi* (approximately 3.6 m) above the nearby floodplain. From the top of the embankments, residential neighborhoods below appeared as if they were located in a deep valley.[27]

This perilous situation was the direct consequence of inadequate dredging. Dredging projects consumed vast sums from the state treasury, yet failure to perform them not only jeopardized lives, farmland, and livestock through catastrophic flooding but also paralyzed the vital transport of grain. For this reason, contemporary officials and engineers advocated for a comprehensive and lasting restoration of the Bian River, hoping to put an end to its recurrent problems once and for all.

The Plan to Divert the Bian River and Its Failure

The first attempt to resolve the problems of the Bian River was made by Xing Yongzhi. As a Doctor of the National University, he submitted a memorial at the end of Emperor Taizong's reign proposing the excavation of a canal from Baigou (near present-day Kaifeng, Henan) to Lüliangkou in Pengcheng, covering a total length of about six hundred li. This canal was intended to facilitate grain transport between the Yangtze

---

[26] Treatise on canals and waterways III: the Bian River. In *History of Song*, vol. 93.

[27] Miscellaneous records II. In *Dream pool essays*, vol. 25.

and Huai rivers.[28] The canal Xing Yongzhi suggested was essentially a restoration of the old Bian Canal route dating back to the Eastern Han dynasty. His plan was that, once this new canal was completed, the Bian River itself could be entirely abandoned. Emperor Taizong had already accepted this proposal and begun the excavation work. However, the project faced strong opposition. Some critics accused Xing Yongzhi of having farmland in Xiangyi and claimed that his real motive was to divert the water away from his own land to avoid flood damage. This controversy stirred public outcry, and as a result, the canal construction was halted.

Ten years later, in the third year of the Jingde reign of Emperor Zhenzong (1006 CE), another proposal was made to excavate a new canal to replace the Bian River. This new canal would divert water from the Guangji River near Dingtao and channel it toward Xuzhou, eventually converging with the Qing River (the Si River). The route appeared to follow, with some modifications, the ancient course of the He River from antiquity. After the canal was successfully excavated, a topographical map was submitted to the capital. Upon reviewing it, Emperor Zhenzong observed that the terrain along the canal varied greatly in elevation, the water flow was shallow, and the route would still have to pass through the treacherous lüliang region. Seeing no clear advantage over the existing Bian River, the emperor ultimately decided to abandon the project.[29]

Later, during the reign of Emperor Shenzong, when Wang Anshi held power, the proposal to replace the Bian River was raised once again. This time, however, the plan differed significantly from the previous two. The proposer argued that the constant silting of the Bian River was due to its reliance on water diverted from the Yellow River, which carried an excessive amount of sediment. To address the issue at its root, he suggested completely blocking off the mouth of the Bian River and instead utilizing the nearby Jingshui River and Suo River as alternative sources. By constructing sluices and dams, water could be stored and regulated, thereby maintaining a navigable water level for boat traffic. Since the riverbed of the Bian River had already risen significantly due to years of sediment accumulation, the proposal recommended abandoning the Bian River altogether and excavating a new canal from Baigou to Pengcheng. This was a far-reaching proposal, and even Emperor Shenzong hesitated to approve it outright. However, Wang Anshi strongly supported the plan, and construction was initiated. Yet after just one year, widespread dissatisfaction arose both within the court and among the general public. The project proved impractical and inconvenient, and it was ultimately abandoned.[30] Following these repeated failures, the idea of replacing the Bian River with an entirely new canal was no longer seriously pursued.

[28] Treatise on canals and waterways IV: the Baigou River. In *History of Song*, vol. 94.

[29] Treatise on canals and waterways IV: the Guangji River. In *History of Song*, vol. 94.

[30] Treatise on canals and waterways IV: the Baigou River. In *History of Song*, vol. 94.

The Regulation of the Bian River Inlet and the Diversion of the Luo River into the Bian River

Long before Wang Anshi advocated the drastic proposal of abandoning the Bian River altogether, the official Guo Zi had already identified the root cause of the river's chronic siltation: the excessive sediment carried by the Yellow River. He argued that unless this problem was resolved at its source, even excavating a new canal would be futile, for it too would inevitably become clogged within a few years.

Guo observed that the mouth of the Luo River lay not far from the Bian inlet and that, in contrast to the Yellow River, the Luo's waters were far clearer. Since the Bian River's dependence on the Yellow River led to continual silting, he proposed diverting the Luo River into the Bian as a fundamental solution. He estimated that the distance from the Luo River's mouth to the Bian inlet was only seventy *li*, a span not prohibitively difficult for canal construction. Unfortunately, Guo died shortly after presenting his proposal, and the plan was shelved.[31]

Later, during Emperor Shenzong's reign, when the ambitious plan to abandon the Bian River had completely collapsed, Guo Zi's proposal was revived. At the time Guo had first suggested it, the Yellow River's main current still pressed tightly against Guangwu Mountain on the river's southern bank. Connecting the Luo River to the Bian under those conditions would have required cutting a canal through the high ground of the riverbank and the mountain slopes—an extremely challenging feat of engineering.

By Shenzong's reign, however, the Yellow River's main current had shifted northward, leaving a broad expanse of open sandbank at the foot of Guangwu Mountain. This presented an ideal opportunity: instead of carving a channel through the mountain, engineers could now build a massive embankment across the exposed riverbed. This embankment effectively blocked the Yellow River's main current from entering the Bian, while diverting the clearer waters of the Luo River into it. During flood periods, if the Bian could not accommodate the full volume of the Luo's flow, the surplus could spill over the embankment and rejoin the Yellow River.[32]

With the successful completion of this project, the Luo River's waters began flowing steadily into the Bian River, greatly reducing the problem of sediment accumulation. From that time onward, the Bian River earned a new name—Qing Bian (the "Clear Bian River")—a testament to the success of this strategic hydraulic reform.

[31] Biography of Guo Zi. In *History of Song*, vol. 326.

[32] Treatise on canals and waterways IV: the Bian River. In *History of Song*, vol. 94.

The Clear and the Turbid Bian Rivers

The plan to divert the Luo River into the Bian Canal was indeed sound, but it still had one significant flaw. At the time, a large embankment was built across the riverbed at the foot of Guangwu Mountain, intended to block the Yellow River from flooding southward. However, two major concerns remained: first, during periods when the Luo River swelled, the Bian Canal might not have the capacity to hold the excess water, raising the issue of where the surplus should go; second, vessels from the Yellow River had long been able to sail directly into the Bian Canal, and officials were eager to preserve this longstanding practice.

To address these concerns, a sluice gate was deliberately left in the embankment so that the Yellow River and the Bian Canal would not be completely severed from each other. Initially, after the diversion, the clear waters of the Luo River entered the Bian Canal, and the waterway earned the name "Clear Bian" (Qing Bian). However, as time passed, the main current of the Yellow River shifted southward and began pressing dangerously close to the embankment. Its muddy waters soon began seeping into the Bian Canal through the sluice gate, and the Clear Bian once again turned into the "Muddy Bian" (Zhuo Bian).

This southward shift of the Yellow River's main channel raised grave concerns. Before the Luo River was diverted, the amount of Yellow River water entering the Bian Canal had always been carefully controlled; when the flow exceeded a certain limit, the Bian mouth would be sealed. Now, although the system was designed to rely on the Luo River, in reality, the Yellow River was still invading the canal. The sluice gate could no longer be freely operated, and as the Yellow River's main current drew closer to the base of the embankment, officials feared that if the dike were to break, the entire Yellow River would pour into the Bian Canal, devastating Kaifeng.

Because of this danger, in the early years of the Yuanyou era (1086–1094) during Emperor Zhezong's reign, the court authorized allowing a controlled amount of Yellow River water to enter the Bian Canal. Yet only a few years later, many came to realize that the earlier fears were not unfounded, and the policy was reversed: only the Luo River was used, and the Yellow River was once again excluded. Some officials even proposed excavating a completely new channel near Guangwu Mountain so as not to rely on the embankment across the riverbed, but such plans were repeatedly delayed and never implemented.[33] Fortunately, by the time the Northern Song court was forced to move south, the Yellow River had not yet fully diverted into the Bian Canal, sparing Kaifeng from a catastrophic flood.

[33] Ibid.

The Regulation of the Huai River and the Han Canal

After the Bian River flows southward, it passes through a section of the Huai River and enters the Han Canal.[34] At that time, there were two notable features of the Huai River and the Han Canal: first, after the Bian River entered the Huai River, the transportation route passed through Hongze Lake, then turned eastward into the Han Canal. The section of the Huai River during this stretch was particularly dangerous, with strong waves that frequently caused the cargo boats to sink. Second, the confluence where the Han Canal entered the Huai River was too swift, making navigation difficult. The first of these issues concerned the rise and decline of the Han Canal. People often questioned why, after the Bian River entered the Huai River, it was necessary to use the Han Canal, which required a long detour, instead of constructing a new canal directly from the point where the Bian River met the Huai River to Yangzhou. These two problems had existed since the Sui and Tang dynasties. At that time, some people had already paid attention to them and attempted to improve the situation, although no satisfactory solutions were achieved.

During the reign of Emperor Taizong, Liu Pan and Qiao Weiyue, who served successively as the Transport Commissioner of the Huainan region, constructed a 60-*li* (approximately 30 km) canal from Chuzhou (modern-day Huai'an in Jiangsu Province) to Huaiyin, which alleviated many of the difficulties.[35] During the reign of Emperor Renzong, the Transport Commissioner Xu Yuan further dug a channel along the Huai River from Huaiyin to Hongze Lake. Later, when Ma Zhongfu served as the Transport Commissioner, he, recognizing that the danger to cargo boats had not been entirely eliminated, excavated a 60-*li* canal known as the Hongze Canal outside Hongze Lake. During the reign of Emperor Shenzong, the Transport Commissioner Jiang Zhiqi continued the work by constructing the Guishan Canal, which connected up to the confluence of the Bian River and the Huai River. The Guishan Canal started near the Guishan Hill close to the confluence and extended to Lake Hongze,

---

[34] During the Song Dynasty, this Han Gou was simply referred to as the canal, without using its old name. According to the *History of Song: Treatise on Canals and Waterways VI, the Southeastern Rivers*, an imperial decree issued in the second year of Xuanhe (1120 AD) stated: "The transportation of grain through the Jianghuai region has a long history. During the Spring and Autumn period, Wu State excavated the Han Canal, which extended northeast to Lake Sheyang and northwest to Mokou. In the Han Dynasty, King Liu Bi of Wu excavated the Han Canal to connect with the sea in Hailing. In the Sui Dynasty, the Han Canal was re-excavated, running from Shanyang to Yangzi, entering the Yangtze River… Now, the canal is shallow and obstructed; it is necessary to investigate the old routes, as well as the current conditions of the river, the surrounding ponds, and water storage areas, to plan for long-term benefits and ensure smooth transportation." It then continues: "At first, due to consecutive years of drought in the Huainan region, the grain transport was blocked, especially in Yangzhou. The emperor ordered an imperial envoy to inspect the situation and considered dredging the canal and the Jianghuai waterways." At the same time, officials' memorials also only referred to the "Huainan Canal," without mentioning the Han Canal. Similarly, during the Song Dynasty, the southern section of the Grand Canal was also referred to simply as the canal, without using the term Southern Canal, which was in line with the renaming of the Han Canal.

[35] Treatise on canals and waterways VI: the Southeastern Rivers I; Biography of Qiao Weiyue. In *History of Song*, vol. 96, 307.

continuing through the Hongze Canal. Thus, the dangers of the long Huai River were largely eliminated.[36]

In the period of Emperor Huizong's Chongning era (1102–1106 AD), a direct canal was begun near the point where the Bian River entered the Huai River, extending directly to Yangzhou. The construction lasted five years and was finally completed. At that time, it was called the Yuming River. It ran from Xuanhua Gate at Zhenzhou (modern-day Yizheng in Jiangsu Province) to the mouth of the Huai River near Sizhou, nearly a straight north–south route.[37] However, this direct canal did not yield significant benefits at the time and soon became silted up. More than a decade later, Tan Zhen, who served as the Regional Commissioner, attempted to reopen the direct canal, but it ultimately failed to be completed.[38]

The Yuming River and Zhenzhou

During the Song dynasty, repeated efforts were made to excavate a canal that would lead directly from the Bian River into the Huai River and continue on to Yangzhou. The eventual outcome of these attempts was the construction of the Yuming River. While on the surface this appeared to be aimed at shortening the water route, in reality, it was primarily intended to accommodate the prosperity of Zhenzhou at the time.

How did this situation arise? The backdrop lies in the decline of Yangzhou. Once a brilliant and prosperous city, Yangzhou had been repeatedly devastated by warfare during the late Tang and Five Dynasties periods, leaving it severely weakened. Yet the region still required a vibrant hub along this critical transportation artery—a place where merchants and travelers could conduct trade and find lodging. It was under these circumstances that Zhenzhou emerged to take Yangzhou's place.

In the Tang dynasty, Zhenzhou was merely a small settlement known as Baisha Town. During the Five Dynasties period, it was elevated to Yingluan Town, and in the early Song it was further promoted to the rank of Jian'an Commandery. In the sixth year of the *Dazhong Xiangfu* reign of Emperor Zhenzong (1013 CE), it was formally designated as Zhenzhou, and in the seventh year of the *Zhenghe* reign under Emperor Huizong (1117 CE), it was granted the prestigious name Yizhen Prefecture.[39]

These successive administrative promotions clearly testify to Zhenzhou's rapid rise to prominence. Its prosperity can be directly attributed to its role as a replacement for Yangzhou as the key commercial hub of the region. In the Sui and Tang periods, Yangzhou, along with Chuzhou (corresponding to modern-day Huaiyin) and Sizhou (modern-day Sixian), had served as the major economic centers along the lower

---

[36] Treatise on canals and waterways VI: the Southeastern Rivers I; Biography of Ma Zhongfu. In *History of Song*, vol. 96, 331.

[37] Treatise on canals and waterways VI: the Southeastern Rivers I. In *History of Song*, vol. 96.

[38] Ibid.

[39] Yizhen county. In *Comprehensive gazetteer of Qing dynasty (Jiaqing edition)*, vol. 96.

reaches of the Bian River and the Han Canal. By the Song dynasty, Zhenzhou had joined—and even surpassed—these historic urban centers in commercial importance.

Scholars and officials of the Song dynasty left numerous accounts describing the prosperity of Zhenzhou. Here we may cite two or three representative examples to offer a glimpse of the broader picture.

Ouyang Xiu once remarked: "Zhen, as a prefecture, lies at the confluence of the southeastern waterways. Thus it was chosen as the administrative seat of the Transport Commissioner for the Jianghuai region, the Liangzhe Circuits, and Jinghu."[40] Hu Su likewise observed: "This is the core region of Yingluan, a grand prefecture on the banks of the Yangtze. …Southward, vessels pass the Five Ranges and sail deep into the Three Xiang regions; westward, they come from the fords of the Ba Gorges; eastward, they anchor at the frontiers of Ou and Min. All routes converge here, and among the surrounding territories, it stands out in prominence. … Ten thousand ships follow one another in endless procession, supplying the capital year after year."[41] Lou Yue went even further: "Zhen's elevation to prefectural status is relatively recent… yet it occupies a critical junction between the Yangtze and Huai rivers. The Grand Canal's command station was established here. Rice shipments from the lakes and rivers are transferred and transported to the capital in quantities that reach tens of millions each year. Although Weiyang, Chuzhou, and Sizhou are all described as prosperous, Zhen remains the foremost among them."[42]

The praises of Ouyang Xiu, Hu Su, and Lou Yue fully encapsulate the reasons behind Zhenzhou's prosperity. As Lou Yue notes, "Weiyang" refers to Yangzhou; Chuzhou corresponds to modern Huai'an County in Jiangsu, located at the mouth where the Han Canal enters the Huai River; Sizhou was situated on the north bank of the Huai River in present-day Xuyi County, Anhui, and served as the mouth where the Bian River entered the Huai. While these places were indeed prosperous, they all fell short when compared with Zhenzhou. This was precisely because, during this period, Zhenzhou had supplanted Yangzhou as the critical juncture where the Han Canal met the Yangtze River.

The excavation of the Yuming River was certainly tied to the prosperity of Zhenzhou; however, even before the Yuming River was constructed, Zhenzhou already had a canal connection to Yangzhou. Not long after Zhenzhou was established as a prefecture, it was already reported that grain transport boats were passing from Zhenzhou into the Huai and Bian Rivers, traversing five weirs and dams along the way—a process that was said to place a heavy burden on the people's labor.[43] This indicates that a canal system linking Zhenzhou and Yangzhou was already in place at the time. Later on, the Zhen–Chu Canal was also dredged,[44] which by necessity passed through Yangzhou. The eventual excavation of the Yuming River further

[40] Ouyang Xiu. Record of the eastern garden in Zhenzhou. In *Collected works of Ouyang Wenzhonggong*, vol. 40.

[41] Hu Su. Record of the water gate in Zhenzhou. In *Collected works of Wengong*, vol. 35.

[42] Lou Yue. Record of the city wall repairs in Zhenzhou. In *Collected works of Gongkui*, vol. 54.

[43] Treatise on canals and waterways VI: the Southeastern Rivers I. In *History of Song*, vol. 96.

[44] Ibid.

contributed to Zhenzhou's flourishing. Had the Yuming River remained navigable and unobstructed for a long time, Zhenzhou might have permanently taken Yangzhou's place as the dominant hub. Yet, that plan ultimately failed. Following a long period of recovery, Yangzhou regained its former prosperity, and as it revived, Zhenzhou gradually fell into decline once more (See Fig. 6.2).

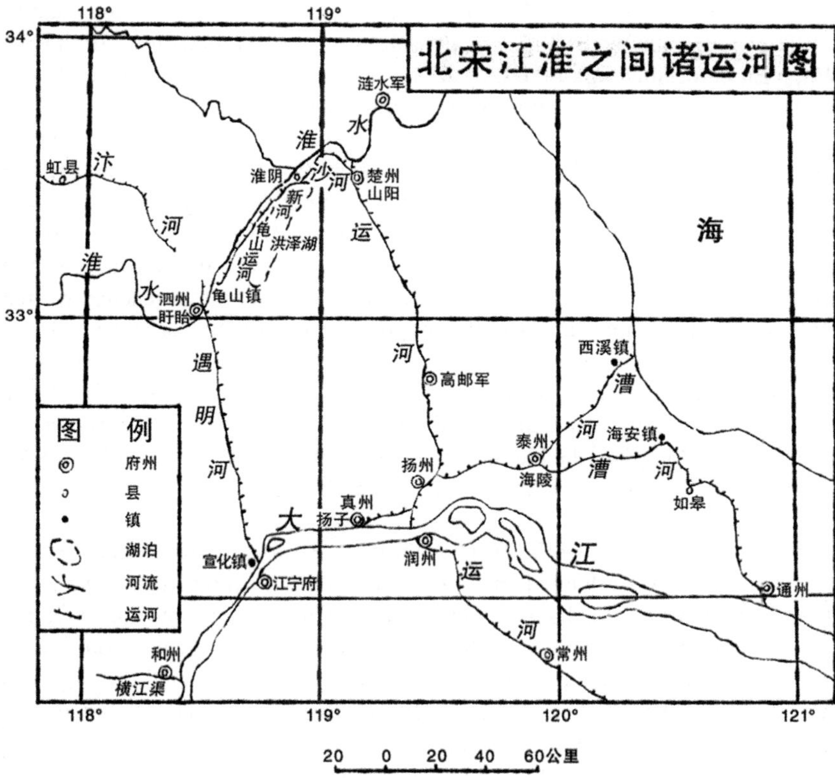

图例 Legend
府州名 Prefectures and Provinces
县 Counties
镇 Towns
湖泊 Lakes
河流 Rivers
运河 Canals

Fig. 6.2 Grain transport canals between the Yangtze and Huai rivers in the northern song dynasty

The Huimin River

Among the four main transport canals of the Song dynasty, the Huimin River was originally known as the Cai River during the Tang period. The Cai River, in turn, was a branch of the Langtang Canal—part of the Hong Canal system—and specifically its offshoot carrying the sandy waters. The Langtang Canal had originally branched off from the main Hong Canal alongside other waterways such as the Bian and Sui rivers. Therefore, even by the Tang era, its water source still relied on the Bian River.

When Emperor Shizong of the Later Zhou undertook dredging works on the Cai River, he too continued to draw water from the Bian River. To the west of the Cai River lay the Min, Wei, and Yi rivers, which originated in Xinzheng (modern-day Xinzheng County, Henan). However, these rivers all flowed southeast into the Ying River and were unrelated to the Cai River. It was not until the first year of the Jianlong reign (960 CE), under Emperor Taizu of Song, that the Min River was diverted from Xinzheng to Kaifeng, where it joined with the Cai River.[45] In fact, the so-called "Min River" was a combined flow of the Zhen and Wei rivers that reached Kaifeng under that collective name. The Min River approached Kaifeng from the southwest, while the Cai River came from the southeast; after the two merged, the Min River became the common designation. In the sixth year of the Kaibao reign (973 CE), the Min River was officially renamed the Huimin River, while the Cai River retained its original name.[46] Although the Huimin River referred specifically to the former Min River, it became common practice to refer to the Cai River as part of the Huimin River as well.

During the Zhidao reign (995–997 CE) of Emperor Taizong, six prefectures transported tribute grain to Kaifeng via the Shitang River and Huimin River: Chen, Ying, Xu (seat in present-day Xuchang County, Henan), Cai (present-day Runan County, Henan), Guang (present-day Huangchuan County, Henan), and Shou (present-day Shou County, Anhui).[47] Among these, Chen Prefecture was located directly along the main course of the Cai River. Ying Prefecture, which bordered the Ying River, lay just downstream of the confluence where the lower Cai River flowed into the Ying. Thus, grain barges from Ying Prefecture could sail upstream along the Ying River and enter the Cai River. Guang and Shou Prefectures, situated along the Huai River, could send their tribute grain via the Huai into the Ying, and then on to the Cai River. Only Xu and Cai Prefectures lay farther west of the Cai River and were likely within the transport network of the Shitang River, which flowed south of present-day Ye County in Henan.[48] North of the Shitang River was the Sha River, which originated from modern-day Lushan County and eventually flowed into the Ru River. East of the Ru lay the Ying River. Although the Ru and Ying were separate waterways, they could be linked at Yancheng in Xu Prefecture (modern Yancheng County, Henan) via

---

[45] Annals of the Later Zhou dynasty V. Hu's commentary notes In *Comprehensive mirror for aid in government*, vol. 294.

[46] Ibid.

[47] Treatise on food and finance I (3): Grain transport. In *History of Song*, vol. 128.

[48] Yuzhou. In *Comprehensive gazetteer of the Yuan dynasty*, vol. 384.

the Shang River,[49] allowing for navigational access between them—a point discussed earlier. In this system, tribute grain barges from Xu Prefecture could take the Shang River into the Ying River and then continue into the Cai River. Barges from Cai Prefecture could descend along the Ru River to the Huai. Alternatively, they could ascend the Ru River to Xu Prefecture, then follow the same route as Xu's grain shipments into the Cai River. If the Huimin River were limited only to the stretch between Xinzheng and Kaifeng, it would be difficult to explain how it served as the conduit for transporting tribute grain from these six prefectures. This demonstrates that the Cai River was commonly and justifiably referred to as part of the Huimin River.

By comparison, the Huimin River was the least prone to silting among the main transport canals of the Song dynasty. Although it still drew water from the Bian River, its primary water source during the Song period was in fact the Min River, formed by the confluence of the Zhen and Wei rivers. In the second year of the Qiande reign (964 CE), Emperor Taizu additionally diverted the Yi River from Changshe (modern-day Xuchang County, Henan) into the Min River.[50] Other minor rivers—such as the Jing and Suo streams flowing from Zhengzhou (modern Zhengzhou, Henan)—also converged into the Huimin River. As these tributaries carried relatively clear water, sediment accumulation in the Huimin River was minimal. However, one concern did arise: since the Huimin River ultimately discharged these waters into the Ying River, during periods of flooding, the combined flow often exceeded the Ying River's capacity. The area around Huaiyang (modern Huaiyang County, Henan) was already low-lying by nature, and in the summer and autumn, the surrounding lands frequently became marshy and inundated.

In the fourth year of the Yuanyou reign (1089 CE) under Emperor Zhezong, Hu Zongyu, the prefect of Chenzhou Prefecture (seat in modern Huaiyang County), proposed following the ancient course of the Bazhang Ditch (Bazhang Gou) to excavate a new canal. This canal would run alongside the Ying River from Chen Prefecture to Shou County (modern Shou County, Anhui), directly entering the Huai River and thereby preventing flood disasters caused by the Ying River's limited capacity.[51] The ancient Bazhang Ditch to which Hu Zongyu referred was a lower-branch distributary of the Honggou canal system. The lower reaches of this system were known as the Sha River, which passed east of Chen (modern Huaiyang County, Henan), and in that stretch, it was called the Baichi Ditch. The waters of this ditch flowed southeast, merging with the Gu River—which came from a marsh west of Chen—and continued southeast to a confluence known as Jiaokou, where they joined the Ying River. At

[49] Northwestern Route of Jingxi. In *Treatise on the Nine Territories in the Yuanfeng Reign*, vol. 1. It states: "At Shangshui in Chenzhou, there are the Ying River and the Shang River. The 'Shang River' is most likely a later name for the Yin River."

[50] Treatise on canals and waterways IV: the Cai River. In *History of Song*, vol. 94. It states: "In the twelfth month of the first year of the Daguan reign (1107 CE) under Emperor Huizong, the Yi River was diverted to flow into the Cai River, following a request by Wu Zeren, the Metropolitan Transport Commissioner of the Capital Region."

[51] Wang Yingling. Historical studies. In *Miscellaneous records from the exhaustive study of the classics*, vol. 16.

this junction stood a large weir, historically called the Baichi Weir (Baichi Ye), or sometimes simply the ancient Baichi embankment. During the Wei and Jin periods, this area still retained substantial water flow.[52] The Baichi Ditch and the Bazhang Ditch were, in fact, the same watercourse.[53] Hu Zongyu's proposed new canal was approximately 800 li (roughly 400 km) long. He completed a substantial portion of it, but the section reaching the Huai River remained unfinished.

In the early years of the Zhenghe reign (1111–1117 CE) under Emperor Huizong, Huo Duanyou, then serving as prefect of Chen Prefecture, once again addressed the problem. He observed that the region around Chen was low-lying and that prolonged rains in autumn often led to crop damage from flooding. Since the previously excavated 800-li canal still fell short of reaching the Huai River, water could not drain efficiently. He thus petitioned to extend the canal by another 200 li, allowing it to reach the Huai fully. This proposal was approved, and the project resumed—eventually leading to full completion.[54] According to Huo Duanyou's proposal, the newly excavated canal began at Xihua, passed through Wanqiu, entered Xiangcheng, and finally reached the Huai River. Xihua and Xiangcheng correspond to present-day Xihua County and Xiangcheng County in Henan, respectively. Wanqiu is the historical name for what is now Huaiyang County. Since Xihua lies to the northwest of Huaiyang—and even farther beyond the historical boundaries of Chen Prefecture—it is likely that Huo Duanyou's canal project encompassed the entire course, including the section originally planned and partially constructed by Hu Zongyu.

The Guangji River and the Tianyuan River

Compared to the Huimin River, the Guangji River was far more susceptible to silting. Originally called the Wuzhang River, it was renamed the Guangji River in the sixth year of the Kaibao reign (973 CE). Like the Huimin River, its water source was drawn from a diversion of the Bian River, but silting along the Guangji River occurred at a much faster rate.

In the sixth year of the Xiande reign of Later Zhou (959 CE), Emperor Shizong ordered the Guangji River to be dredged. Yet, barely two years later, in the second year of the Jianlong reign under Emperor Taizu of the Song dynasty (961 CE), it

---

[52] Commentary on canals. In *Water classic*.

[53] Chenzhou. In *Essentials of historical geography*, vol. 47.

[54] Biography of Huo Duanyou. In *History of Song*, vol. 354.

again required dredging.[55] During this second operation, not only was the riverbed of the Guangji River reinforced, but a new water source was also created.

Previously, the Guangji River relied solely on the diverted waters of the Bian River, a situation that both reduced the water volume available to the Bian and exacerbated silting in the Guangji. To address this problem during the Jianlong dredging, engineers excavated a new canal to divert water from the Jing River at Xingyang, routing it through Zhongmou County to Kaifeng. At Kaifeng, an elevated aqueduct was constructed over the Bian River, allowing the diverted water to pass above the Bian and discharge eastward into the Guangji River. This new canal from the Jing River was called the Jinshui River at the time, and also went by the name Tianyuan River.[56] Because the waters of the Tianyuan River carried relatively little sediment, the Guangji River became less vulnerable to silting once it began to receive this cleaner flow. However, the aqueduct conveying the Tianyuan River necessarily spanned across the Bian River, which inevitably hindered navigation on the Bian itself. For this reason, the aqueduct was sometimes dismantled or rendered inoperable, which in turn cut off the Tianyuan River's contribution to the Guangji. As a result, canal transport along the Guangji River in the Song dynasty was frequently interrupted, likely due to this very issue.

The Yellow River's Transport Routes

Among the four main transport arteries, the Yellow River was a natural waterway and thus quite distinct from the other three canals. The term "Yellow River transport route" here refers specifically to the section from Biankou to Tongguan, rather than the river's entire course.

As early as the Qin, Han, Sui, and Tang dynasties, this stretch of the Yellow River, together with the adjacent canal network, formed a crucial segment of the empire's grain transportation system. The river, however, was notorious for its numerous shoals and treacherous currents, and incidents of boats sinking and cargo being lost occurred frequently.

During the Song dynasty, however, such issues were largely absent. This was because the capital was established at Kaifeng, allowing grain shipments from the Guanzhong region (around present-day Xi'an) to travel downstream—unlike in the Qin, Han, Sui, and Tang dynasties, when shipments had to navigate upstream against the current. Moreover, the volume of tribute grain transported from Guanzhong

[55] Biography of Chen Chengzhao. In *History of Song*, vol. 261. It states: "Emperor Taizu, recognizing Cheng Zhao's expertise in hydraulic works, appointed him to oversee the construction of the Huimin River and the Wuzhang River in order to facilitate grain transport. The people of the capital benefited from this. In the second year of the Jianlong reign (961 CE), the rivers were completed." Biography of Liu Zai. In *History of Song*, vol. 262. It states: "In the early Song period, the Wuzhang River was dredged, extending from Chenqiao to the western border of Cao Prefecture. The court appointed Zai to supervise the project."

[56] Treatise on canals and waterways IV: the Jinshui River. In *History of Song*, vol. 94.

during the Song was relatively small. In the later period, when the state launched campaigns against Western Xia, the grain from Guanzhong was mostly requisitioned locally for military use and was no longer shipped to Kaifeng.

The Evolution of the Yongji Canal

In the Song dynasty, the Yongji Canal was not included among the primary transportation routes. Although it had not become entirely silted up since the late Tang and Five Dynasties period, it was no longer fully navigable. The northern terminus of the Yongji Canal, originally excavated by Emperor Yang of Sui, once extended as far as Zhuojun. However, beginning in the Five Dynasties period, Khitan influence steadily advanced southward, and the Later Jin handed over the Sixteen Prefectures of Yan and Yun to them. As a result, the northern part of what is now Hebei Province fell beyond the reach of the Central Plain dynasties. Although Emperor Shizong of the Later Zhou launched northern campaigns and managed to reclaim some territory, his premature death cut short any lasting success.

After the founding of the Song dynasty, the northern frontier remained largely unchanged from the situation at the time of Emperor Shizong's death, with the Baigou River serving as the boundary between the two realms. This river was commonly known at the time as the "Boundary River." Because of this shift in the frontier, the northern terminus of the Yongji Canal could now only extend as far as present-day Qing County, where it joined the Boundary River before flowing into the sea. As for its southern terminus—originally constructed by Emperor Yang of the Sui dynasty to link the Qin River with the Yellow River—by the end of the Tang and Five Dynasties period it had also become blocked and impassable. By the early Song period, what remained of the Yongji Canal was therefore only a truncated middle section, which at the time was commonly referred to as the Imperial River (Yuhe).

The Regulation of the Upper Reaches of the Yuhe

In the Song dynasty, the so-called *main transport routes* referred to the waterways used to convey tribute grain from various regions to the capital. The Imperial River (*Yuhe*), by contrast, was primarily used to deliver military provisions to frontier areas and therefore was not counted among these main transport arteries.

By this time, the Yuhe no longer benefited from the waters of the Qin River and relied solely on the Baimen Springs in Gongcheng County of Weizhou (Weizhou's seat was in present-day Ji County, Henan, and Gongcheng in modern Hui County, Henan). Below Jixian, the spring-fed stream became deep enough to carry large vessels. In the early Song period, military provisions from the capital to Hebei were first transported via the Bian River into the Yellow River, then either to Liyang (northeast of present-day Xun County, Henan) or to Maling Crossing (south of modern

Daming County, Hebei). From there, they were transferred by cart to the banks of the Yuhe and reloaded onto boats for further transport.

In the eighth year of the Xining reign (1075), Emperor Shenzong adopted the proposal of Cheng Fang and Liu Hui to excavate the old course of the Sha River in Weizhou, diverting Yellow River water northward into the Yuhe at Wanggongsao (east of Jixian, then close to the Yellow River and not far from the Yuhe). Cheng and Liu cited five benefits for this project:

1. Flood prevention at Wanggongsao—This location was a weak point on the Yellow River's banks, often requiring costly reinforcement to prevent breaches. Diverting water into the new channel would relieve pressure on the embankments at Wanggongsao and, by reducing the Yellow River's flow volume, mitigate the risk of breaches elsewhere.
2. Safer navigation—Boats leaving the Bian River could enter the Sha River directly, avoiding the dangerous currents and waves of the Yellow River.
3. Regulation of Yellow River overflow—The Sha River could act as a buffer, easing Yellow River flooding by channeling excess water into the Yuhe.
4. Prevention of siltation in the Yuhe—If the Yuhe experienced high water levels, pre-installed sluice gates (*doumen*) could be operated as needed to prevent destructive overflow and sediment accumulation.
5. Improved transport for Dezhou and Bozhou—Vessels from these prefectures (modern Ling County and Liaocheng County, Shandong) would be spared the hazards of navigating several hundred *li* of the Yellow River.

Because Wanggongsao was close to the Yuhe and the old Sha River course could be reused, the work was completed swiftly, in just one month.[57]

Before this project, when the southern end of the Yuhe was blocked, grain shipments to Hebei had to be detoured through Liyang or Maling Crossing for overland transfer. The new channel eliminated these delays. Cheng and Liu's claim that the vessels from De and Bo could now avoid "several hundred *li* of the Yellow River" referenced the situation following the Yellow River's breach at Shanghu (near present-day Puyang County, Hebei) in the eighth year of the Qingli reign (1048, under Emperor Renzong). One branch of the river flowed north into the Yuhe, while another flowed east into the Majia River. The eastern branch passed through Dezhou and Bozhou, enabling boats to follow it upstream, pass Maling Crossing, and reach Daming Prefecture (present-day Daming County, Hebei), where they could enter the canal network. This was the strategic context for Cheng and Liu's plan.

There remains an issue that requires further clarification. The plan proposed by Cheng Fang and Liu Hui was premised on the existence of an already silted-up former course of the Sha River in Weizhou, which they intended to reutilize. However, questions remain as to how this Sha River came into being, and where exactly its starting and ending points were—these matters have yet to be conclusively resolved. Wanggongsao lay directly along the banks of the Yellow River, and the distance northward to Qing River was not particularly great; theoretically, there should have

[57] Treatise on canals and waterways IV: the Yuhe. In *History of Song*, vol. 95.

been no other watercourse between them. This raises the possibility that the so-called Sha River's old course was, in fact, the former channel dug by Emperor Yang of the Sui to divert the Qin River eastward. Yet Emperor Yang's Yongji Canal did not only draw on the Qin River; it also made use of the Qing River's course downstream in order to merge with the Qi River. The Qin River, being sediment-laden, may have left deposits in its former eastern flow path once it ceased to flow that way, which may explain the formation of "accumulated sand". However, while the Qin River no longer flows eastward, the Qing River has continued to flow unimpeded, making it less likely for a silted riverbed to have formed there. Therefore, the matter warrants further investigation and consideration.

Among those who opposed the proposal by Cheng Fang and Liu Hui, Wen Yanbo was the most adamant. At the time, Wen served as the Pacification Commissioner of Daming and genuinely feared that the diversion of Yellow River water into the Yuhe would cause flooding in the Daming region.[58] From a practical standpoint, although this newly proposed canal was indeed convenient, it had a critical flaw. In general, when a canal draws directly from the Yellow River, the greatest danger lies in the main current of the river surging into the canal. The new canal in Weizhou, which was designed to divert Yellow River water into the Yuhe, relied on nothing more than a single large embankment at the intake point to resist the force of the river's main flow. Relying solely on this embankment to hold back such a powerful current was extremely difficult. As it turned out, Wen Yanbo's concerns were fully justified. Not long after the canal was completed, the Yellow River broke into the Yuhe. This resulted not only in flooding in the Daming prefecture, but also brought widespread devastation along the banks of the Yuhe.

The Influence of the Yellow River on the Yuhe

Among the various factors affecting transport on the Yuhe during the Song dynasty, none was more significant than the repeated course changes of the Yellow River. The incident where the Yellow River flowed into the Yuhe via the Sha River near Weizhou was, in fact, only a minor occurrence. As early as the reign of Emperor Zhenzong (Dazhong Xiangfu period), the Yellow River had already breached near present-day Xun County and flowed northward, merging with the Yuhe to reach the sea.[59] That breach was soon sealed, and its damage was relatively limited.

However, in the eighth year of Qingli under Emperor Renzong (1048 CE), the Yellow River breached again at Shanghu (northeast of modern Puyang, Henan), splitting into two courses: one flowed east to join the Majia River and passed through Bozhou and Dezhou into the sea; the other flowed north, passing north

[58] Ibid.

[59] Li Tao. Reign of Zhenzong. In *Extended continuation to the comprehensive mirror for aid in government*, vol. 76. Annals of Emperor Zhenzong III. In *History of Song*, vol. 8. This event was recorded in the eighth month of the fourth year of the Dazhong Xiangfu era (1011 CE).

of Weizhou (i.e., Daming Prefecture), through Enzhou (modern Qinghe County, Hebei), Jizhou (modern Ji County, Hebei), and Qianning Commandery (modern Qing County, Hebei) before reaching the sea. Along the way, it also passed through Shenzhou (south of modern southern Shen County, Hebei) and Yingzhou (modern Hejian County, Hebei).[60] From Qianning Commandery onwards, the Yellow River entered what had originally been the Yongji Canal; in essence, the river was now using the canal's path to flow northward. Later, in the fourth year of Yuanfeng under Emperor Shenzong (1081 CE), the Yellow River breached again at Xiaowusao (southwest of modern Puyang County, Henan). It flowed north from Weizhou and joined the Yuhe near Enzhou, then branched off through Jizhou and Yingzhou before rejoining the Yuhe further northeast to reach the sea.[61]

Again in the second year of Yuanfu under Emperor Zhezong (1099 CE), the Yellow River breached at Neihuangkou (modern Neihuang County, Henan). This time, its eastern flow ceased entirely, and the entire river diverted north. The new course passed through Ping'en County in Mingzhou (south of modern Qiu County, Hebei), Julu County in Xingzhou (modern Julu County, Hebei), Hengshui County in Jizhou (west of modern Hengshui County, Hebei), Wuqiang County in Shenzhou (southwest of modern Wuqiang County, Hebei), and Leshou County in Yingzhou (modern Xian County, Hebei), before once again merging with the Yuhe and reaching the sea.[62] Though this breach formed a new river course, much of it still flowed into the Yuhe en route to the sea. For example, in the fifth year of Zhenghe under Emperor Huizong (1115 CE), a breach occurred at Zaoqiangsao in Jizhou (modern Zaoqiang County, Hebei). In the seventh year, officials petitioned to repair the main channel at Ninghua

[60] Treatise on canals and waterways: the Yellow River I. In *History of Song*, vol. 91. It states: "In the first year of the Xining reign of Emperor Shenzong (June 1068), the Yellow River overflowed the Wulan Dike in Enzhou, and also breached the Zaoqiang Embankment in Jizhou, diverting northward into Yingzhou. In July, it again overflowed the Leshou Embankment in Yingzhou... The Assistant Director of the Directorate of Waterways, Li Lizhi, petitioned to construct 367 li of new dikes across the prefectures of En, Ji, Shen, Ying, and others to contain the river."

[61] Treatise on canals and waterways II: the Yellow River II. In *History of Song*, vol. 92. It states: "In the fourth year of the Yuanfeng reign (1081 CE), the Xiaowu Embankment suffered another major breach, allowing the Yellow River to pour into the Yuhe, placing Enzhou in grave danger."It is also recorded that Li Lizhi stated: "The areas of Nanle, Guantao, Zongcheng, Weixian, Qiankou, Yongji, Yan'an Town, and Jingcheng Town in Yingzhou lie between the two embankments of the Yellow River." Commentary: Nanle corresponds to modern Nanle County in Hebei; Guantao is present-day Guantao County, Shandong; Zongcheng was located south of today's Qinghe County, Hebei; Wei County corresponds to northwest of modern Daming County, Hebei; Qiankou and Yongji were situated north of present-day Guantao County; the exact location of Yan'an Town is uncertain; Jingcheng Town was in the southeast of modern Hejian County, Hebei. Among these, Guantao, Zongcheng, Weixian, Qiankou, and Yongji were all situated near the Yu River, showing that at that time, the Yellow River and the Yu River shared the same channel in these regions. The *Treatise* further records that, in the seventh year of Yuanfeng (1084 CE), Wang Lingtu of Jizhou reported: "The main current of the Yellow River has become dispersed and sluggish, with no strong, concentrated flow within the riverbed."This indicates that, downstream of Enzhou, the Yellow River had branched off into the Yu River, flowing through Jizhou and Yingzhou, before once again rejoining the Yellow River.

[62] Treatise on canals and waterways III: the Yellow River II. In *History of Song*, vol. 93.

Town in Enzhou. In the third year of Xuanhe (1121 CE), another breach occurred at Qinghesao—while this was nominally the Yellow River, it was in effect the Yuhe, as Zaoqiangsao lies just east of the Yuhe, and the newer course of the Yellow River at the time was still some distance away. Both Ninghua and Qinghe counties were located within Enzhou, directly along the path of the Yuhe. As for the repairs to Qingzhou's waterways in the second year of Daguan (1108 CE), they concerned the river course formed by the confluence of the new Yellow River (flowing down from Xing, Ji, Shen, and Ying prefectures) with the Yuhe.[63] With such frequent course changes and repeated breaches into the Yuhe, the transport capacity of the Yuhe inevitably suffered severe decline (See Fig. 6.3).

The Canal System Centered on Kaifeng

As previously mentioned, one of the defining features of the canal system during the Song dynasty—setting it apart from those of other dynasties—was that its major canals radiated outward with Kaifeng as the central hub. This pattern applied not only to the canals already constructed at the founding of the dynasty, but also to many of those later developed. For instance, during the reign of Emperor Taizong, there was a plan to excavate a new canal connecting Baigou to Pengcheng; and in the reign of Emperor Zhenzong, work began on a new canal between Dingtao and Xuzhou—both projects were conceived within this framework. Likewise, at the beginning of the Taiping Xingguo era (976–983), Cheng Nengxian, then the Transport Commissioner of the Western Capital (Xijing), proposed excavating the ancient Bai River, effectively creating a canal route from Nanyang to the capital. Though Cheng's proposed canal was never completed, it nonetheless held considerable significance in terms of planning and strategic vision.

Proposals for Excavating the Ancient Bai River and the Xiang-Han Grain Canal

Originally, the Song dynasty's transport waterways were organized as follows: the Yellow River connected westward to the Guanzhong region; the Guangji River linked eastward to the Qi and Lu regions; the Huimin River extended southward to the Chen and Ying areas; the Bian River connected southeastward to the JiangHuai, and Taihu regions; and the Imperial River reached various locations in Hebei to the north. However, places such as Nanyang (present-day Nanyang City, Henan), Xiangzhou (administered from present-day Xiangfan City, Hubei), and Dengzhou (administered from present-day Deng County, Henan) were unable to connect with the capital via navigable waterways. Grain tribute from Xiang (now Changsha, Hunan), Tan (now

[63] Ibid.

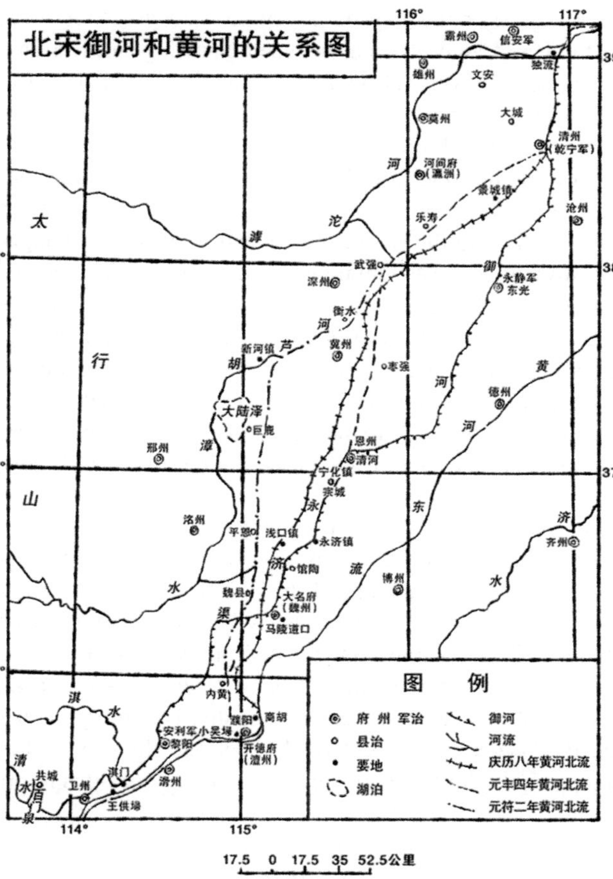

图例 - Legend

府州军治 Military Administration under Prefectures and Provinces

县治 County Seats

要地 Strategic Locations

湖泊 Lakes

御河 Imperial River

河流 Rivers

庆历八年黄河北流

Yellow River' s Northward Course in the 8th Year of the Qingli Era (1048)

元丰四年黄河北流

图例 - Legend

府州军治 Military Administration under Prefectures and Provinces

县治 County Seats

要地 Strategic Locations

湖泊 Lakes

御河 Imperial River

河流 Rivers

庆历八年黄河北流

Yellow River' s Northward Course in the 8th Year of the Qingli Era (1048)

元丰四年黄河北流

Fig. 6.3 The relationship between the imperial river and the yellow river during the northern song dynasty

Xiangtan, Hunan), and Lingnan could only be transported downstream along the Yangtze River to enter the Han Canal. A canal between Nanyang and Kaifeng was feasible and has been discussed previously. Cheng Nengxian proposed building a dam at Xiaxiangkou, north of Nanyang County, to divert water from the Bai River into the Shitang and Sha Rivers, merging with the Cai River and reaching the capital—thus facilitating the transport of grain tribute from Xiang and Tan. The construction area covered Bowang, Luoqu, Shaozhe Mountain, and Fangcheng. Bowang lies northeast of Xiaxiangkou; Luoqu is further northeast of Bowang and southwest of Fangcheng (modern Fangcheng County). Shaozhe Mountain is located north of Fangcheng. The terrain north and south of Fangcheng consists mostly of mountainous land, with Fangcheng itself situated in a relatively lower area. Since the Spring and Autumn and Warring States periods, this region has served as a passage between the Central Plains and the Jingxiang region to the south and beyond, and today remains a route for north–south highways. For this project, tens of thousands of laborers were conscripted locally. Although the canal was excavated, water could not be successfully diverted into it.[64]

This failure was clearly influenced by the topography. The canal needed to pass through Fangcheng County; although this area was a known north–south passage, its elevation remained relatively high. A tributary of the Zhe River flowed from north to south here, naturally sloping downward. Present-day Nanyang City stands at an elevation of approximately 127 m, while Fangcheng County rises above 200 m. Although nearly a millennium has passed since Cheng Nengxian's proposal, the topographic conditions would not have changed drastically. Diverting water eastward into this canal faced considerable difficulty. Although the elevation difference was not extreme, overcoming it with the engineering technology of that era proved challenging.

Although Cheng Nengxian's proposal ultimately failed, contemporaries did not completely abandon the idea. Twelve years later, in the first year of the Duangong era under Emperor Taizong (988 CE), two court officials—Yan Wenxun and Miao Zhong—jointly petitioned to reopen the transport route along the ancient Bai River. At the same time, they also proposed excavating a grain transport canal east of the city of Jiangling, connecting Shizi Mouth (Shizikou) to the Han River. These plans were remarkably ambitious. Had they been fully realized, the north–south waterway trunk line that had once existed prior to the Western Jin dynasty might have been restored. In practice, the canal east of Jiangling connecting to Shizikou and the Han River was successfully completed, but the transport route along the ancient Bai River once again ended in failure (See Fig. 6.4).[65]

The transport canal excavated by Yan Wenxun and Miao Zhong in the city of Jiangling was likely a restoration of the Yangkou route originally established by Du Yu during the Western Jin dynasty, though it differed slightly in the location where it entered the Han River. The canal constructed by Yan Wenxun and Miao Zhong entered

[64] Annals of Taizong I; Treatise on canals and waterways IV: the Bai River; Biography of Li Jilong; Biography of Wang Wenbao. In *History of Song*, vol. 4, 94, 257, 274.

[65] Treatise on canals and waterways IV: the Bai River. In *History of Song*, vol. 94.

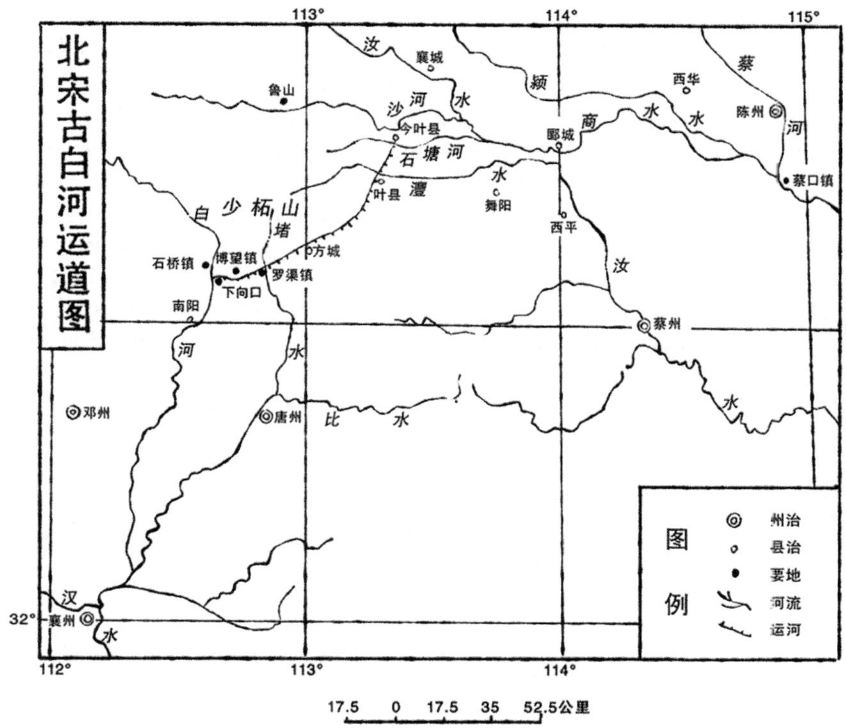

图例	- Legend	州治	Prefecture Seats
县治	County Seats	河流	Rivers
要地	Strategic Locations	运河	Canals

Fig. 6.4 The ancient Bai river transport route in the northern song

the Han River at Shizi Mouth (Shizikou). Yue Shi once wrote: "In Qianjiang County of Jingzhou, Mount Han lies twenty li north of the county seat; from Changshou County to Shizi Mouth, the route passes through the county's territory."[66] Lion's Mouth is believed to be identical with Shizi Kou. Changshou County corresponds to present-day Zhongxiang County in Hubei, which lies to the north of Qianjiang. This would place Shizi Kou approximately twenty *li* north of Qianjiang. During the Song dynasty, Qianjiang County was located in the northwest part of present-day Qianjiang, and Shizi Kou lay even farther to the northwest. Yangkou, too, was situated northwest of modern Qianjiang.[67] The precise spatial relationship between Yangkou and Shizi Kou—whether the latter lay to the left or right of the former—remains

[66] Jingzhou. In *Universal geographic record of the Taiping era*, vol. 146.

[67] Yang Shoujing. In *Illustrated commentary on the water classic.*

unclear. However, the two sites could not have been far apart. In other words, the canal excavated by Yan Wenxun and Miao Zhong was based on the precedent set by Du Yu, with only a slight alteration at the point where it entered the Han River. Later, during the final years of Emperor Zhenzong's Tianxi reign (1017–1021), the Assistant Director of the Ministry of Revenue, Li Yiyu, again "dredged the ancient canal and cleared the mouth at Xiakou to facilitate the transport of tax grain".[68] The so-called "ancient canal" refers to the one constructed by Yan Wenxun and Miao Zhong, which in turn followed Du Yu's earlier Yangkou route. As previously discussed, Yangkou was also known as Xiakou. Thus, although the canal dredged by Li Yiyu was essentially the same as that of Yan and Miao, it differed slightly in its course. In practice, it restored Du Yu's original plan, re-entering the Han River at Yangkou rather than Shizi Kou. It is also worth mentioning that during the Five Dynasties period, Gao Conghui of the Southern Ping State diverted water from the Baikai River into the city from the west of Jiangling and channeled it into the transport canal.[69] The Baikai River was a minor watercourse outside Jiangling. Gao Conghui's canal was quite short, simply diverting water from outside the city into the urban area where it connected with the transport canal. Its primary function was to facilitate the delivery of grain tribute to the granaries on the west side of the city.

Minor Canals Between the Yangtze and Huai Rivers

In addition to major arterial canals, many smaller-scale canals were also excavated during the Song Dynasty. The earliest of these is likely the Hengjiang Canal, opened during the Kaibao era of Emperor Taizu (968–976 CE) in Hezhou (present-day Hexian, Anhui Province). This canal was built primarily to facilitate military logistics during the campaign against the Southern Tang dynasty.[70]

Later, Xu Di, who served as Deputy Commissioner for Transport and Supply in the Huainan, Jiangzhhe, and Jinghu Circuits, excavated another canal between Haian and Rugao counties in Taizhou (modern-day Tai County, Jiangsu Province). Haian was then a market town (zhen), and Rugao remains a county today.[71] Xu Di lived during the reign of Emperor Renzong.[72] *The Treatise on the Nine Territories in the Yuanfeng Reign* records a canal that existed within Hailing County in Taizhou,[73] presumed to be the one excavated by Xu. Xu also undertook the management of the

[68] Jiangling prefecture I: Scenery I – Transport canal. In *Topographical records of the imperial realm*, vol. 64.

[69] Jiangling prefecture I: Scenery I – Transport canal. In *Topographical records of the imperial realm*, vol. 64, citing "Chronicles of prefectures and counties of the imperial dynasty."

[70] Biography of Li Fu. In *History of Song*, vol. 270.

[71] Biography of Xu Di. In *History of Song*, vol. 300.

[72] The barbarians. In *History of Song*, vol. 493.

[73] Huannan lu. In *Treatise on the nine territories in the Yuanfeng reign*, vol. 5.

Xixi River in Taizhou.[74] Xixi Town, located in the northeast of Taizhou,[75] was likely connected to this river. The canal Xu constructed extended eastward from Taizhou, passing through Haian and reaching Rugao. South of Rugao, however, no further canal excavations from the Northern Song period are documented.

In the fifth year of the Xianchun reign of Emperor Duzong (1269 CE), Li Tingzhi, then Commissioner for the Lianghuai region, excavated a 40-*li* stretch of canal to reach the Jinsha and Yuqing depots, and simultaneously dredged several other canals.[76] Jinsha Depot is now the seat of administration for present-day Nantong County, while Yuqing Depot was likely located to its east.

Contemporary historical maps record the existence of Yudong Depot and Yuxi Depot. It is likely that Yuqing Depot was located somewhere between the two. The distance between Yudong and Yuxi alone is more than forty li, not to mention Jinsha Depot, which is even further. It is presumed that Li Tingzhi's canal construction began in Tongzhou. The stated "forty li" does not indicate a continuous excavation but rather the total length of the canal sections under construction. East of Yuqing Depot lies Lusi Depot. Fan Zhongyan had once repaired the sea embankment beginning at Lusi Depot,[77] which indicates that, by this time, Lusi had already become solid land. Since Lusi was not far west of Yuqing, it is plausible that Li Tingzhi's canal extended to Lusi Depot.

The Commissioner for the Lianghuai region was stationed in Yangzhou at the time, and the canals leading to Jinsha and Yuqing were primarily intended for the transport of salt. The prosperity of Yangzhou depended heavily on the salt trade. Salt transported from Tongzhou to Yangzhou would necessarily pass through Rugao and Haian, suggesting that a navigable canal already existed between Tongzhou and Rugao. Since Li Tingzhi also dredged other canals, it is reasonable to assume that the stretch from Tongzhou to Rugao was included among them. Prior to this, in the fifth year of Emperor Guangzong's Shaoxi reign (1194 CE), Chen Sunzhi, Supervisor of Huai East, had also excavated a canal linking Hailing in Taizhou to Taixing in Yangzhou, which connected to the Yangtze River.[78]

The canal system extending from Yangzhou through Taizhou, Haian, Rugao, and Tongzhou, then further eastward to the Jinsha, Yuqing, and Lusi depots, along with the canal running from Taizhou to Taixing and the Yangtze River, comprised the southernmost branch among the smaller canals located east of the main canal connecting Yangzhou and Chuzhou (administrative center in present-day Huai'an, Jiangsu).

Opposite this southern branch was another canal route running from Gaoyou through Xinghua to Yancheng (all counties bearing these same names today). This route was also constructed under Chen Sunzhi's direction in the fifth year of the Shaoxi era. At that time, the area between Gaoyou and Chuzhou was largely submerged by vast marshes and dense reed beds—a natural consequence of canal

[74] Biography of Xu Di. In *History of Song*, vol. 300.

[75] Huannan lu. In *Treatise on the nine territories in the Yuanfeng reign*, vol. 5.

[76] Biography of Li Tingzhi. In *History of Song*, vol. 42.

[77] Tongzhou. In *Essentials of historical geography*, vol. 23.

[78] Canals and waterways VII: Southeastern rivers II. In *History of Song*, vol. 97.

blockages caused by the Jurchen occupation north of the Huai River. In response, Chen Sunzhi petitioned to construct embankments encircling the vast marshes in the area and to excavate a new canal, approximately 240 li in length, running alongside the dikes from Gaoyou and Xinghua to Yancheng in order to facilitate navigation. From Yancheng, he further proposed extending the canal eastward to reach the sea.[79] Along its course, thirteen *shida*—masonry sluice structures built of brick and stone—were erected to guard against scouring by currents and wave action. One such sluice, named Guanghui, was situated three li east of Yancheng County and was later referred to as the Baiboqiu Canal.[80]

Before Chen's renovation of the Gaoyou–Xinghua waterway, a smaller canal branched south from the main Yangzhou–Chuzhou canal, passing through Fancha Town (situated between Gaoyou and Taizhou), then turning south to reach Taizhou. This canal was later called the Yunyan River (Salt Transport River). Tradition holds that it was originally excavated by Liu Bi, the Prince of Wu, during the Western Han dynasty. While Liu Bi may have constructed a canal connecting to the granaries in Hailing, it likely did not extend as far as this point. It is more plausible that this was an ancient watercourse of uncertain origin. In the ninth year of Emperor Shenzong's Xining reign (1076 CE), Transport Commissioner Wang Zijing undertook repairs on this canal.[81] Later, a canal was built north of Fancha Town to link with the Gaoyou–Xinghua canal, thereby connecting the southern and northern branches of the minor canal network. However, this northern canal's construction may not have been completed until as late as the Ming dynasty.

Two additional minor canals merit mention: the Baita River near Yangzhou and the Tonglian River near Chuzhou. The Baita River is located sixty *li* northeast of Yangzhou, connecting southward to the Yangtze River and northward to the Grand Canal. It is traditionally said to have been excavated by Chen Xuan in the seventh year of the Xuande reign of the Ming Emperor Xuanzong (1432 CE).[82] However, as this canal is already recorded in the *Topographical Records of The Imperial Realm*,[83] it is unlikely to have been first constructed as late as the Ming dynasty. More plausibly, the canal had long existed but had become silted over time, and Chen Xuan merely undertook its dredging. The Tonglian River lies northwest of present-day Lianshui County. It connects the Huai River and the Lian River, following the course of the Zhijia River near Chuzhou, which was re-excavated to form this canal.[84] This region originally featured a new transport canal constructed during the reign of Empress Wu Zetian of the Tang dynasty, and the Tonglian River likely follows the remnants of this earlier canal.

---

[79] Canals and waterways VII: Southeastern rivers II. In *History of Song*, vol. 97.

[80] Huai'an prefecture. In *Essentials of historical geography*, vol. 22.

[81] Taizhoui. In *Essentials of historical geography*, vol. 23.

[82] Yangzhou prefecture. In *Essentials of historical geography*, vol. 23.

[83] Yangzhou. In *Topographical records of the imperial realm*, vol. 37. It refers to the river as Baita River.

[84] Canals and waterways VI: Southeastern rivers I. In *History of the Song*, vol. 96; also see *Wu Juhou's biography*, vol. 343.

Minor Canals South of the Yangtze River

By the end of the Xuanhe reign of Emperor Huizong (circa 1126 CE), three canals were excavated along the southern bank of the Yangtze River to mitigate the dangers of stormy navigation on the river. One of these was located near Chizhou (modern-day Guichi County in Anhui Province), where a canal was dug from the sandy mouth of the Chezhu River into the Duwu River.[85] Duwu, also known as the Chikou River, lies west of Chizhou city. In this region west of Chizhou, the Yangtze flows northward before turning east. Along this northward segment, submerged rocks were common along the eastern bank, while the western bank was dotted with sandbars—both of which posed significant hazards to navigation.

Following the excavation of this transportation canal, vessels could re-enter the Yangtze River at Chikou, thereby avoiding the perilous 200-*li* stretch of stormy river passage. Another canal, the Jing'an River in Jiangning Prefecture (present-day Nanjing), was completed in the sixth year of the Xuanhe reign (1124 CE). This canal diverted water from the Yangtze at Jing'an Town, passed through the Qingshajia area, followed the northern bank, and entered the new canal at Yizheng. This route provided a safe passage of over 80 *li*, thus circumventing approximately 150 *li* of hazardous open river navigation.[86]

The reason for constructing this new canal was the discovery of remains of an ancient transport canal, which similarly drew water eastward from the Yangtze. Remnants of this old canal were visible at Xiashu Port,[87] located north of Jurong County in present-day Xiashu Town, through which the Shanghai–Nanjing railway now runs. The exact date of the ancient canal's construction is unknown, but it is clear that canal routes running along the riverbank to avoid stormy river travel have existed since early times. A third route involved dredging the course of an ancient canal east of the Yangtze. This began at the riverside at Wuhu (modern-day Wuhu City, Anhui Province) and extended via Xuanxi and Lishui to Zhenjiang, where it rejoined the Yangtze. This route offered an alternative to a hazardous 600 *li* journey along the river.[88] Lishui received inflow from several lakes including Danyang and Gucheng, and discharged into Jingxi, eventually entering Tai Lake. The location of Xuanxi is uncertain, but it likely lay between the Wuhu county seat and Danyang Lake. This canal, commonly known as the Ancient River East of the Yangtze, roughly

---

[85] *The History of Song (vol. 96): Canals and waterways records:* "Lu Zongyuan further memorialized... Currently, there exists over 400 li of sandy terrain at the Chezhou River mouth on the eastern bank. If dredged, these waters could be channeled into Du Lake..." However, no lake exists along the Yangtze riverbanks in Chizhou. *The Essentials of historical geography, Volume 27*, under the "Yangtze River" entry for Guichi County, Chizhou Prefecture, cites this location as "Du Wu" instead, clarifying that the Du Wu River lies west of the county seat. This latter account should be considered authoritative.

[86] Jiangning prefecture, citing Songji lue. In *Comprehensive gazetteer of Qing dynasty (Jiaqing edition)*, vol. 73.

[87] Jiangning prefecture, citing Jiankang zhi. In *Comprehensive gazetteer of Qing dynasty (Jiaqing edition)*, vol. 73.

[88] Canals and waterways VI: Southeastern rivers I. In *History of Song*, vol. 96.

corresponds to the canal attributed to Wu Zixu during the Spring and Autumn period, which was said to be the "River of Jiang" referred to in the *Tribute of Yu* section of the Book of Documents according to traditional exegetes. The version dredged during the Song dynasty, however, deviated slightly from Wu Zixu's original route. In Wu's time, the capital of the State of Wu was located in Gusu (present-day Suzhou, Jiangsu Province), and his canal likely led directly from Lishui to Tai Lake.

In contrast, the Song dynasty intended for the canal to cross the Yangtze and connect with the Han Canal. Thus, the route from Liyang had to bend northward after passing Lishui to reach Runzhou (modern Zhenjiang, Jiangsu Province). Although Changdang Lake lies between Liyang and Runzhou, no direct waterway seems to have existed there at the time. The project to dredge the Ancient River East of the Yangtze began in the seventh year of the Xuanhe reign (September 1125 CE). By the second year of the Jingkang reign (February 1127 CE), Jurchen forces had entered the capital Bianjing (Kaifeng), and military operations continued in the lower Yangtze and Tai Lake region. It appears that this canal project was ultimately not completed. In the Southern Song period, however, additional smaller canals were gradually constructed in the region, partially realizing the intended eastern section of the Ancient River. In the ninth year of Emperor Xiaozong's Chunxi reign (1182 CE), a canal named Baihe Stream was excavated southwest of Changzhou (modern Changzhou City), and another canal, the Xili River, was opened to the south of the city.[89] Baihe Stream diverted waters from the Tianhuangdang marshes north of Jintan County, encircled the northeast of the county, and then split into two branches—one flowing southeast into Lake Ge, and the other flowing eastward to merge with the Jiangnan Canal near Changzhou.

The Xili River originated in Yixing County (which still bears the same name today), where it diverted water from the Jingxi River northward, passed east of Ge Lake, and continued north to join the Jiangnan Canal near Changzhou. This waterway was later referred to as the Western Canal (Xiyunhe).[90] It is worth noting that, according to records from the Northern Song period, a canal already existed in Yixing County at the time.[91] The county's principal waterway, the Jingxi River, was a natural stream and not part of the canal system. As Yixing County fell under the jurisdiction of Changzhou, it is likely that the existing canal was intended to connect Yixing with Changzhou. Among the possible routes, the Xili River would have been the most direct. If this reasoning holds, the Xili River must have been excavated prior to the Southern Song period. It is possible that the river underwent further dredging during the Southern Song, which may explain the discrepancies found in later records. As for the Jintan River, which connected Liyang and Danyang counties, it was not formally opened until the reign of Emperor Lizong (r. 1225–1264 CE). The Jintan River channeled water from the Jingxi River in Liyang County northward to Jintan County, then continued northward via the Erdu River (also known as the Qili River) to join the Jiangnan Canal near Danyang. This waterway was later known as the Jintan

89 Canals and waterways VII: Southeastern rivers II. In *History of Song*, vol. 97.

90 Changzhou prefecture. In *Essentials of historical geography*, vol. 25.

91 Liangzhe lu. In *Treatise on the nine territories in the Yuanfeng reign*, vol. 5.

Canal,[92] or in modern terms, the Jinli Transport Canal (Jinli Caohé). Near Danyang County, another waterway—the Jiuqu River—extended westward to the Jiangnan Canal and northeastward to the Yangtze River. Also known as the Xin River, it was once described as a waterway "through which tides formerly flowed, facilitating grain transport."[93] Both the Jiuqu River and the Jintan River were connected to the Jiangnan Canal, and their points of convergence lay in close proximity. It is likely that the two canals were constructed during the same general period.

To the west of Changzhou City stood the Lietang River. Further west was the Mengdu River, said to have been constructed by Meng Jian during the Tang dynasty. The Lietang River was dredged under the administration of Prefect Li Jiayan during the Shaoxing reign of Emperor Gaozong (1131–1162 CE) and later became known as the Victory New Canal (Desheng Xinhe) during the Ming dynasty.[94] The canal drew water from the Jiangnan Canal 18 li west of Changzhou City and carried it northward to the Yangtze River, with a total length of 43 li. To the east of the Lietang River lay the canal of Jiangyin County. This waterway began at Huangtian Port in the north of Jiangyin and flowed southward to Wuxi County, where it connected with the Grand Canal. It was reportedly dredged during the Huangyou reign of Emperor Renzong (1049–1053 CE), and further dredging was undertaken during the Yongle reign of the Ming Emperor Chengzu (r. 1402–1424).[95] This canal has remained navigable and unobstructed up to the present day.

As for the Zhihe Pond in Changshu County, it was originally the Yuanhe Pond constructed during the Tang dynasty, and by this time it remained fully navigable and functional.[96]

Additionally, two other canals on the southern shore of Tai Lake deserve mention: the Wuxing Pond and the Di Pond. The Wuxing Pond was located in Gui'an County of Huzhou Prefecture.[97] Gui'an County was the seat of Huzhou Prefecture and corresponds to present-day Huzhou City in Zhejiang Province. According to tradition, the Wuxing Pond was constructed under the supervision of Prefect Shen Youzhi and was initially intended for irrigation purposes.[98] Shen Youzhi was an official during the Liu Song dynasty. If this account is reliable, then the Wuxing Pond had already demonstrated utility by the Southern and Northern Dynasties period.[99]

[92] Zhenjiang prefecture. In *Comprehensive gazetteer of Qing dynasty (Jiaqing edition)*, vol. 90.

[93] Zhenjiang prefecture. In *Comprehensive gazetteer of Qing dynasty (Jiaqing edition)*, vol. 90.

[94] Changzhou prefecture. In *Essentials of historical geography*, vol. 25.

[95] Ibid.

[96] Suzhou prefecture. In *Essentials of historical geography*, vol. 24.

[97] Liangzhe Circuit. In *Treatise on the nine territories in the Yuanfeng reign*, vol. 5.

[98] Huzhou. In *Yuanhe maps and records of prefectures and counties*, vol. 25.

[99] Biography of Shen Youzhi. In *Book of Song*, vol. 74. "In the fourth year of the Taishi reign, Shen Youzhi was appointed governor of Wuxing but declined the post." Since Shen Youzhi never actually served as governor of Wuxing, he could not have overseen the construction of the Wuxing Pond; the story is likely apocryphal.

To the south of present-day Zhenze County in Jiangsu lies the Di Pond Canal, said to have been excavated by Yu Di, the Tang dynasty prefect of Huzhou.[100] Another tradition attributes its construction to the Qingli reign of the Song dynasty.[101] This canal ran from Nanxun Town to Pingwang Town, where it joined the Nantang River—the canal running through northern Jiaxing County.

During the Ming and Qing dynasties, the canal network in Huzhou Prefecture continued to thrive. Its upper reaches drew from the Yubu Stream and Tiao Stream, and it flowed eastward through the towns of Nanxun, Zhenze, and Pingwang before joining Jiaxing's canal system. This route largely followed the course of the ancient Wuxing and Di Ponds.[102] This serves to confirm that although both ponds were originally developed for irrigation purposes, their scale was substantial enough to accommodate boat traffic, and they thus came to be regarded in later times as formal canals (See Fig. 6.5).

Expansion of the Eastern Zhejiang Canal

During the Yuanhe reign of the Tang dynasty, Meng Jian not only constructed the Mengdu Canal in Changzhou but also excavated a transportation canal (known as Yundaotang) in Yuezhou. This canal was located to the north of the city of Yuezhou, though its exact starting and ending points remain unknown.

By the Song dynasty, both Xiaoshan and Shangyu (both retaining their historical names today) were known to have canal systems.[103] Xiaoshan lies to the west of Yuezhou and connects further westward to the Qiantang River. Shangyu is located to the east of Yuezhou and is connected farther east to the Yuyao River. The lower reaches of the Yuyao River flow into the Yin River, which runs beneath the city of Qingyuan Prefecture—modern-day Ningbo in Zhejiang Province.

Whether both of these canal segments were initially constructed by Meng Jian is uncertain; definitive attribution remains difficult. However, by the Song period, the general structure and scale of the Eastern Zhejiang Canal network had clearly taken shape.

[100] Huzhou prefecture. In *Essentials of historical geography*, vol. 91.

[101] Suzhou prefecture. In *Comprehensive gazetteer of Qing dynasty (Jiaqing edition)*, vol. 77.

[102] Huzhou prefecture. In *Comprehensive gazetteer of Qing dynasty (Jiaqing edition)*, vol. 289.

[103] Liangzhe lu. In *Treatise on the nine territories in the Yuanfeng reign*, vol. 5.

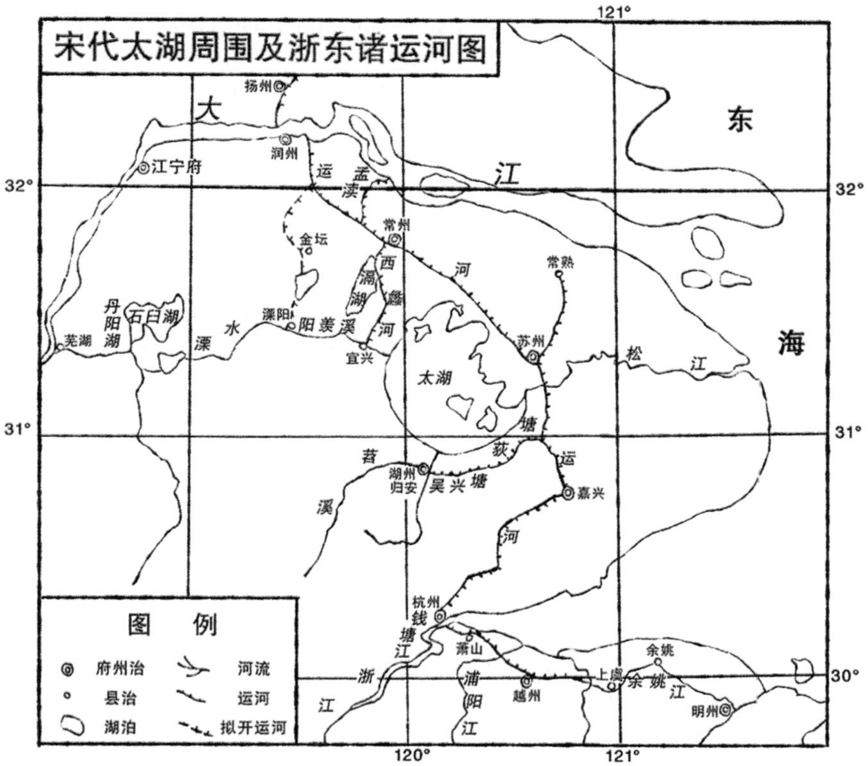

图例 - Legend	河流 Rivers
府州治 Prefecture and Province Seats	运河 Canals
县治 County Seats	拟开运河 Planned Canals
湖泊 Lakes	

Fig. 6.5 Canals surrounding Tai lake and the eastern Zhejiang transport system in the song dynasty

Minor Canals and Reservoirs in Hebei

During the Song dynasty, the main waterway in Hebei was the Imperial River (*Yuhe*), yet this river was often disrupted by the Yellow River's encroachments, rendering navigation unreliable. Nevertheless, the Song government spared no effort in developing Hebei's water infrastructure, constructing several minor canals in addition to the *Yuhe*. Though these canals may seem insignificant from a modern perspective, they carried great strategic value at the time. The Northern Song's most persistent external threat came from the Khitan-led Liao dynasty—a nomadic power that repeatedly launched swift raids across the northern frontier. With their superior cavalry, the

Liao forces could strike suddenly and retreat with equal speed, leaving Song forces unable to intercept or pursue them effectively.

Unlike the Qin and Han dynasties, which had built the Great Wall across mountainous terrain, the Song-Liao border lay on the flat plains of Hebei, offering no natural barriers for constructing walls or fortresses. The Song thus devised an alternative defensive measure: the construction of a series of reservoirs (*beitang*) along the Baigou River, the de facto boundary between the two states. These reservoirs stretched discontinuously from the vicinity of present-day Xushui County eastward through Xiong County and Bazhou, terminating near Qing County. Roughly aligned in an east–west direction, they formed a linear system that functioned as a "Great Wall of water" across the plains. Because of their vastness and depth, the reservoirs were not only effective in impeding enemy cavalry but also served secondary functions, particularly in facilitating grain transport to supply military outposts along the frontier.

The reservoirs constructed during this period were limited by topographical constraints and could not extend to the eastern foothills of the Taihang Mountains. West of present-day Xushui County and Baoding City, the terrain gradually rises in elevation, making the construction of such reservoirs virtually impossible in those areas.

As early as the sixth year of the Taiping Xingguo reign of Emperor Taizong (981 CE), waters from the Xu River and the Jiju River were diverted eastward into the Baigou River to facilitate grain transport in the region south of the frontier passes (*guannan*). Subsequently, in the fourth year of the Xianping reign of Emperor Zhenzong (1001 CE), a channel was excavated from Jingrong Garrison (located south of modern Yixian County, later renamed Ansu Garrison), directing water from the Bao River eastward, south of the Great Wall, into Xiongzhou (modern Xiong County, Hebei). The following year (1002 CE), a new channel was opened, again diverting Bao River water from Jingrong Garrison eastward to Shun'an Garrison (modern Gaoyang County), and from there westward into Weilu Garrison (modern Xushui County). All these waterways served the purpose of transporting grain.[104]

The term *guannan* refers to the area south of Waqiao Pass, where garrisons such as Jingrong, Shun'an, and Weilu were situated—precisely the region where the border reservoir system (*tangluo*) was located. The successful operation of grain transport in *guannan* demonstrates the navigability of the *tangluo* system. The artificial channels opened along the Xu River, Jiju River, and Bao River were designed to compensate for the system's inability to reach the higher ground along the eastern flank of the Taihang range.

The Jiju River was often referred to as Jiju Spring, located west of modern Dingxing County in Hebei. Its waters were notably abundant. In the first year of the Jingde reign (1004 CE), Yan Chenghan diverted water from the Zhaobin Weir in Baozhou (modern Baoding) into Jiju Spring to reduce the need for manual towing of

[104] Treatise on canals and waterways V. In *History of Song*, vol. 95.

boats,[105] demonstrating the substantial flow volume of the spring. The outflow from Jiju Spring could thus rightfully be referred to as the Jiju River.

The Xu and Jiju rivers discharged downstream into the Yi River, which then flowed into the Baigou River. Also in 1004 CE, Yan Chenghan excavated a canal east of Jiashan (northeast of present-day Quyang County), diverting waters from the Tang River—also known as the Kou River. This channel extended 32 *li* to Dingzhou (modern Ding County), where it was divided into a canal that ran a further 62 *li* eastward to Puyin County (modern Anguo County). The canal then joined the Sha River, passed through Bianwu Marsh, and ultimately entered the Boundary River (Baigou River), thus facilitating waterborne grain transport.[106] Jiashan lay northeast of modern Quyang County, with the Tang River flowing beneath its slopes. At that time, the administrative seat of Dingzhou was located at modern Ding County. Puyin corresponds to present-day Anguo County. The Sha River flowed from the northwest and joined the Tang River, while Bianwu Marsh lay north of modern Gaoyang County. These details suggest that Yan Chenghan's canal ran along the left bank of the Tang River, passing near Dingzhou and through Bianwu Marsh, thereby serving as a substitute for the Tang River in transporting grain. This canal also connected directly to the Baigou River, supplementing the limitations of the *tangluo* system just as the Xu and Jiju river diversions had done.

Beyond these reservoirs and channels, additional canals were also constructed between the Imperial River (*Yuhe*) and the eastern edge of the Taihang Mountains to strengthen the northern transport network. In the second year of the Chunhua reign (991 CE), under Emperor Taizong, the so-called "New River" was excavated from Xinzhai Town in Shenzhou (modern Shen County). A branch of the Hulu River (modern Fuyang River) was diverted along this new channel for a distance of 200 *li*, reaching Changshan (modern Zhengding County), thus opening a new route for grain transport.[107] This canal effectively connected the waters of the Hutuo and Hulu rivers.

A minor issue arises here regarding the geography. Shenzhou lies to the east of Changshan and is therefore situated at a lower elevation. The Hutuo River flows from Changshan through Shenzhou. However, the canal opened in the second year of the Chunhua reign (991 CE) did not divert water from the Hutuo River into the Hulu River, but rather redirected water from the Hulu River into the Hutuo—a reversal that may reflect an error in the historical record.[108] This newly constructed canal was likely the longest of the east–west canals running across the plains to the east of the Taihang Mountains at the time. Later, in the fourth year of the Xianping reign of Emperor

[105] Treatise on canals and waterways V. In *History of Song*, vol. 95.

[106] Ibid.

[107] Ibid.

[108] According to the *History of Song: Treatise on Rivers and Canals*: "During Emperor Shenzong's Xining reign, the palace eunuch Cheng Fang proposed dredging and diverting water into the old course of the New River. An imperial edict ordered local officials to investigate. Lin Shen, assistant magistrate of Yongjing Commandery (modern Dongguang County, Hebei), and Zhang Yanju, magistrate of Dongguang County, reported that the New River's terrain was elevated and might damage farmland. Cheng Fang, however, asserted that the terrain was highly favorable and

Zhenzong (1001 CE), another canal was excavated to divert water from the Hutuo River at Changshan Town in Zhenzhou into the Xiao River, thereby establishing a water route to Zhaozhou (seat located in present-day Zhao County, Hebei).[109] This canal, too, could be used for grain transport. These projects underscore the suitability of the eastern plains of the Taihang Mountains for canal construction.

The Destruction of the Central Plains Canal System

From the perspective of the canal system as a whole, the network in the Song dynasty was more orderly and integrated than in any previous era. Were it not for external threats, the transport of tribute grain via waterways would have been extremely efficient. However, the hostile neighboring states did not permit the Song to enjoy such convenience for long. The successive rise of the Liao and Jin regimes rendered the northern frontier perpetually unstable. In the end, the Jin forces breached Kaifeng, occupied large parts of the Central Plains, and Emperor Gaozong was driven to retreat to the southeastern region, clinging to a fractured domain and struggling to sustain a precarious situation.

Amid such chaos and displacement, the canals—once maintained through the collective efforts of the state—became a source of peril. Particularly during Emperor Gaozong's hasty southward flight, the Jin took advantage of the canal routes to pursue him by boat. In order to prevent the enemy from advancing further, Gaozong issued an edict to destroy the canal infrastructure. This entailed not only demolishing the embankments along the waterways but also dismantling the weirs and dams.[110] The primary target of this sabotage was a section of the Han Canal, which was the most strategically vital. As for the Bian River further north, the Song court no longer had the capacity to reach or maintain it, and thus had no choice but to let it silt up and fall into disuse.

presented no such risk. The court then dispatched Liu Kuai and Li Zhigong to verify the situation, who ultimately corroborated Cheng Fang's assessment. As a result, Lin Shen and others were demoted for their erroneous judgment."

[109] Treatise on canals and waterways V. In *History of Song*, vol. 95.

[110] Treatise on canals and waterways VII. In *History of Song*, vol. 97. "At the beginning of the Shaoxing reign of Emperor Gaozong, as the Jin forces ravaged Huainan and had not yet retreated, an edict was issued to burn and destroy the sluices at the port of Yangzhou's Wantou, at Jiangyan in Taizhou, and the Baipu weir in Tongzhou, along with other sluices, ordering local officials to open and burn them to prevent enemy ships from passing. Additionally, the pacification command ordered the demolition of the sluices at Zhen and Yang, as well as the Gongtang in Zhenzhou, to prevent enemy access to the canal system." Under such rigorous anti-enemy measures, the canal system suffered severe destruction.

The Southeastern Canal System

As the turmoil gradually subsided, the Southern Song established Lin'an (modern Hangzhou, Zhejiang Province) as its temporary capital and revitalized the grain transport system in the southeastern region. Fortunately, since Emperor Yang of Sui excavated the Jiangnan Canal, the grain transport routes in the southeast had been continuously maintained without interruption. Lu You's journey into Sichuan, taking the Jiangnan Canal, provides a firsthand account confirming this. According to Lu You, the canal originally constructed by Emperor Yang was approximately 800 *li* in length, with a width of about ten *zhang*, flanked by embankments resembling mountain ridges, formed by the accumulated earth at the time. Lu You also remarked that "the reason the court could station troops at Qiantang was because of this canal."[111] This statement reveals the importance attached to the canal by officials and commoners alike. Given such a foundation and the government's attention, only periodic dredging was needed to keep the canal navigable. During this period, several canal sections were also excavated in western and eastern Zhejiang, forming a small but functional transportation network.

After the Jin forces ceased their southern incursions, peace gradually returned to the Jianghuai region, and the former course of the Han Canal was once again dredged and brought back into use. This revival involved more than simple dredging. During the Shaoxi reign of Emperor Guangzong (1190–1194), Chen Sunzhi, the regional inspector of Huai East, submitted a proposal stating: "Between Gaoyou (modern Gaoyou County, Jiangsu) and Chuzhou, the area is covered in expansive marshes and lakes, densely filled with reeds and aquatic plants. It is advisable to construct embankments and weirs for the purposes of water retention and drainage, so as to guard against both flooding and drought." With imperial approval, a stretch of embankments totaling 360 *li* was constructed,[112] extending from Jiangdu County in Yangzhou (modern Yangzhou City, Jiangsu) to Huaiyin County in Chuzhou (now southwest of modern Qingjiang City). These efforts restored the functionality of this ancient canal. In addition, Chen Sunzhi undertook the construction of a new canal running from Gaoyou and Xinghua to Yancheng (all in present-day Jiangsu), as previously discussed. This new waterway significantly enhanced the Southern Song's inland water transport network.

---

[111] Lu You, *Record of Entering Shu*. In this work, Lu You remarks: "From Jingkou to Qiantang, there was no navigable canal prior to the Liang and Chen dynasties." This statement is of questionable reliability. Here, it is cited solely as a contemporary observation of the canal conditions at the time.

[112] Treatise on canals and waterways VII: Southeastern rivers II. In *History of Song*, vol. 97.

Canals Under Jin Rule

While the Southern Song centered its transportation network around Lin'an, the Jin dynasty established a smaller network centered on Daxing (modern Beijing), utilizing the Imperial Canal to transport grain from the Central Plains. The Jin made use not only of the section south of Bai Canal at the Song-Liao border but also restored the lower reaches of the Sui dynasty's Yongji Canal. Except for a silting section near the Qinshui River's mouth, the rest of the Yongji Canal was nearly fully navigable again.

One notable engineering feat was related to the Yongji Canal's original construction under Emperor Yang, which required reaching Ji County in Zhuo Commandery. This involved a section along the lower reaches of the Lei River—today's Yongding River—which frequently shifts its course. After the Jin restored the Yongji Canal, this section was replaced by a waterway along the lower Lu River, extending all the way to Lu County (modern Tongzhou, Beijing). The Jin also excavated a canal connecting Lu County to their capital Daxing,[113] foreshadowing the Yuan dynasty's later construction of the Tonghui Canal.

The Silting of the Bian River

Following the standoff between the Song and Jin dynasties, the Grand Canal effectively split into two independent systems, each developing along separate trajectories. The Bian River, however, silted up rapidly after the turmoil of the Jingkang Incident. Even after diplomatic relations were reestablished between the two states, envoys continued to travel along the old course of the Bian River—though by that time, the river had already ceased to flow.

Lou Yue, who traveled north on a diplomatic mission during the reign of Emperor Xiaozong of the Southern Song, provided one of the clearest and most reliable accounts of the river's condition. He recorded: "On the second day of the twelfth month in the fifth year of the Qiandao reign (1169), under clear skies and wind, we traveled eighty *li* eastward and made an early stop at Hong County… After lunch, we rode on horseback another eighty *li* and lodged at Lingbi. A few *li* farther on, we found the Bian River had dried up…On the third day, the jia-shen day (of the lunar calendar), the weather was clear. We traveled sixty li by carriage and made an early stop at Jing'an Town, then continued another sixty li to lodge in Suzhou. Since leaving Sizhou, we had been following the course of the Bian River. By this point, the river had become so thoroughly silted that its bed was nearly level with

[113] According to *History of the Jin Dynasty, vol. 110: Biography of Han Yu*, "During the Taihe reign period, he proposed the excavation of a canal from Tongzhou to the Lu River to facilitate grain transport to the capital." This refers to the project of opening the canal.

the surrounding banks—so much so that carts and horses could traverse it with ease, and even houses had been built directly upon its former channel."[114]

Later, when Lou Yue returned from his diplomatic mission to the Jin court, he once again followed the same route and recorded further observations: "On the twentieth day of the first month in the sixth year of the Qiandao reign (1170), we traveled sixty li by carriage and reached Yongqiu County… After another sixty *li*, we gradually found ourselves traveling within the riverbed of the Bian River… On the twenty-fourth, we lodged again at Suzhou. The bottom of the Bian River is now extensively planted with wheat."[115] Compared to the Bian River's heyday—when sails crowded the sky and boat prows jostled along the banks—the contrast is nothing short of astonishing.

With the silting of the Bian River, the Han Canal naturally suffered as well. When the Bian River remained navigable, Chuzhou at the northern end of the Han Canal and Sizhou at the southern end of the Bian River were prosperous cities along the Grand Canal—surpassed only by Zhenzhou and Yangzhou. Yet by the reign of Emperor Ningzong of the Song dynasty, Yue Ke had already remarked: "Huaiyin in Chuzhou, once a town on the banks of the transport canal, is now desolate among the marshlands and settlements. During the northern campaign of the Kaixi reign (1205–1207), my boat passed through this area."[116]

Huaiyin lies only a few dozen li from Shanyang County, the seat of Chuzhou's prefecture. If Shanyang could be regarded as occupying the northern terminus of the Han Canal, then Huaiyin—controlling the critical confluence of the Han Canal and the Huai River—was of even greater importance. How startling, then, that such a key locality had fallen into ruin in so short a time!

Although Chuzhou retained its administrative status as a prefecture, it was no longer able to prosper. In truth, from the turmoil of the Jingkang Incident to the fifth year of the Qiandao reign was a span of merely forty-four years. Even by the Kaixi reign, only seventy or eighty years had passed—yet the transformation had already been so drastic! Fan Chengda also recorded similar observations,[117] though his account is comparatively less detailed.

Since the beginning of the Southern Song, the Bian River has never been restored. The extensive efforts invested in its construction and maintenance over the course of the Sui, Tang, and Northern Song dynasties were ultimately rendered futile.

---

[114] Lou Yue. A record of the Northern journey (Beixing rilu). In *Collected works of Gong Kui*, vol. 111.

[115] Ibid.

[116] Yue Ke. Temple of Huaiyin. In *History of the pillow*, vol. 12.

[117] Fan Chengda. *Notes on holding the reins (Lanpei lu).*

Chapter 7
The Construction and Decline of the Grand Canal

Transformation of the Canal Network's Main Course

By the Yuan dynasty, the configuration and function of the Grand Canal had undergone marked changes from those of earlier periods. Much of the canal system from former dynasties had become choked with silt and rendered unusable. The surviving segments—such as the Han Canal between the Yangtze and Huai Rivers, the Imperial Canal in Hebei, and the Jiangnan Canal near Tai Lake—had become fragmented and disconnected, unable to constitute an integrated transportation network. Although the Yuan dynasty still made use of these vestiges of earlier waterways, their role and significance had fundamentally shifted.

The principal reason for the divergence of the Yuan canal system from those of earlier dynasties lay in the relocation of the imperial capital. In previous dynasties—so long as the realm remained unified and was not experiencing division or temporary retreat—the capital had invariably been situated in Chang'an, Luoyang, or Kaifeng. At the same time, the centers of economic wealth were located in the eastern and southeastern regions. Consequently, the canal system primarily followed an east–west orientation, forming the main arteries of inland transportation. Although some canals extended in a north–south direction, these were originally designed to serve as auxiliary routes that supported the east–west trunk lines; without this supporting function, their significance would have been greatly diminished.

With the advent of the Yuan dynasty, however, the selection of the capital underwent a radical departure from precedent: it was relocated to Dadu (present-day Beijing), thereby departing from the longstanding pattern of capitals located in Chang'an, Luoyang, or Kaifeng. This constituted one of the most consequential shifts in Chinese imperial history. The change in capital location necessitated a corresponding reorientation of the canal system. Although the Yuan canal was still intended to connect the capital with the affluent southeastern regions, the fact that the capital now lay in the far north meant that the canal had to adopt a north–south

N. Shi, *A Historical Geography Research of Canals in China*,
https://doi.org/10.1007/978-981-95-3517-0_7

axis, in contrast to the formerly dominant east–west orientation. If one takes modern-day Xuzhou as a geographical reference point, one finds that the principal course of the canal had turned from an east–west alignment to a north–south one—almost a ninety-degree rotation. This dramatic shift in orientation underscores a deeper historical transformation: the decline of the economic prominence of the Guanzhong and Central Plains regions, and the corresponding rise of the North as the new center of gravity in China's political and economic landscape.

In addition to changes in orientation, Yuan canals also differed from earlier ones in terms of construction techniques. One of the defining features of Yuan canal engineering was the widespread use of gated canals (*zhahe*, or lock canals). These were constructed in areas with significant differences in water elevation. The gates helped impound water to raise the level in lower-lying stretches, thereby allowing boats to gradually ascend to higher levels. Conversely, releasing the water through the gates enabled vessels to descend. This technique was especially effective in regions with uneven terrain and rapid water flow.

Although this technique was not invented during the Yuan dynasty—it had already been used in the mid-Tang period—it was not widely adopted until later. For instance, Li Bo of the Tang dynasty built sluice gates (*doumen*) along the Lingqu Canal connecting the Xiang and Li rivers, enabling the passage of grain transport boats.[1] These early sluice gates can be considered a rudimentary form of canal locks, and were particularly useful given the uneven topography between the two rivers. However, due to the predominance of east–west canals at the time—which, owing to China's natural topography of being higher in the northwest and lower in the southeast, allowed for gravity-fed water flow—there was little need to apply lock technology extensively.

In the Song dynasty, the use of gated canals gradually became more common, though primarily confined to the Han Canal and the Jiangnan Canal, and not implemented across the entire system. It was during the Yuan dynasty, with its reliance on north–south canal routes, that lock technology became indispensable. Unlike the earlier canals, which could rely on gravity for water flow, the new south-to-north orientation necessitated more advanced hydraulic engineering, and the gated canal system emerged as the key solution to this problem.

The Relationship Between the Grand Canal and the Yellow River in the Yuan Dynasty

China's major canals have historically been inextricably linked to the Yellow River—either directly utilizing segments of the river or drawing on its water supply. The frequent breaches and course changes of the Yellow River's lower reaches often had a significant impact on canal transportation, as discussed previously. During the Western Han, a breach in the Yellow River seriously affected the Hong Canal, even

[1] Treatise on geography VII, part I. In *New book of Tang*, vol. 43.

leading to the siltation of some sections. Similarly, in the Northern Song, the shifting of the Yellow River's course caused interruptions in transport on the Imperial Canal. In the Yuan dynasty, the Grand Canal ran in a north–south direction and had to cross the Yellow River in the middle, making its relationship with the river even more crucial. This was particularly true in the early Yuan period, when the canal had not yet been successfully completed and transport routes still relied on segments of the Yellow River. As a result, careful attention to the state of the river became essential.

From the beginning of recorded history up through the Northern Song period, the Yellow River had altered its course on several occasions, yet it consistently emptied into the Bohai Sea to the northeast. A significant shift occurred in the early Southern Song period. In the second year of the Jianyan reign (1128) of Emperor Gaozong, Du Chong, the military governor of the Eastern Capital (Kaifeng), deliberately breached the Yellow River's embankments to divert its waters into the Huai River via the Si River, with the strategic aim of halting the advance of the Jin army.[2] This act marked the beginning of the Yellow River's gradual southward shift during the Southern Song.

In fact, the Yellow River had previously overflowed into the Si and Huai Rivers as early as the Western Han dynasty. At that time, the river burst through its levees at Huzi in Puyang (modern-day Puyang County, Henan), causing its waters to divert southward. Over a decade later, Emperor Wu of Han personally oversaw the closure of the breach and the reestablishment of the river's original course, thereby curbing its southern inundation.

Following Du Chong's deliberate breaching of the riverbanks—an act taken amidst the exigencies of wartime—there was neither the initiative nor the capacity to repair the breach. Consequently, not only did the original opening remain unsealed, but new breaches continued to occur, ultimately rendering the restoration of the former river course impossible.

In the twenty-seventh year of the Dading reign of Emperor Shizong of Jin (1187), an imperial decree was issued, ordering the administrative officials of four prefectures, sixteen sub-prefectures, and forty-four counties along the river to assume responsibility for riverine defense.[3] Of these forty-four counties, several were situated upstream of Guangwu Mountain in Xingyang—areas belonging to the river's middle reaches—where no course alterations were likely to occur, and may thus be excluded from the present discussion.

South of Guangwu Mountain, the Yellow River traversed over thirty counties, which may be grouped into three main routes: The principal current of the Yellow River passed through Zuochen, Changyuan, Jiyin (modern Heze County, Shandong), Shanfu, Yucheng, Feng County, Pei County, Xiao County, and Pengcheng (modern Xuzhou). A second cluster of counties, lying north of the main channel, included

[2] The annals of Emperor Gaozong, part II. In *History of Song*, vol. 25. According to *A Chronological Record of Events Since the Jianyan Era*, it is recorded: "Du Chong, Military Commissioner of Eastern Capital, upon learning of the Jin army's approach, diverted the Yellow River into the Qing River to impede the enemy." The Qing River here refers to the Si River.

[3] Treatise on canals and waterways: Yellow River. In *History of Jin*, vol. 27.

Xinxiang, Ji County, Wei County (modern Qi County, Henan), Baima (modern Hua County, Henan), Puyang, Yuncheng, Jiaxiang, and Jinxiang. The third group, situated to the south of the main stream, comprised Yanjin, Fengqiu, Xiangfu (modern Kaifeng County), Kaifeng, Chenliu, Qixian, Ningling, and Songcheng (modern Shangqiu County, Henan).

Some historical accounts suggest that in the fifth year of the Mingchang reign of Emperor Zhangzong of Jin (1194), the Yellow River burst its banks at Yangwu (east of modern Yuanyang County, Henan), flowing eastward into Liangshan Lake (beneath Mount Liang in present-day Liangshan County, Shandong, now silted), and then splitting into two branches: the northern branch entered the Bohai Sea via the North Qing River, and the southern branch flowed into the Huai River via the South Qing River, ultimately entering the East China Sea in conjunction with the Huai's waters.[4]

In reality, however, there was no breach at Yangwu in 1194. At the time, Tian Li, a vice commissioner of the Waterworks Bureau, had proposed diverting the river northward through the wall village levee into the old course of Liangshan Lake, thereby restoring the former bifurcation into the North and South Qing Rivers. The court, deeming his proposal impractical, dismissed it without implementation.[5] This confirms that no such diversion took place.

In the late Jin and early Yuan periods, the Yellow River did in fact experience two major artificial breaches. The first occurred in the ninth year of the Zhengda reign of Emperor Aizong of Jin (1232), when Mongol forces besieged Guide (modern Shangqiu County, Henan). They deliberately broke the levees, diverting floodwaters through the city's southwest into the old course of the Sui River.[6] The second breach took place in the sixth year of Emperor Taizong of the Yuan (1234), when Southern Song forces under Zhao Kui attacked Bian (Kaifeng), and the Mongols opened the levees at Sunjin Swamp, north of Xiangfu County, flooding the city.[7] This flow may have continued southward into the Guo River, although this is not explicitly recorded in historical sources.

Subsequent changes likely occurred. For example, in the twenty-third year of Emperor Shizu's Zhiyuan reign (1286), the Yellow River breached its banks in fifteen locations across Kaifeng, Xiangfu, Chenliu, Qi, Taikang, Tongxu, Yanling, Fugou, Weichuan, Weishi, Yangwu, Yanjin, Zhongmou, Yuanwu, and Suizhou.[8] The numerous and scattered locations of these breaches reflect the increasing fragmentation of the river's course, with the divergence occurring below Yuanwu and Yangwu.

---

[4] Supplementary discussion on successive historical river course changes. In *Detailed commentary on the tribute of Yu*, vol. 13, part 2.

[5] Treatise on canals and waterways: Yellow River. In *History of Jin*, vol. 27.

[6] Biography of Shi Zhan's Daughter Lu Huan. In *History of Jin*, vol. 116.

[7] Supplementary discussion on successive historical river course changes. In *Detailed commentary on the tribute of Yu*, vol. 13, part 2.

[8] *History of Yuan (vol. 14): Annals of Emperor Shizu* states: The last location referred to as Muzhou is erroneous, as no Muzhou existed in the Yellow River basin at that time. It is likely a miswriting of Suizhou.

Among the diverging routes, one flowed through Zhongmou, Weishi, Weichuan, Yanling, and Fugou to enter the Huai River via the Ying River. Another passed through Xiangfu, Kaifeng, Tongxu, and Taikang, entering the Huai via the Guo River. A third followed the path through Chenliu, Qi County, and Sui County, utilizing the ancient Han-dynasty Bian Canal route to reach the Si River.

Subsequent flood events further shaped these paths. In the first year of the Dade reign of Emperor Chengzong (1297), the river overflowed at Linying and Yancheng[9]; in the second year of the Huangqing reign of Emperor Renzong (1313), it burst its banks at Chenzhou; and in the third year of the Yanyou reign (1316), it overflowed at Taihe. All these incidents occurred along the branch entering the Huai River via the Ying River. Additional breaches at Luyi in 1297 and at Bozhou in 1313 occurred along the branch entering the Huai through the Guo River.

In the 25th year of the Zhiyuan era of Emperor Shizu (1288), the Yellow River breached at Xiangyi (present-day Sui County, Henan), overflowing in Suiyang (present-day Shangqiu, Henan) and Kaocheng (present-day Lankao, Henan).[10] During the 2nd year of Emperor Chengzong's Yuanzhen reign (1296 CE), the river broke through at Ningling.[11] In the first year of the Dade era, it flooded Guide, Xuzhou, Suqian, and Suining.[12] In the 2nd year of Emperor Yingzong's Zhizhi reign (1322), the river overflowed at Yifeng (present-day Lankao, Henan).[13] These events all occurred along the branch flowing from Bian (Kaifeng) into the Si River. Later, after some smaller changes, by the 11th year of Emperor Shundi's Zhizheng reign (1351), Jia Lu successfully managed river control, directing the Yellow River to flow north of Fengqiu and Kaifeng, along the border of present-day Henan and Shandong provinces, eastward to Xuzhou, where it joined the Si River, which then joined the Huai River—generally following the eastern Han Dynasty's old course of the Bian Canal.[14]

During the Ming Dynasty, the Yellow River experienced persistent disasters and occasional alterations to its course; however, the main riverbed generally remained along the eastern Han Bian Canal route. It was not until the fifth year of the reign of Emperor Wen Zong of the Qing Dynasty, Xianfeng (1855 CE), when the Yellow River breached at Tongwaxiang and changed course once again.[15] The current course of the Yellow River is the result of this mid-19th-century alteration.

---

[9] Treatise on the five elements I. In *History of Yuan,* vol. 50. The entries for the 2nd year of the Huangqing era (1313), 3rd year of the Yanyou era (1316), and 1st year of the Dade era (1297) follow the same pattern.

[10] Annals of Emperor Shizu XII. In *History of Yuan,* vol. 15.

[11] Annals of Emperor Chengzong II. In *History of Yuan,* vol. 19.

[12] Treatise on the five elements I. In *History of Yuan,* vol. 50.

[13] Annals of Emperor Yingzong II. In *History of Yuan,* vol. 28.

[14] Annals of Emperor Shundi V. In *History of Yuan,* vol. 42.

[15] Treatise on canals and waterways II: The Canal. In *History of Qing*, vol. 108.

The Yuan Dynasty's Waterway System

Before the Yuan Dynasty's conquest of the Jin, the Yuan already controlled the lands north of the Yellow River, including both ends of the Imperial Canal (*Yuhe*). The canal remained navigable under the Jin, and naturally, after acquiring Hebei, the Yuan continued to utilize it. In the 3rd year of Emperor Shizu's Zhiyuan reign (1266), the Director of the Metropolitan Waterways Bureau reported: "The canal extends over two thousand *li*; transportation of grain and goods by official and private vessels is of great benefit. Since the outbreak of war, maintenance has been neglected. From Qingzhou southwards to Jingzhou northwards, more than thirty sections of embankments and riverbanks have collapsed, and fifteen li of river channel are silted. In the Gui-si year… dredging and repairs were carried out, allowing navigation to resume, but now it has been over thirty years again."[16]

The Gui-si year corresponds to the 5th year of the reign of Ögedei Khan (1265), which coincides with the 2nd year of Jin Emperor Aizong's Tianxing reign and the 6th year of Song Emperor Lizong's Shaoding reign. Qingzhou corresponds to present-day Qing County, Hebei, and Jingzhou to Jing County, Hebei; the distance between them is not great. The relatively limited damage—only fifteen *li* of silted waterways and collapsed embankments—indicates the *Yuhe* was largely navigable. After the *Yuhe* became separated from the Qin River, its upper reaches only extended upstream to the Baimen Spring in present-day Hui County, Henan.

The waters of Baimen Spring, as well as the Qi and Huan Rivers, were all notably clear, thus there was little concern over siltation. From the Jin to the Yuan periods, the Zhang River and Hutuo River both flowed northeastward on the eastern slopes of the Taihang Mountains, meeting the Yuhe only below present-day Qing County. Although the Zhang and Hutuo rivers carried substantial sediment, their confluence with the *Yuhe* was near the coast, so their impact was relatively minor. It is particularly important to note that during the Northern Song, due to the Yellow River's frequent breaches and course changes, the *Yuhe* often suffered disruptions from incursions of Yellow River water, causing occasional blockages. However, from the early Southern Song onward, as the Yellow River shifted southward, the *Yuhe* became more freely navigable and continued to benefit from this situation through the Yuan period.[17]

After the Yuan Emperor Shizu conquered the Song in the 16th year of the Zhiyuan era (1279), all territory south of the Yangtze River and Huai River came under Yuan control. The fertile regions of the Yangtze River Delta and Taihu Lake basin, which had been the economic heartland of the Southern Song for over a century, produced large quantities of grain and other resources. These goods needed to be transported north to the Yuan capital, Dadu (present-day Beijing). Since the north–south canal

[16] Treatise on canals and waterways I. In *History of Yuan*, vol. 64.

[17] The modern scholar Zhu Xie, in his *Selected Historical Materials on China's Canals*, lists the Guangji Canal first among Yuan dynasty canal records. The Guangji Canal, located in the Huaimeng Circuit, diverted water from the Qin River to connect with the [Yellow] River. This canal is documented in *History of Yuan*, vol. 65: "Treatise on Canals and Waterways, Part 2," and served solely as an agricultural irrigation channel, unrelated to transportation.

system had not yet been constructed, transportation relied heavily on the Yellow River. At that time, the Yellow River had several branches flowing into the Huai River—via the Ying River, the Guo River, and the old Eastern Han Bian Canal route through the Si River. Of these, the Guo-Huai branch offered a relatively moderate shipping distance.

During the Song-Jin standoff, as the Bian River became silted, traffic on the Han Canal also declined, although not completely ceased. After the Yuan began transporting grain from the southeast, the Han Canal remained the sole navigable waterway for transport. Grain collected from the southeast was shipped into the Han Canal, then transferred via the Huai River into the Yellow River, moving upstream to Zhongluan (a town located southwest of present-day Fengqiu County on the north bank of the Yellow River in Henan). From Zhongluan, the goods were transported overland to Qimen (a town south of present-day Qi County, Henan) on the bank of the *Yuhe*, and then floated down the Yuhe all the way to Dadu.[18] Though this route relied mostly on waterways, its detours and required land transport made it time-consuming and labor-intensive, falling short of an ideal shipping route. Therefore, it was in great need of improvement.

The Jizhou River and Related Transport Routes

In the 18th year of the Zhiyuan era (1281), efforts commenced to excavate a canal aimed at restoring the ancient Si River waterway. The intention was that after grain barges exited the Han Canal, they would no longer need to ascend the Yellow River via the Huai River, but instead would navigate the Huai directly into the Si River.

The Si River originates from present-day Sixian County in Shandong Province, flowing westward to present-day Jining, then turning southward. At that time, grain barges traveling up the Si River could only reach as far north as Lu Bridge, located south of Jizhou City (modern Jining City), without having to follow the Si River's eastward bend. Accordingly, near Jizhou City, a canal was dug northward to connect the Si River and the Wen River. This newly excavated canal was approximately 150 li in length and came to be known as the Jizhou River.[19] The Wen River flows northward and joins the Daqing River northwest of Xucheng County (present-day Dongping County, Shandong), which was formerly a channel of the ancient Ji River.

The opening of the Jizhou River resembles the canal works of the Eastern Jin period by Huan Wen and Liu Yu, who excavated the Huangong Canal to connect the Si and Ji Rivers. The Jizhou River similarly connected the Si River and the Wen River. Because the Wen River actually connects to the Daqing River—and the Daqing River corresponds to the ancient Ji River—the projects are analogous. The primary

---

[18] Treatise on food and finance I. In *History of Yuan*, vol. 93.

[19] Treatise on canals and waterways II. In *History of Yuan*, vol. 65; Waterways and ditches VI. In *Essentials of historical geography*, vol. 129.

difference is geographic: the Huangong Canal lay to the west, while the Jizhou River lay more to the east.

The Huangong Canal followed the eastern edge of the Juye Marsh northward, drawing water from the Xuexun Marsh. The southern section also drew from the Huang River flowing out of Juye Marsh. These waters naturally flowed into the Si River. Its northern section received floodwaters from Xuexun Marsh, which entered Juye Marsh at a point near the northern flow of the Ji River. Due to the relatively flat terrain along Juye Marsh, with little north–south elevation difference, the waters of Xuexun Marsh could split north and south without difficulty.

By the time the Jizhou River was dug, however, Juye Marsh had gradually silted up, the Xuexun Marsh had likely disappeared, and the Huang River was no longer flowing south. Thus, the Huangong Canal could scarcely be restored.

The Wen River originally had a tributary called the Guang River, which flowed southwest from Gangcheng, northeast of present-day Ningyang County, joining the southern flow of the Zhu River near present-day Jining. During the Eastern Jin, the general Xun Xian, in his northern campaign against Former Yan, is said to have excavated a canal from the Guang River to the Wen River, reaching Dong'e. Navigating northward on the Guang River would require moving against the current, which was likely challenging as there is no record of locks on this canal, implying manual towing was necessary.

When the Yuan excavated the Jizhou River, they avoided the detour via the Guang River and instead proceeded directly to the Wen River. This choice likely aimed to circumvent the difficulties posed by upstream navigation against the current.

Due to the inability to utilize favorable conditions such as the Huangong Canal and the deliberate avoidance of the difficulties caused by upstream navigation against the Guang River's current, the decision was made to rely on the Si and Wen Rivers. Although the upper reaches of both the Si and Wen Rivers flow from east to west, their lower courses diverge: the Si River bends southward, while the Wen River bends northward. The increasing distance and elevation difference between these two rivers naturally made water diversion challenging.

To address this problem, sluices and weirs were installed along the waterways to adjust water levels as needed, facilitating the passage of vessels. To divert Wen River water into the Guang River, the Gangcheng Sluice was constructed thirty-four li northeast of Ningyang County, allowing Wen River water to flow into the Guang River channel beneath Jizhou City. Similarly, to control the Si River, the Yanzhou Sluice was built outside Yanzhou City[20] to channel the Si River southwestward toward Jizhou City. At Jizhou City, the Jizhou Sluice connected the Wen and Si Rivers.

The Wen River not only converged with the Si River but also turned northward. The Jizhou Sluice complex consisted of three gates—upper, middle, and lower—spaced approximately three li between the upper and middle gates and two li between the middle and lower gates. These sluices enabled the Wen and Si Rivers to interconnect while allowing each to maintain separate flow paths. The northward-flowing Wen River was called the Jizhou River. Along this river were three sluices: Kaihe,

[20] Treatise on canals and waterways: Yanzhou Sluice. In *History of Yuan*, vol. 64.

Anshan, and Shouzhang. Kaihe Sluice was located thirty *li* southwest of Wenshang County, Anshan Sluice was situated below Anshan, and Shouzhang Sluice lay north of Shouzhang County (modern Liangshan County). The distances between these sluices were eight li from Shouzhang to Anshan, eighty-five *li* from Anshan to Kaihe, and 120 *li* from Kaihe to Jizhou Sluice. The Jizhou River flowed north of Jizhou City and converged with the Daqing River just beyond Shouzhang Sluice.[21]

After joining the Daqing River, waters naturally flowed into the sea. Thus, the transport route proceeded along the Jizhou River, then the Daqing River to Lijin County, where it entered the Bohai Sea. After traversing a maritime route, it reached Zhigu and joined the Imperial Canal (*Yuhe*). This route was more convenient than the earlier circuitous path via the old course of the Guo River, a tributary of the Yellow River. However, the detour via the sea involved considerable complexity. The junction of the Jizhou and Daqing Rivers was less than 200 *li* from the Imperial Canal. Then why was there no direct canal linking the Jizhou and Imperial Canals, and instead a detour around the Bohai Sea? Possibly, such a plan had not yet been conceived, or the idea of digging a new canal here had not arisen.

The opening of the Jizhou River established a navigable connection with the Daqing River, which in turn flowed into the sea, thereby facilitating transportation compared to previous routes. However, this route also had certain drawbacks: the tidal fluctuations at the river's estuary continually affected the ingress and egress of vessels, often damaging ships. Moreover, the estuary was frequently obstructed by sediment accumulation, which further impeded smooth navigation. Consequently, the Jizhou River remained operational for only four years and ceased transportation activities by the 24th year of Zhiyuan (1287).[22] Some sources suggest that the abandonment of the Jizhou River occurred in the 31st year of Zhiyuan (1294),[23] but this assertion is

[21] Treatise on canals and waterways I: Huitong River. In *History of Yuan*, vol. 64.

[22] Treatise on food and finance I. In *History of Yuan*, vol. 93.

[23] *History of the Yuan, Volume 65, Treatise on Canals II: The Jizhou River*, records the historical development of this important waterway. However, in the *New History of the Yuan Dynasty, Volume 53, Treatise on Canals II, the Jizhou River* is categorized among those canals that were abandoned shortly after their excavation—an assessment that appears to involve a degree of textual confusion. It is true that in the 21st year of the Zhiyuan reign (1284 CE), Emperor Shizu ordered the suspension of transportation along the Jizhou River. However, this suspension applied only to the section running along the Daqing River to Lijin, where it emptied into the sea. The segment between Rencheng and Xuchang remained in use. The reason for suspending the former section was the buildup of tidal silt at the river's mouth, which impeded navigation; this did not significantly affect the functioning of the river system as a whole. In fact, the same volume of the *New History of the Yuan Dynasty*, under the entry for the Huitong Canal, clearly notes: "At the beginning of the Yuan, the Wen River was diverted into the Si River to facilitate grain transport. The Wen then joined the Guang, Si, and Yi Rivers, though it had not yet been diverted northward. In the 20th year of Zhiyuan (1283), a new canal was excavated from Jining, directing the waters of the Wen and Si northwest to Anmin Mountain in Xucheng, where they entered the old course of the Ji River, though they still had not yet reached the Imperial Canal. In this year [i.e., the 26th year of Zhiyuan], another canal was opened from Anmin Mountain southwestward to Linqing, whereby the waters of the Wen and Si were finally connected to the Imperial Canal." Since the Jizhou River drew upon the waters of the Wen and Si Rivers—and these waters were ultimately routed through the Huitong Canal into the Imperial Canal—it is clear that the Jizhou River remained operational. Thus, the claim in the New

inaccurate. By the 26th year of Zhiyuan (1289), the Huitong River had already been completed.[24] After the completion of the Canal, the Jizhou River's significance was enhanced rather than diminished, making claims of its abandonment untenable.

The Huitong River

Once grain and other goods had been transported via the Jizhou River, they faced two possible routes northward: either shipment by sea through the Daqing River, or overland transport from Shouzhang (in the northwest of present-day Liangshan County) to Linqing, where the goods would then enter the Imperial Canal (*Yuhe*). This overland segment stretched over 200 *li*, making it not only longer than the Zhongluan–Qimen route but also more challenging, as it traversed low-lying terrain prone to waterlogging and impassability during the summer and autumn rainy seasons. The difficulty of overland transport in such conditions posed significant obstacles for both public and private freight operations.

In the 26th year of the Zhiyuan reign (1289 CE), Han Zhonghui, magistrate of Shouzhang County, and Bian Yuan, a clerk of the Bureau of Astronomy, independently submitted memorials to the throne proposing the excavation of a canal to replace this inefficient land route. The new waterway, subsequently named the Huitong River,[25] extended over 250 *li*, running from Anshan (also known as Anminshan, located west of present-day Dongping County, at the confluence of the Wen and Ji Rivers) to Linqing. Its primary water source remained the Wen River, thereby functioning as a direct continuation of the Jizhou River system.

Originally, the term Huitong River referred specifically to the section of the canal running from Anshan to Linqing. Over time, however, its scope expanded to include the Jizhou River, and eventually even the section of the Si River south of Pei County. In the third year of the Zhizhi reign of Emperor Yingzong (1323), a deputy official of the Water Transport Office memorialized: "At Jingou and Gutou, east of Pei County along the Huitong River, the terrain is elevated. During periods of low water, the depth becomes insufficient and navigation is obstructed. The provincial authorities have already approved the construction of two overflow weirs."[26]

Following the excavation of the Jizhou River, sluice gates were installed along its course to regulate water levels, as discussed above. Although the terrain along the original route of the Huitong River was relatively level, additional gates were also constructed to facilitate smoother navigation. North of Linqing County stood the Huitongzhen Sluice; 152 *li* to its south was the Lihaiwu Sluice, which lay within

History of the Yuan Dynasty that it was abandoned does not accurately reflect the historical reality of the time.

[24] Treatise on canals and waterways I: Huitong River. In *History of Yuan*, vol. 64.

[25] Ibid.

[26] Ibid.

the southern part of Liaocheng County. Twelve li further south stood the Zhoujiadian Sluice, followed in sequence by the Qiji Sluice, A'cheng Sluice, and Jingmen Sluice—all situated between the counties of Liaocheng and Shuzhang. South of Jingmen Sluice were the various gates originally belonging to the Jizhou River system.

Further downstream, south of Jizhou City, were the sluices at Zhaocun, Shifo, Xindian, Shijiadian, Zaolin, Mengyangbo, Jingou, Gutou, Sanchakou, and Tushan. Most of these were one-to-one gate-site correspondences, though a few comprised multiple gates at a single named location. For example, the Huitongzhen Sluice actually consisted of three separate gates, known as the Head Gate, the Central Gate, and the Narrow Boat Gate. In total, there were thirty-one sluice gates spanning from north of Linqing to south of Pei County. According to the *History of the Yuan: Treatise on Canals and Waterways*, "Construction on the Huitong River began in the first month of the twenty-sixth year of the Zhiyuan reign (1289 CE), on the day Jihai, starting from southwest of Anshan in Xucheng and terminating at the Imperial Canal (Yuhe) in Linqing, covering a distance of more than 250 *li*. Along the route, thirty-one sluice gates were built to accommodate variations in elevation, regulate long-distance flow, and manage the storage and release of water." This description, however, is somewhat imprecise, as not all of the thirty-one gates were located between Anshan and Linqing. Nonetheless, it illustrates that the name Huitong River eventually came to refer to a broader system, extending beyond its original boundaries (See Fig. 7.1).

The Tonghui River

With the successful excavation of the Huitong River, the main canal system of the Yuan dynasty was largely complete. However, there remained a short yet critically important segment of canal—the Tonghui River. Although its name resembles that of the Huitong River and the two are often confused, they are in fact geographically distant and functionally distinct.

As previously discussed, when Emperor Yang of the Sui dynasty constructed the Yongji Canal, his intent was to connect to Ji County—the administrative seat of Zhuo Commandery at the time—which corresponds to present-day Beijing. The canal followed the Sanggan River upstream to the south of Ji County. During the Jin dynasty, the waterway was rerouted to follow the Lu River (modern-day Tonghui River), reaching Lu County (modern Tongzhou District, Beijing). This change was prompted by the turbid nature of the Sanggan River (known as the Lugou River during the Jin period), which made it unsuitable for navigation. However, a short land route still separated Lu County from Zhongdu (the Jin capital, present-day Beijing), posing logistical difficulties. During the Taihe era of Emperor Zhangzong (1201–1208 CE), Han Yu completed a canal connecting these points.[27]

[27] Treatise on canals and waterways: Grain canals. In *History of Jin*, vol. 27. Biography of Han Yu. In *History of Jin*, vol. 110.

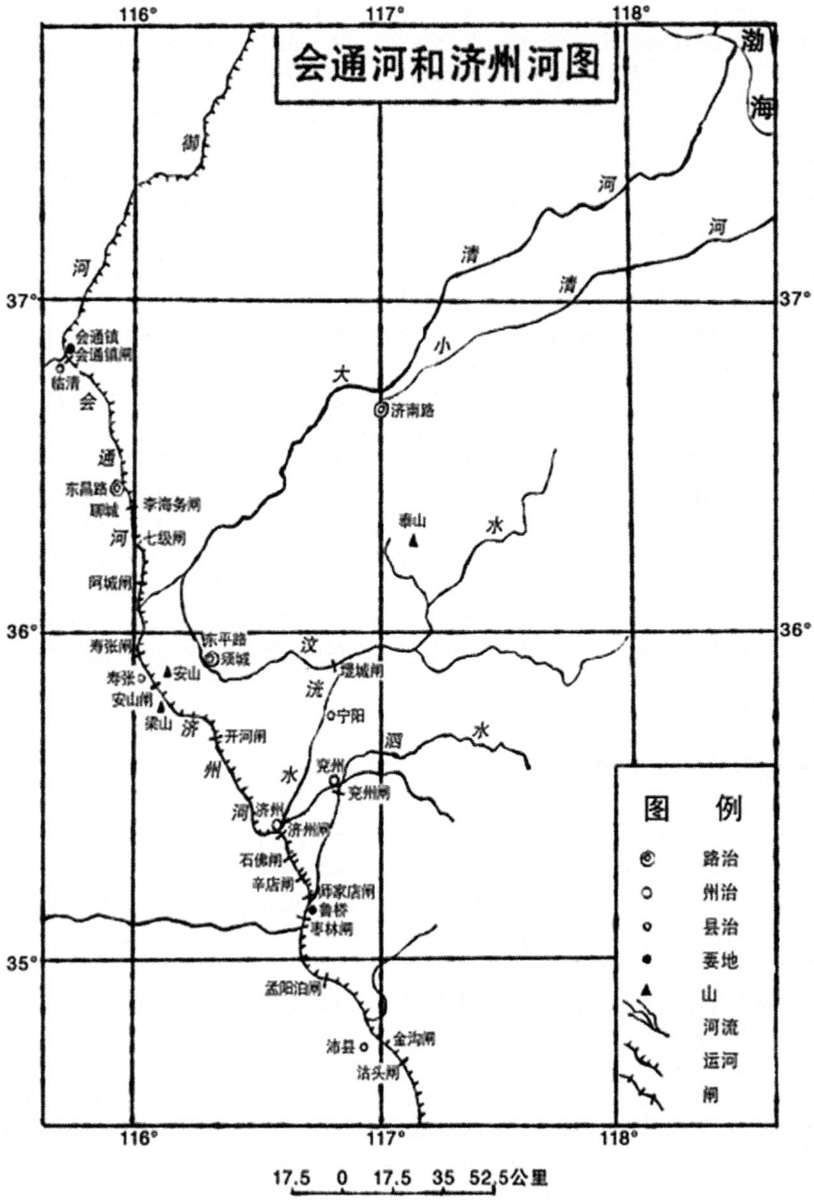

路治 Circuit Seats
州治 Prefecture Seats
县治 County Seats
要地 Strategic Locations
山 Mountains
河流 Rivers
运河 Canals
闸 Sluices

Fig. 7.1 The Huitong and Jizhou rivers

After the Jin capital was relocated to Kaifeng near the dynasty's end, this canal likely fell into disuse and eventually became silted up. In the third year of the Zhongtong reign of Shizu (1262), Guo Shoujing submitted a memorial to the Yuan government outlining six major waterworks proposals. The first item referenced the old transport canal of Zhongdu, which extended eastward to Tongzhou (whose seat was Lu County), and recommended diverting water from Yunquanshan to make the canal navigable. Guo estimated that this would save 60,000 strings of copper cash annually in cartage expenses.[28]

Guo's memorial suggests that Han Yu's canal had long been abandoned by the time, well before the Yuan conquest of the Song dynasty. However, his proposal seems to have been initially rejected, and grain transported from southeastern regions to Lu County continued to require costly and labor-intensive overland transit. Not until the twenty-eighth year of the Zhiyuan reign (1291) did Guo Shoujing, now serving as Director of the Waterworks Bureau, again submit a proposal to restore the canal. This time it was approved, and work commenced the following year. By the thirtieth year of Zhiyuan (1293), the canal was completed and named the Tonghui River.

The newly constructed canal stretched 164 *li* and featured ten hydraulic installations, each equipped with a pair of sluice gates—an upper and a lower—totaling twenty gates altogether. The ten sites were as follows:

1. Guangyuan Sluice—Located downstream of the diversion channel from Wengshanbo, west of the city.
2. Xicheng Sluice (later renamed Huichuan Sluice)—Situated northwest of Heyimen Gate, near modern-day Gaoliang Bridge outside Xizhimen.
3. Zhaozong Sluice—South of the Wanyi Treasury, which was located within the north water gate of Heyimen and north of Gaoliang Bridge, near the west city wall of Dadu (Yuan capital).
4. Haizi Sluice (later renamed Chengqing Sluice)—Located beneath the Wanning Bridge within the city.
5. Wenming Sluice—The upper gate was southeast of the Lizzhengmen Water Gate (modern-day area just east of Zhengyi Road North). The lower gate was approximately one *li* southwest of Wenmingmen Gate. Its remains have been discovered buried deep underground in what is now Taijichang Second Alley.
6. Weicun Sluice (later renamed Huihe Sluice)—Southeast of Wenmingmen Gate.
7. Jidong Sluice (later renamed Qingfeng Sluice)—Located in Wangjiazhuang, southeast of Dadu.
8. Jiaoting Sluice (later renamed Pingjin Sluice) was located at Yinwangzhuang, approximately twenty-five *li* southeast of the capital city of Dadu.
9. Yangyin Sluice (later renamed Pujì Sluice) was situated thirty *li* southeast of Dadu.

[28] Biography of Guo Shoujing. In *History of Yuan*, vol. 164.

10. Tongzhou Sluice (later renamed Tongliu Sluice) had its upper gate outside the west gate of Tongzhou, and its lower gate outside the south gate of the same city.[29]

The Tonghui River constructed by Guo Shoujing differed in certain respects from the earlier transport canal opened by Han Yu. Han Yu had utilized water from Yuquanshan, which flowed southeast into the Gaoliang River, and from there was channeled eastward to Lu County. In contrast, Guo Shoujing sourced water from the Shenshan Spring in Baifu Village, Changping County, which he supplemented with waters from Shuangta Spring, Yuhe, Yimu Spring, and Yuquan Spring, all located to the northwest of present-day Beijing.[30]

To collect and channel these waters, Guo Shoujing constructed a sixty-li-long dike, known as the Baifu Weir, northwest of the Yuan capital, Dadu.[31] The spring waters passed through the weir into the Western Water Gate, entered the city to form Jishuitan (modern-day Shichahai in Beijing), and from there were diverted along the former canal route of Han Yu toward Gaolizhuang in Tongzhou (southeast of modern Tongzhou District), where the canal joined the Bai River. Consequently, Jishuitan became a busy hub where riverboats and seafaring vessels arriving via the Tonghui River congregated and moored.

Although Guo Shoujing used a different water source, much of the canal's alignment and features remained consistent with those of Han Yu's earlier work. By the time the Tonghui River was constructed, Han Yu's original canal had already silted up and become level ground. Nevertheless, the excavation of the new canal made it possible to reuse bricks and stones from the previous canal's sluices and dams. The extent to which the new works matched the earlier layout in both form and function was impressive—and far from easily achieved.[32]

The Yuan capital of Dadu lay west of the Lu River, whereas the Lugou River (now the Yongding River, historically also known as the Hun River (Muddy River) flowed south of Dadu. In terms of proximity, the Lugou River was actually closer to Dadu. However, the decision not to use the Lugou River but rather the more distant Lu River was motivated by sound reasoning: the Lu River had a gentler flow, and the mountain springs feeding the Tonghui River were notably clear, whereas the Lugou's current was muddier and swifter.

In the late Yuan period, efforts were made to abandon Guo Shoujing's hydraulic system in favor of utilizing the Lugou River, but these ultimately ended in failure. The proposed rerouting of the canal system occurred in the second year of the Zhizheng reign (1342), during the reign of Emperor Huizong. That year, a plan was advanced to excavate a new canal running from Gaolizhuang in Tongzhou to the Tieban Gorge

---

[29] Annals of Shizu XIV; Treatise on canals and waterways I: Tonghui River, Haizi bank, Shuangta River, Baifuweng Mountain. In *History of Yuan*, vol. 17, 64. Hou Renzhi. The capital city of Yuan Dadu and Ming-Qing Beijing. In *The theory and practice of historical geography*.

[30] Treatise on canals and waterways I: Tonghui River. In *History of Yuan*, vol. 64.

[31] Hou Renzhi. *Bufang collection* (p. 26). Beijing Renmin Press.

[32] Annals of Shizu XIV; Treatise on canals and waterways I: Tonghui River; Biography of Guo Shoujing. In *History of Yuan*, vol. 16, 64, 164.

at Jinkou, an ancient watergate in the Western Hills (*Xishan*), intended to replace the Tonghui River. The proposed canal stretched over 120 *li*, with a depth of five *zhang* and a width of fifteen *zhang*. It was designed to divert water from Jinkou eastward to Gaolizhuang, where it would merge with the Imperial Canal.

At the time, the ruling authorities approved the excavation of the canal. However, Vice Chancellor Xu Youren raised objections on several grounds. First, in the second year of the Dade reign (1298) under Emperor Chengzong, a major flood of the Hun River (present-day Yongding River) caused severe damage to the local population. In response, sluice boards were installed to block the Jinkou Gate, and over the following five years, the river channel above Jinkou was entirely dammed with sand, gravel, and earth—indicating the river's force and danger. Secondly, from Lugou Bridge to the river's point of confluence, not a single fishing boat has ever been seen navigating upstream or downstream—clear evidence that the river is unfit for navigation. Moreover, while Tongzhou lies forty li from Dadu, Lugou is only twenty li away. If the river had indeed been navigable, why would a port not have been established at Lugou, where everything would be far more convenient? Why, instead, would one resort to a location twice as distant, such as Tongzhou? Third, the Western Hills supplied torrents of water that were rapid and forceful. During the rainy seasons of summer and autumn, these waters could swell and overflow, potentially causing great floods. Such disasters, he argued, could not be prevented in advance. Moreover, traces of ancient canals in the area—initiated and then abandoned during the Jin dynasty—stood as further testimony to the infeasibility of such projects. Fourth, the varying topography of the area posed an additional challenge. Without the construction of sluices, water would become too shallow and sluggish for navigation. Yet if sluices were built to regulate the flow, sediment would quickly accumulate due to the muddy and turbulent waters, necessitating constant dredging—monthly, even daily. This endless maintenance would prove unsustainable. Xu concluded that the original decision to source water from the Baifu springs—as done during the construction of the Huitong River—instead of relying on the torrents of the Western Hills, was entirely justified. Nonetheless, the authorities disregarded his advice and proceeded with the project, which ultimately ended in failure.[33]

Xu Youren's remarks make the matter abundantly clear: the waters of the Lugou River are, in the final analysis, too laden with silt and too swift in current. As such, sluice gates could neither be installed nor omitted—rendering the river fundamentally unsuitable for navigation. This is precisely why Guo Shoujing refrained from utilizing its waters; and it is likewise the reason that the Ming and Qing dynasties, following Guo's precedent, never attempted to alter the established course (See Fig. 7.2).

[33] Treatise on canals and waterways III: Jinkou River. In *History of Yuan*, vol. 66.

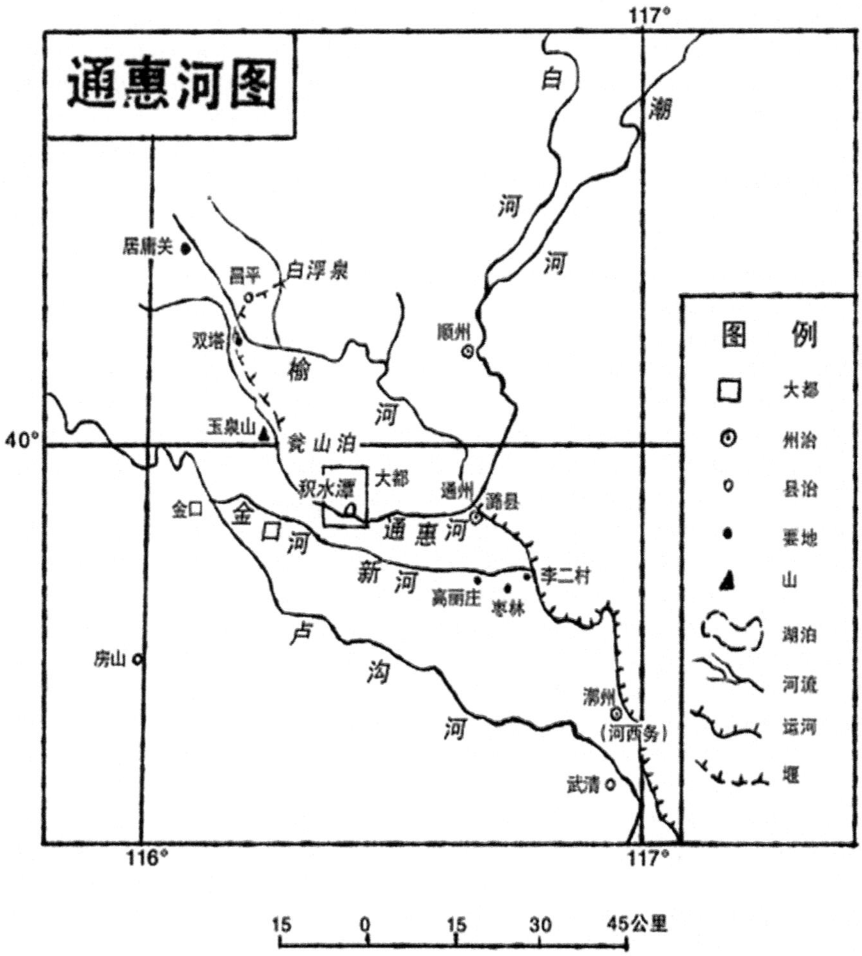

图例 - Legend
大都 Dadu
州治 Prefecture Seats
县治 County Seats
要地 Strategic Locations
山 Mountains
湖泊 Lakes
河流 Rivers
运河 Canals
堰 Weirs

Fig. 7.2 The Tonghui River

Dredging and Regulation of the Jiangnan Canal

Since its construction under Emperor Yang of the Sui Dynasty, the Jiangnan Canal had long remained the most stable and navigable among all sections of the Grand Canal. Nevertheless, during the Yuan dynasty, it too underwent periods of dredging and maintenance. Two major engineering efforts were particularly noteworthy: The first involved dredging a 131-*li* stretch of waterway from Zhenjiang Circuit (with its administrative seat in present-day Zhenjiang, Jiangsu) to Lücheng Dam (located southeast of present-day Danyang, Jiangsu). The second was the regulation of Lian Lake (located in present-day Danyang), which served as the principal water source for this section of the canal. As Lian Lake frequently retained excess water, even a one-inch release could raise the canal's water level by as much as one foot—provided the channel was not obstructed. For this reason, the regulation of Lian Lake was considered a matter of critical importance in the upkeep of the Jiangnan Canal at the time.[34]

During the Yuan dynasty, efforts were also undertaken to manage the Longshan River located outside the city of Hangzhou (present-day Hangzhou, Zhejiang Province). Although this short river was not officially part of the Jiangnan Canal system, it had served since the Song dynasty as the primary channel through which riverine grain transport entered the city from the Qiantang River. In the first year of the Zhida reign (1308 CE) under Emperor Wuzong of Yuan, the river underwent dredging and repair, thereby becoming effectively connected to the main canal system.

The Completion of the Grand Canal

With the successful excavation of the Tonghui River, it became possible to travel by canal from Dadu (present-day Beijing) all the way to Hangzhou on the Qiantang River. This Grand Canal thus constituted a continuous north–south waterway. Although it was shorter than the north–south water route established in the early Western Jin dynasty—when Du Yu opened the Yangkou channel and successfully connected Ji County (in the north) to the South China Sea—the Yuan dynasty canal traversed some of the most prosperous regions in the country. As a result, its practical effectiveness was even greater. From its completion during the Yuan dynasty, the Grand Canal was preserved and maintained throughout the Ming and Qing dynasties, with only minor adjustments made to certain sections. Its overall structure and scale remained largely unchanged.

[34] Treatise on canals and waterways II: Lian lake; Longshan waterways. In *History of Yuan*, vol. 66.

Names of the Canal Sections in the Yuan Dynasty

Although the north–south waterway was generally referred to as the "Grand Canal" during the Yuan dynasty, each section of it had its own unique history of construction and, accordingly, a distinct name that varied by region. Some of these names have already been mentioned earlier, but further clarification is warranted. The section from Dadu to Tongzhou was naturally called the Tonghui River, though it was also known by alternative names such as the Futong River or the Ba River. The segment from Tongzhou to Zhigu (modern Tianjin) followed the Bai River, a natural watercourse which, though not artificially excavated in origin, underwent various technical improvements to facilitate transport.

From Zhigu to Linqing lay what had once been known as the *Yuhe* or "Imperial River." Although the Yuhe originally extended from Mount Sumen in Huixian (Henan province) and was still navigable along its full length, only the segment north of Linqing functioned as an integral part of the Yuan dynasty's north–south canal system.[35] South of Linqing were the Huitong River and the Jizhou River, both of which were key segments of the Grand Canal. In later periods, the name Huitong River came to refer not only to its original section but also to the Jizhou River and, further south, the canal segment extending as far as Pei County. Continuing southward to Yangzhou, the canal was known as the Yangzhou Canal. South of the Yangtze River, the section stretching from Zhenjiang to Lücheng Dam near Changzhou was referred to as the Zhenjiang Canal.[36] As for the section of the canal stretching south from Lücheng Dam to Hangzhou, it appears that during the Yuan period it was generally referred to simply as "the Canal," with no other specific designation found in contemporary sources.[37]

River Transport and Maritime Transport

Although both the Yuan court and officials were actively committed to the excavation of canals, the dynasty's grain transport could not rely solely on inland waterways. The Yuan dynasty's canal system primarily utilized gated waterways, in which the regulation of water levels through sluice gates required considerable time, resulting in frequent delays for grain barges. Moreover, the volume of water within these canals was limited, and when the cargo capacity of vessels exceeded prescribed limits, navigation became increasingly difficult.

Throughout the Yuan period, successive regulations were enacted to strictly limit vessel tonnage. From the inception of canal construction in the early Yuan, only

---

[35] Treatise on canals and waterways I; In *History of Yuan*, vol. 64; Treatise on canals and waterways II; In *New history of Yuan*, vol. 53.

[36] Treatise on canals and waterways II: Yangzhou canal- Lian lake. In *History of Yuan*, vol. 65.

[37] Treatise on canals and waterways II. In *History of Yuan*, vol. 65. The canal section south of Lücheng Dam is not specifically discussed nor given a distinct name.

boats of 150 *liao* (料, a traditional Chinese unit of cargo measurement, with 1 *liao* approximately equal to 60 kg) were permitted to operate on the waterways. This regulation remained unchanged for an extended period, though in practice, it was often ignored. In the second year of the Yanyou reign (1315), provincial officials submitted complaints that the shallowness of the canals and the congestion caused by large vessels were hindering other ships from passing. Although the regulations were in place, powerful elites and wealthy merchants, driven by profit, constructed vessels of 300, 400, or even 500 *liao* to navigate the canals, thus obstructing official and civilian traffic.

In response, auxiliary sluices were constructed beside the Gutou Sluice (located south of present-day Pei County, Jiangsu), and similar installations were established near Linqing. Ships carrying over 200 *liao* were prohibited from entering the canals via these points.[38] These restrictions were especially stringent for the Jizhou River section, where the terrain presented significant elevation differences, necessitating the construction of many locks, which further complicated navigation.

Nevertheless, such regulations failed to eliminate abuse entirely. In the fourth year of the Taiding reign (1327), censors of the Yushitai (Censorate) again reported: "The narrow gates established in the north and south each measure nine feet wide. Only vessels under 200 *liao*, with a beam of no more than eight *chi* five *cun* can enter. Yet commoners, insatiable in their pursuit of profit, circumvent these restrictions by modifying their ships—reducing the beam while lengthening the hull—to create vessels of 80 to 90 *chi*, or even 100 *chi*, reaching capacities of 500 to 600 *liao*. Once inside the gates, these vessels cannot turn, frequently run aground in shallow sections, and obstruct the passage of other boats. This problem arises because the narrow sluices only limit beam width, not vessel length."

According to official regulations, vessels passing through canal locks were strictly limited in size: the beam could not exceed eight *chi* five *cun*, and the maximum length was set at six *zhang* five *chi*. These dimensions accommodated ships capable of carrying up to 200 *liao*. To enforce these limits, it was stipulated that stone markers be erected at the foot of narrow locks, and any vessel exceeding the prescribed length would be denied entry.[39] Such restrictions were undoubtedly stringent.

Despite the construction of the canal system, the actual volume of tribute grain transported via the Grand Canal to Dadu (present-day Beijing) remained limited. Most of the imperial grain shipments continued to rely heavily on maritime transport. Since the Yuan dynasty initiated official sea transport, the annual volume of seaborne grain steadily increased. In the 20th year of the Zhiyuan reign (1283), over 40,000 *dan* of grain were delivered. Twelve years later, by the end of the Zhiyuan era (1294), this figure rose to more than 500,000 *dan*. By the end of Emperor Chengzong's reign (11th year of the Dade era, 1307), the total reached approximately 1.64 million *dan*. Under Emperor Wuzong (4th year of the Zhida era, 1311), it further increased to about 2.77 million *dan*. During Emperor Renzong's reign (7th year of the Yanyou era, 1320 CE), the volume reached 3.24 million *dan*, and by the 2nd year of the Tianli

[38] Treatise on canals and waterways I: Huitong River. In *History of Yuan*, vol. 64.

[39] Ibid.

era under Emperor Wenzong (1329), it had grown to approximately 3.34 million *dan*, likely the highest level achieved during the Yuan dynasty.[40]

This continuous rise in maritime transport volumes clearly reflects the weakness of canal-based grain transport. In fact, by the late Zhiyuan period, river transport had already reached its limits and was suspended on two occasions: first in the 20th year of Zhiyuan (1283) and again in the 24th year (1287). By the earlier date, the Jizhou Canal—linking the Si and Wen rivers—had already been completed. As noted previously, though this canal connected the two river systems, grain shipments still needed to enter the sea via the Daqing River. At its estuary, cargo vessels had to wait for high tide to proceed, during which they were susceptible to damage. Meanwhile, sea vessels arriving from the south had already reached the destination ahead of them, rendering river transport relatively insignificant. Given these operational difficulties, river transport was suspended.

In the 24th year of Zhiyuan (1287), the Yuan government established the Xingquan Fusi (Maritime Transport Office), tasked specifically with managing sea transport operations. Two additional Wanhufu (Commandant Offices) were created, bringing the total to four. That same year, grain shipments via the Dongping River were terminated.[41] The so-called "Dongping River" referred in fact to the Jizhou River, which passed through Dongping (present-day Dongping County, Shandong), hence the alternate name. Although these two suspensions of the Jizhou River were relatively brief and occurred prior to the excavation of the Huitong River, they nevertheless indicate the broader trend of declining canal-based grain transport during the Yuan period.

Maritime Transport

The origins of maritime transport in China did not begin with the Yuan dynasty. As early as the Spring and Autumn period, there were already records of its use. In the tenth year of Duke Ai of Lu (485 BCE), the state of Wu attacked the southern border of Qi. Xu Cheng led a naval force intending to enter Qi via the sea but was defeated by Qi forces, forcing the Wu army to retreat.[42] During the conference at Huangchi between Wu and Jin (present-day southwest Fengqiu County, Henan), the state of Yue sent troops by sea and ascended the Huai River in an effort to cut off Wu's return route.[43] This may be the earliest known documentation of maritime operations. Coastal inhabitants, being well acquainted with watercraft, naturally sought to exploit the advantages of sea transport. Du Fu, in his poem *Going Beyond the Passes*, described Tang dynasty maritime activity as follows:

---

[40] Treatise on food and finance I: Maritime transport. In *History of Yuan*, vol. 93.

[41] Ibid.

[42] *Zuo commentary*, 10th year of Duke Ai.

[43] Discourses of the state of Wu. In *Discourses of the states*.

"Yuyang, land of warriors bold, Drums beat, flutes and reed pipes rolled.
Cloud-like sails traverse the Liao Sea wide,
Grain and rice from Wu's east side.
Robes of Yue, silks of Chu shine bright,
Reflecting the courtiers' splendid might."[44]

In another poem, *Former Wandering*, he wrote:

"Warfare long in You and Yan,
Supplying troops was hardship then.
From Wu came grain and silk supplies,
O'er stormy seas to Penglai skies."[45]

Such large-scale maritime transportation—from the southeast across the sea to reach You and Yan—was already practiced during the Tang dynasty.

However, it was the Yuan dynasty that first systematically developed maritime transport as a national enterprise. The Yuan initially employed maritime routes during their early campaigns in the Jiangnan region. At that time, captured Song archives and goods were shipped by sea to Dadu (present-day Beijing), although grain transport still relied on inland waterways. When river-based grain transport proved insufficient, efforts were made to revive maritime shipping (in 1282, the 19th year of the Zhiyuan reign under Emperor Shizu).

The earliest Yuan maritime route began at Liujiagang (now called Liu River, located northeast of Taicang County in Jiangsu Province), entering the sea and proceeding northward along the coast. In 1292, due to the many dangers posed by the coastal route, a new, offshore route was established. This route began at Liujiagang, continued through Chengjiaosha, turned past Shazui, proceeded to Sansha Yangzijing, crossed Biandansha and Dahong, passed Wanli Changtan, then entered the open sea to Qingshuiyang, via Heishuiyang, to Chengshan, through Liujia Island, and then to the islands of Zhifu and Shamen, finally traversing the Laizhou Sea and entering Jiehekou. This offshore route was slightly shorter than the coastal one.

One year later, a third and even more direct route was charted: it also began at Liujiagang, passed through Sansha near Chongming Prefecture into open sea, then proceeded eastward through Heishui Ocean, reached Chengshan, turned west to Liujia Island and Shamen Island, crossed the Laizhou Sea, and entered Jiehekou. This final route was the shortest and most expedient; under favorable winds, it could be completed in just over ten days. However, the sea conditions on this path were extremely rough, with frequent shipwrecks and loss of life.

However, these issues were not the primary concern of the government at the time. What worried the authorities most was the potential loss of the grain tribute during transport. Subsequently, regulations were enacted stipulating that any lost

[44] *Categorized and annotated collection of Du Gongbu's poetry*, vol. 15.
[45] *Categorized and annotated collection of Du Gongbu's poetry*, vol. 19.

grain should be compensated by the transport officials. Despite these challenges, maritime transport continued unabated[46] almost until the fall of the Yuan dynasty.

The Jiaolai Canal

In the early period of maritime transport, the Jiaolai Canal was also excavated. Also known as the Jiaodong River, it was dug by diverting the Jiao River. The Jiao River originates southwest of Jiaozhou (the administrative center of present-day Jiao County, Shandong), flows northward to the northeast of Jiaozhou, and then splits into two branches flowing north and south. The southern branch enters the sea at Mawankou in Jiaozhou, while the northern branch flows into the sea at Haicangkou in Ye County.

Maritime transport at the time required ships to sail around Chengshan, a route that was long and perilous due to turbulent seas. Thus, there was a strong need to excavate a canal crossing the Shandong Peninsula. Following the recommendation of Yao Yan, excavation began in the 17th year of the Zhiyuan era (1280).[47]

After the successful completion of the canal, its transport capacity was considerable. In the 22nd year of Zhiyuan (1285), it was decreed that one million dan of grain tribute would be transported annually from the Jianghuai region, distributed as follows: 100,000 *dan* by sea, 300,000 *dan* via the Jizhou River, and 600,000 *dan* solely through the Jiaolai Canal.

Although the Jiaolai Canal's transport capacity was substantial, its maintenance and operation costs were high. As a result, in that same year, grain transport via the Jiaolai Canal was discontinued.[48] The cessation, however, did not seem to be thorough, for three years later, maritime transport officials again petitioned for its abolition.[49] Subsequently, the canal was never reopened. This likely resulted from the shift in maritime routes—from entering the sea at Liujia Port directly to Chengshan—abandoning the coastal route northward.

In the Ming dynasty, there were occasional mentions of this waterway, and several attempts were made to restore it, but none succeeded (See Fig. 7.3).

The Jialu River

In addition to the aforementioned canals, the Yuan dynasty also saw the construction of the Jialu River, a significant waterway named after the official Jialu, who proposed and oversaw its excavation. The project likely took place between the 11th and 13th

[46] Treatise on food and finance I: Maritime transport. In *History of Yuan*, vol. 93.

[47] Annals of Shizu VIII. In *History of Yuan*, vol. 11.

[48] Annals of Shizu X. In *History of Yuan*, vol. 13.

[49] Annals of Shizu XII. In *History of Yuan*, vol. 15.

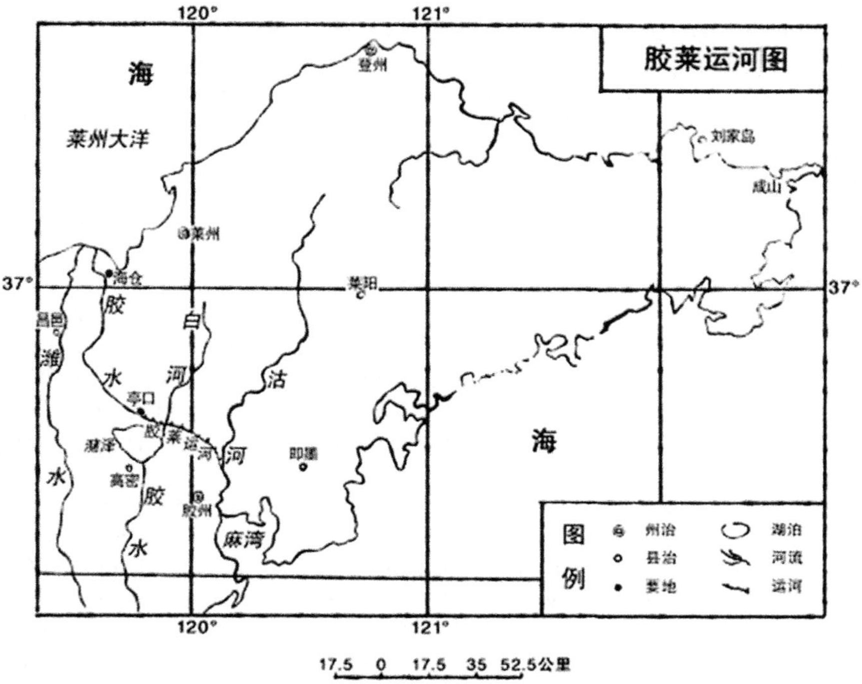

图例 - Legend
州治 Prefecture Seats
县治 County Seats
要地 Strategic Locations
湖泊 Lakes
河流 Rivers
运河 Canals

Fig. 7.3 The Jiaolai Canal

years of the Zhizheng reign (1351–1353) under Emperor Shun, during which Jialu served as Minister of Works and was charged with managing riverine defenses. As Jialu passed away in the 13th year, the canal must have been completed before that time, although the precise date of its commencement remains uncertain.[50]

The upper source of the Jialu River can be traced back to the Jingshui River, which itself has three sources: the western source originates from Shengshuiyu in Mi County, Henan; the middle source begins at the Warm Spring and Ice Spring in Xingyang County, Henan; and the eastern source springs from Jiuxian Temple in Zhengzhou. These three sources converge at Jingshui Town, forming the Jingshui

[50] Biography of Jia Lu. In *History of Yuan*, vol. 187.

River. To the north, it further receives the Xu and Suo Rivers, together called the Shuangqiao River.[51]

Originally, this river flowed from Zhongmou County to Kaifeng City. The Jialu Canal diverts water west of Zhuxian Town in Xiangfu County (modern Kaifeng County), then flows south of Weishi County, and reaches Fugou County (all retaining their historical names today), where it joins the Wei River. Downstream, it merges into the Ying River at Xihua County.[52] At that time, only the section from Zhengzhou to Zhuxian Town was referred to as the Jialu River.[53]

The modern usage of the name "Jialu River" is no longer confined to this section but also includes the segment where it joins the Wei River. Originally, the Jialu River flowed through Zhuxian Town and, when passing Weishi County, it was closer to the neighboring Tongxu County to the east. Currently, however, it runs nearer to Weishi County and farther from Tongxu County, as well as farther from Zhuxian Town. This rerouting had already taken place by the Qing dynasty during the Jiaqing reign (1796–1820) (See Fig. 7.4).[54]

Rivers named Jialu did not only refer to the stretch between Zhengzhou and Zhuxian Town. There was also a river called Jialu southwest of Cao County in Shandong Province, extending between Huanglinggang and the vicinity of Shan County.[55] These two Jialu Rivers were likely excavated around the same time.

At that time, the Yellow River had broken through at Baimao in Cao County, flooding surrounding counties, advancing northward into Anshan, and breaching into the Grand Canal, severely damaging the north–south grain transport. Jialu was tasked with river management and excavated this waterway to channel the Yellow River's flow properly. Consequently, later generations collectively referred to these waterways as the Jialu River, although they were neither connected nor related in any way.

Reflecting on the Jialu River between Zhengzhou and Zhuxian Town calls to mind the Song dynasty's Tianyuan River and Huimin River. The Jialu River's course overlaps the old channel of the Tianyuan River, which flowed eastward into the Bian River and then into the Guangji River. The Jialu River, however, bends southward into the old course of the Song dynasty's Huimin River. By then, the Yellow River had turned southward, and the Guangji River had long ceased to exist. Although the Huimin River downstream still received water from the Yi and Wei Rivers, its volume was too small to accommodate boat traffic, partly due to the silting of the Bian River upstream. The excavation of the Jialu River restored the former scale of the Huimin River and allowed it to remain an important waterway.

[51] Yellow River. In *A Mirror on River Management*, vol. 17, citing *Travels across the four seas*.

[52] Kaifeng prefecture. In *Comprehensive gazetteer of Qing dynasty (Jiaqing edition)*, vol. 186.

[53] Yellow River. In *A Mirror on River Management*, vol. 17, citing *Travels across the four seas*.

[54] Chenzhou prefecture. In *Comprehensive gazetteer of Qing dynasty (Jiaqing edition)*, vol. 191.

[55] Yellow River. In *A Mirror on River Management*, vol. 17, citing *Brush notes from Gushan*.

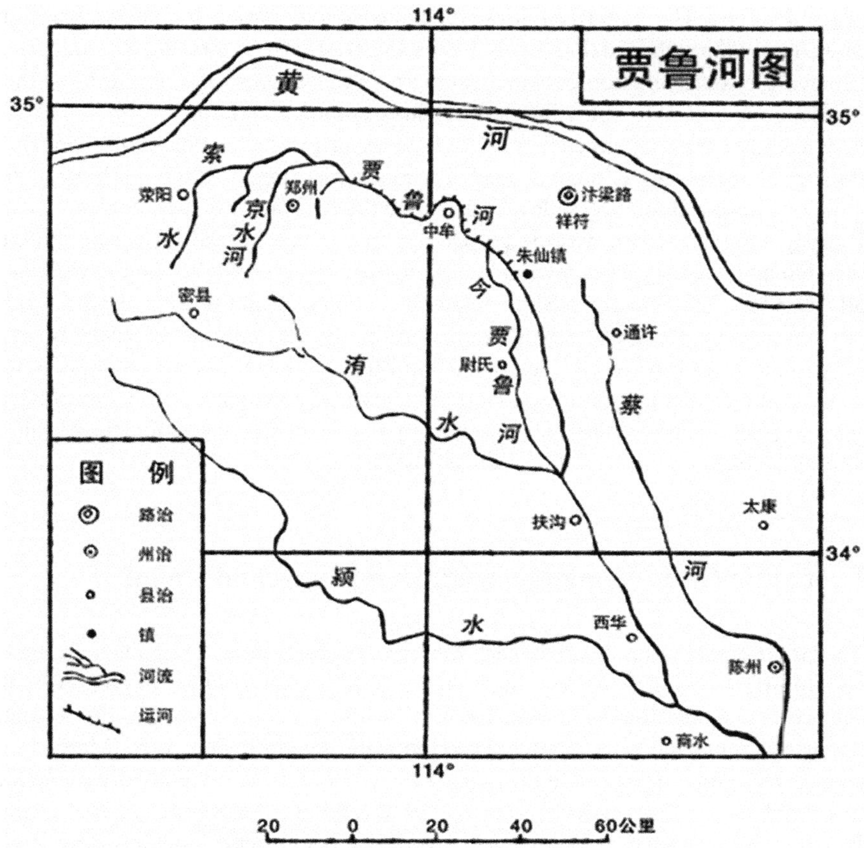

图例 - Legend
路治 Circuit Seats
州治 Prefecture Seats
县治 County Seats
镇 Towns
河流 Rivers
运河 Canals

Fig. 7.4 The Jialu River

The Jialu River and Zhuxian Town

The smooth navigability of the Jialu River directly contributed to the prosperity of Zhuxian Town, and the town's rise and fall were closely tied to the condition of the river itself. In earlier times, when the Huimin River maintained a full flow, grain and goods transported from regions such as Chen and Cai, as well as from various parts of Huainan, could travel directly via the Huimin River into the Bian River and reach the city of Kaifeng. The Jialu River, however, adjusted to the water sources available and flowed only through Zhuxian Town, bypassing Kaifeng. As a result, Zhuxian Town became a critical junction between southern river transport and northern overland routes, gradually developing into a flourishing commercial hub.

One might ask why, given that the Jialu River originated upstream in Zhengzhou, prosperity did not center there instead. The answer is simple: although Zhengzhou is a major transportation hub today, at that time the regional transport center was still Kaifeng. Zhuxian Town was the closest location along the Jialu River to Kaifeng, which made it strategically favorable. As the upper reaches of the Jialu River gradually silted up, Zhuxian Town began to decline, and its former prosperity was eventually replaced by Zhoujiakou to the south. Originally, the name "Jialu River" referred only to the section between Zhengzhou and Zhuxian Town, but over time, common usage came to include the lower section—previously known as the Huimin River. As the upper reaches of the river became impassable, the name "Jialu River" shifted to refer primarily to the lower stretch below Zhuxian Town.

Economic Hubs Along the Yuan Dynasty Grand Canal

The Grand Canal of the Yuan dynasty ran north to south, connecting Dadu (modern-day Beijing) to Hangzhou. Crossing both the Yellow River and the Yangtze River, the canal linked the political center of the north with the economic center of the south. Along this route, numerous regional economic hubs naturally emerged. Marco Polo, during his journey from Dadu to Yangzhou and Quanzhou, traveled along much of this canal route.[56] He left vivid accounts of several cities along the way—Linqing, Yanzhou, Pizhou, Huai'an, Baoying, Gaoyou, Zhenzhou, Guazhou, Yangzhou, Zhenjiang, Changzhou, and Suzhou. He noted the abundance of goods, the prosperity of commerce, and especially the busy traffic of boats traversing the canal. Such records offer valuable glimpses into the economic vibrancy of Grand Canal cities during the Yuan dynasty.

However, not all canal cities developed equally. For instance, Baoying and Gaoyou, located between Huai'an and Yangzhou, likely had more limited room for development. Guazhou, merely a river port, faced stiff competition from nearby economic centers—Zhenzhou to the west, Yangzhou to the north, and Zhenjiang across the river. Guazhou was notable mainly because it stockpiled large quantities of grain for shipment to Khanbaliq (i.e., Dadu).

Marco Polo does not mention Jizhou or Xuzhou by name but does refer to "Linzhou" south of Dongping and "Xizhou" near Pizhou, suggesting possible transliteration inconsistencies. Jizhou lay at the southern end of the Jizhou River, where the Wen River was obstructed by the Gangcheng Dam and flowed southwest to join the Si River. Xuzhou marked the junction where the Si River entered the

[56] Marco Polo. *The travels of Marco Polo, chapters 130–151* (Feng Chengjun, Trans.). Beijing: Zhonghua Book Company.

Yellow River. Given the variance in vessel sizes and shipping conditions, trade and travel often paused at these locations, likely spurring local economic development.

At the southern end of the Grand Canal, Marco Polo provided detailed descriptions of Hangzhou. Not only was Hangzhou located near Suzhou and the fertile Tai Lake region, but it had also served as the capital of the Southern Song dynasty. Having suffered relatively limited destruction during the Yuan conquest, Hangzhou remained a vibrant and prosperous city. According to Marco Polo, the city's streets were lined with countless shops, in addition to ten large markets and innumerable smaller ones. Each major market was surrounded by tall buildings housing various merchants. These large markets held fairs three times a week, each attracting some 40,000 to 50,000 people. Such prosperity was indeed rare for the time.

According to The Travels of Marco Polo, his southbound journey passed through Hejian Prefecture and then Jingzhou. This suggests that he did not take the eastern route from Dadu along the Tonghui River and then into the Imperial Canal. Along this western route, Tongzhou and Hexiwu stood out as major commercial hubs. Tongzhou controlled the confluence of the Tonghui and Bai Rivers, making it a key node in grain transport. Further south was Hexiwu, which, much like Zhuxian Town on the Jialu River, served as a point of convergence between overland and waterborne transport. Hexiwu's strategic role was clearly recognized: in the 25th year of Zhiyuan (1288), the Yuan government established two grain transport offices—one in the capital and one in the provinces, the latter being located in Hexiwu and charged with managing the reception and dispatch of sea-transported grain.[57] From Hexiwu southward lies present-day Tianjin, then still a modest river town known as Haijin under Jinghai County. At that time, it marked the confluence of the Bai and Imperial Canals but had not yet emerged as a major maritime terminus (See Fig. 7.5).[58]

The Calamities of the Grand Canal During the Yuan–Ming Transition

During the transition from the Yuan to the Ming dynasty, the Grand Canal—running vertically through the empire from north to south—suffered several major catastrophes that ultimately rendered it impassable.[59] During the reign of Emperor Shun of the Yuan, the Yellow River repeatedly burst its banks, inundating over a thousand *li* between Jining, Cao (present-day Cao County, Shandong), and Yun (present-day Yuncheng County, Shandong), with floodwaters flowing northeastward into the canal.[60] Given the canal's shallow channel, how could it possibly withstand such repeated devastation? In the eleventh year of the Zhizheng reign (1351), Jia Lu was

---

[57] Treatise on food and finance I: Maritime transport. In *History of Yuan*, vol. 93.

[58] Tianjin prefecture I. In *Comprehensive gazetteer of Qing dynasty (Jiaqing edition)*, vol. 24.

[59] Treatise on canals and waterways III: Grand Canal, part 1. In *History of Ming*, vol. 85.

[60] Annals of Emperor Shun IV; Treatise on canals and waterways III: Yellow River. In *History of Yuan*, vol. 41, 66.

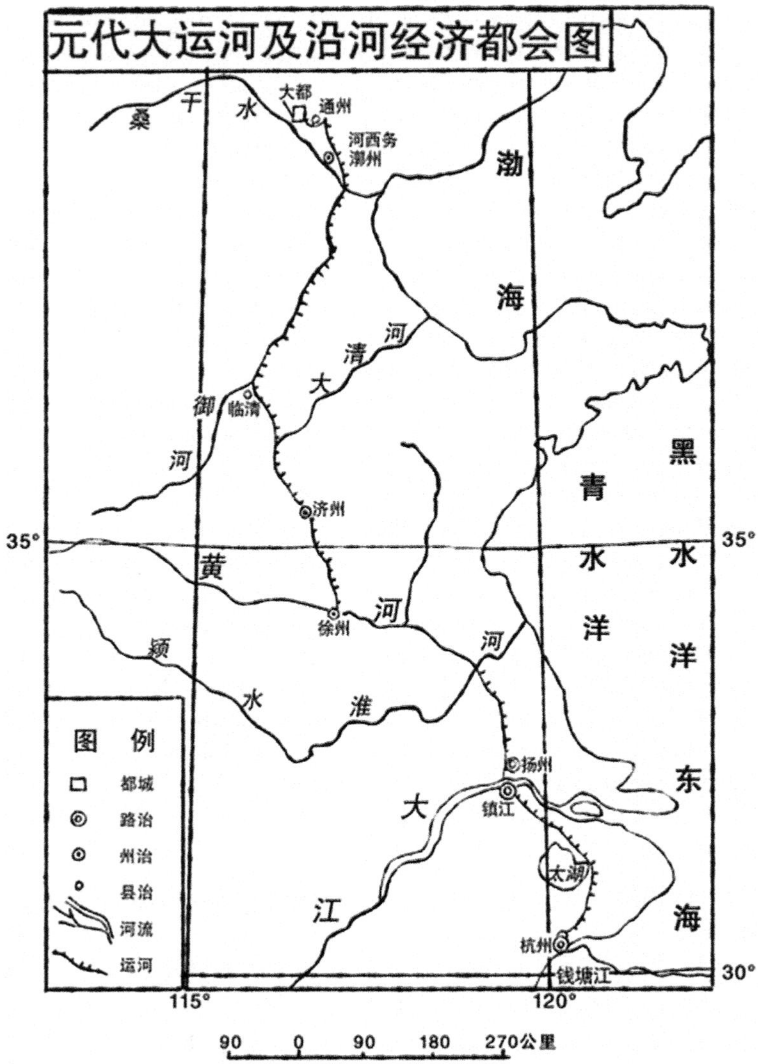

图例 - Legend
都城 Capital Cities
路治 Circuit Seats
州治 Prefecture Seats
县治 County Seats
河流 Rivers
运河 Canals

Fig. 7.5 Economic Hubs along the Yuan-era Grand Canal

appointed to direct river control works and was tasked with redirecting the Yellow River to its former course—from Huanglinggang (on the northern bank of the former Yellow River, southwest of present-day Cao County, Shandong) to Hazhikou in Guide Prefecture (present-day east of Shangqiu County, Henan).[61] At that time, merely sealing the breaches of the Yellow River was a daunting task in itself; there was little capacity to spare for dredging a canal that was already on the verge of losing its function.

Although the Yellow River was nominally restored to its old course, flooding disasters continued to occur intermittently. In the twenty-second year of Zhizheng (1362), the river breached its banks at Fan County[62] (southeast of present-day Fan County, Henan). In the following year (1363), it broke through again at Shouzhang County in Dongping[63] (now northwest of Liangshan County, Shandong). In 1365, further breaches occurred at Xiaoliukou within the counties of Xucheng, Dong'e, and Pingyin, also in Dongping, with the floodwaters reaching as far as the Qing River.[64] Xucheng corresponds to present-day Dongping County; Dong'e lies to the south of modern Dong'e County; Pingyin remains the same in name and location in present-day Shandong.

These breaches indicate that the Yellow River flowed through the administrative boundaries of these counties at the time. In the twenty-sixth year of Zhizheng (1366), the Yellow River shifted northward, wreaking havoc from the upper reaches in Dongming, Cao, and Pu to the lower reaches in Jining.[65] Dongming is located south of present-day Dongming County, Henan; the seat of Caozhou corresponds to modern Heze County, Shandong; the administrative center of Puzhou was situated south of present-day Juancheng County, Shandong; and the seat of Jining Circuit was located in present-day Juye County, Shandong. The term "shifted north of the river" is somewhat ambiguous, and the precise location of the shift remains unclear. However, records show that in the same year the Yellow River overflowed in the region of Jining Circuit.[66] As Fan County and Shouzhang County were located to the northwest and northeast of Jining respectively, it is reasonable to infer that Jining Circuit lay along the river's path and was thus subject to flooding. Therefore, it seems that no dramatic course change occurred in 1366, but the Yellow River's destructive overflows persisted nonetheless. It is said that the old courses of the Yellow River can be found approximately 35 *li* west of present-day Dingtao County, 5 *li* east of Heze County, 25 *li* west of Yuncheng County, and 60 *li* east of the former Puzhou (now

[61] Annals of Emperor Shun V; Treatise on canals and waterways III: Yellow River. In *History of Yuan*, vol. 41, 66.

[62] Annals of Emperor Shun IX. In *History of Yuan*, vol. 46.

[63] Treatise on five elements II. In *History of Yuan*, vol. 51.

[64] Ibid.

[65] Ibid.

[66] *Complete collection of illustrations and writings from the earliest to current times*, vol. 233, citing Shandong Gazetteer.

north of Juancheng County). These are considered remnants of the river's course during that period.[67]

The following listing of locations where the Yellow River flowed after the twenty-second year of Emperor Shun's Zhizheng era is intended to illustrate the canal routes for grain transport during the early Ming dynasty. In the first year of the Hongwu reign (1368) under the Ming Taizu, the Yellow River broke through at Shuanghekou in Caozhou, flowing southeast into Yutai County (southwest of present-day Yutai County).[68] Shuanghekou is located approximately 5 *li* east of present-day Heze County,[69] exactly corresponding to the area through which the Yellow River flowed at the end of the Yuan dynasty. Yutai County, situated adjacent to the canal, inevitably received floodwaters emanating from the Yellow River's breach, further exacerbating the silting problems that had already endangered the canal's navigability.

---

[67] Caozhou prefecture. In *Comprehensive gazetteer of Qing dynasty* (Jiaqing edition), vol. 181. This source records: "After the Jin and Yuan dynasties, the Yellow River flowed from the territory of present-day Yifeng County, passed through Cao County, then northeast past the southeast of Dingtao County and the southeast of the prefectural seat, then again northeast past the western edge of Yuncheng County. From there, it flowed eastward, sixty li past Puzhou (present-day north of Juancheng County), then turned north past the west of Fan County, and finally northeast into Yanggu County under Yanzhou Prefecture." Yifeng County corresponds to the northeastern part of present-day Lankao County in Henan; the prefectural seat of Caozhou is located in what is now Heze City, Shandong. However, this depiction of the Yellow River's post–Jin and Yuan course is exceedingly general. In reality, the river's path during the Jin and Yuan periods was highly variable and cannot be reduced to a single fixed channel. For instance, during the Jin dynasty, breaches were recorded in both Cao and Shan prefectures (*Gu Jin Tushu Jicheng*, vol. 232, citing the *Gazetteer of Yanzhou Prefecture*), and similar incidents occurred between Cao and Puzhou (*History of Jin* vol. 97, "Biography of Kang Yuanbi"), all of which suggest river routes that diverged considerably from those described in the *Comprehensive gazetteer of Qing dynasty*. In the Yuan dynasty, Jiyin County—the seat of Caozhou—experienced multiple breaches, as well as dike-repair initiatives. These include embankment work in 1325 (Taiding 2, *History of Yuan*, vol. 29, "Annals of Emperor Taiding"); a breach in 1330 (*History of Yuan*, vol. 65, "Treatise on Waterways II: Yellow River"); further breaches in 1334 (*Gu Jin Tushu Jicheng*, citing *Xu Wenxian Tongkao*), 1339 (*Gu Jin Tushu Jicheng*, citing the *Gazetteer of Huai'an Prefecture*), and 1345 (*History of Yuan*, vol. 41, "Annals of Shundi"). These incidents, all concentrated in Jiyin alone, offer only limited insight into the river's broader trajectory. Additional accounts further complicate the picture. *History of Yuan*, vol. 50, "Treatise on the Five Elements I," states that in 1324 (Taiding 1), flooding occurred in Chuqiu County (under Caozhou) and in Puyang County (under Kaizhou). Vol. 37 ("Annals of Emperor Wenzong") reports a dike collapse in Chuqiu in 1332, while vol. 41 ("Annals of Shundi") records a breach in Caozhou in 1344. Kaizhou was then based in what is today Puyang County, Henan; Caozhou's seat was in Jiyin; and Chuqiu County corresponds to the area southeast of modern-day Cao County. These three counties form a northwest-to-southeast axis, suggesting that the Yellow River at the time likely followed this orientation—flowing from the northwest toward the southeast. This trajectory contradicts the course outlined in the *Comprehensive gazetteer of Qing dynasty*. Thus, the *Comprehensive gazetteer of Qing dynasty*'s reference to "after the Jin and Yuan" appears to be a generalized statement, and in substance, it seems to reflect a specific Yellow River course from the late Yuan period, particularly following the *Zhizheng* reign era.

[68] Treatise on canals and waterways I: Yellow River, part 1. In *History of Ming*, vol. 83.

[69] Caozhou prefecture. In *Comprehensive gazetteer of Qing dynasty* (Jiaqing edition), vol. 181. At that time, the Yellow River breached at Shuanghekou in Caozhou, flowing southeast; the canal built by Xu Da started at Tachangkou, and did not extend as far as Shuanghekou.

In that year, General Xu Da led a northern military campaign, which made it impossible for the grain transport boats to travel upstream along the canal. Consequently, a new river channel was opened from Tachangkou, located forty *li* south of Yutai County, to Nailaopo, twenty *li* west of Jining, thereby connecting the Si River to the Ji Canal for transportation purposes.[70] Tachangkou was situated on the then Yellow River bank, so this newly excavated channel extended northward from there to Nailaopo, linking with the Si River.

Xu Da's river excavation in this region did not involve the reuse of the former canal route north of Jizhou, but rather the opening of a newly diverted channel of the Yellow River. The primary objective of this project was to ensure the transport of grain to provision the Liang and Jin regions.[71] As the Yellow River flowed southeast past Caozhou, it became possible to navigate upstream along its course, delivering grain to Caozhou, from where it was redistributed to various destinations within Liang and Jin. Since this stretch of the Yellow River had only recently taken shape, certain areas inevitably suffered from insufficient water depth, while others were prone to obstruction. In response, laborers were frequently dispatched to Juye, Cao, and Puyang to dredge and clear the riverbed in order to keep the grain transport route navigable.[72] The repeated impediments encountered vividly reflect the arduous nature of grain transport during this period.

Given these challenges, the grain supplies intended for the garrisons in northeastern regions had to be transported by sea, following the precedent established by the Yuan dynasty. As early as the first year of the Hongwu reign (1368), watermen were recruited to manage the maritime granaries at Laizhou (modern Ye Couty, Shandong) to provision Yongping Guard (present-day Lulong County, Hebei). Subsequently, sea transport was also organized to supply Beiping and Liaodong, becoming an established practice.[73]

However, the misfortunes of the canal did not end there. In the 24th year of the Hongwu reign (1391), the Yellow River breached its banks at Yuanwu (present-day Yuanyang County, Henan), causing a massive flood that flowed southeast past the east of Kaifeng, through Chen Prefecture (administered from present-day Huaiyang County, Henan), Xiangcheng, Taihe, Ying Prefecture (administered from present-day Fuyang, Anhui), Yingshang, and finally reached Shouzhou (administered from present-day Shou Couty, Anhui) at Zhengyang Town, where it emptied into the Huai River.[74] Another flood branch broke out from the old two river mouths near Caozhou and Yuncheng, flooding through Anshan in Dongping, severely damaging the canal.

Anshan in Dongping was a strategic hub of the Huitong River; however, after it was inundated by floodwaters, the river gradually ceased to flow.[75] The major

[70] Caozhou prefecture. In *Comprehensive gazetteer of Qing dynasty* (Jiaqing edition), vol. 181.

[71] Treatise on canals and waterways III: Grand Canal, part 1. In *History of Ming*, vol. 85.

[72] Huang Zhe. Preface to the poem "river's turbulence". In *Collected poems of successive dynasties I*, vol. 21.

[73] Treatise on food and finance III: grain transport. In *History of Ming*, vol. 79.

[74] Treatise on canals and waterways I: Yellow River, part 1. In *History of Ming*, vol. 83.

[75] Treatise on canals and waterways III: Grand Canal, part 1. In *History of Ming*, vol. 85.

southward shift of the Yellow River's main current also had a profound impact on the Grand Canal. During the Yuan dynasty, the Yellow River followed the course of the ancient Bian Canal from the Eastern Han period, flowing southeastward; thus, the section of the canal between Xuzhou and Huaiyin essentially relied on the Yellow River's waterway.

In the early Ming dynasty, Xu Da opened a breach at Tachangkou, allowing the canal to borrow the Yellow River's course and extend further north into the territory of Yutai County. While the Yellow River channel was far from ideal for transportation, its use saved the labor that would otherwise have been required to excavate a new canal. As the Yellow River's main flow shifted further south, its old course became blocked, and its former beds filled with sediment. Naturally, the canal could no longer make use of this section of the river.

Fortunately, with the Ming capital established in Jiangning (modern-day Nanjing), located in the prosperous southeast, there was no longer a pressing need to transport large quantities of grain northward. As a result, the already silted sections of the Grand Canal were simply left to continue their decline.

The Yanzhi River

Ming Taizu (the Hongwu Emperor) established the capital at Jiangning, an ancient city that had served as a capital since the Six Dynasties period and was located in the prosperous southeastern region of China. There, the existing transportation system—centered around the Yangtze River and the canal networks of Huainan and Jiangnan—formed a coherent and efficient waterway system. Such a network posed no major obstacles to grain transport. Nonetheless, Ming Taizu sought to further optimize this infrastructure by constructing a short canal west of Lishui County: the Yanzhi River, which connected Shijiu Lake to the Qinhuai River.[76]

Although modest in length, this canal greatly enhanced the efficiency of grain transportation. Previously, grain from the Liangzhe region was partly shipped along canals east of Tai Lake to Zhenjiang, then transferred to the Yangtze River for direct shipment to the capital. However, most of it was transported into Tai Lake, passed through Dongba into Shijiu Lake, and from there was hauled overland to the capital. Compared to this arduous overland transport, the new canal provided a much more convenient alternative.

Within Jiangning itself, the upper reaches of the Qinhuai River lay separated from Shijiu Lake by only a low ridge known as Yanzhi Ridge. During the Hongwu reign, this ridge was cut through to create the Yanzhi Canal. Once completed, grain from Liangzhe could be transported directly to the capital via a continuous water route.

---

[76] According to the *History of Ming (vol. 132): Biography of Li Xin*, "In the 26th year of Hongwu (1393), he supervised officials to open the Yanzhi River at Lishui, extending westward to the Yangtze River and eastward to the Liangzhe region, facilitating grain transport. The river was completed, and the people greatly benefited from it."

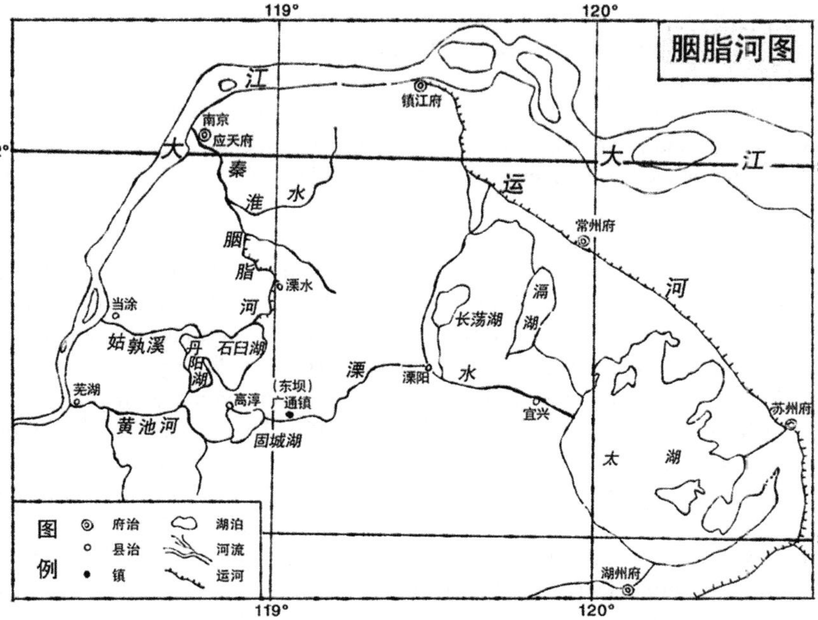

图例 - Legend
府治 Prefecture Seats
县治 County Seats
镇 Towns
湖泊 Lakes
河流 Rivers
运河 Canals

Fig. 7.6 The Yanzhi River

Even rice and millet from the upper reaches of the Yangtze River could be rerouted: either via Wuhu into the Huangchi River and eastward into Danyang Lake, or via Dangtu into Gushu Creek and onward to Shijiu Lake. From there, all routes could converge through the Yanzhi River and continue to Nanjing.

This canal thus opened a highly convenient transport corridor. However, after Emperor Chengzu relocated the capital to Beijing, the strategic importance of the Yanzhi River quickly waned (See Fig. 7.6).

The Canal System After the Capital Relocation to Beijing

When Emperor Chengzu relocated the capital to Beijing, the canal system underwent another major transformation. By this time, the Huitong River had already silted up, and although maritime transport was still in use, it remained limited in scale. As a result, a strategy was adopted that echoed the early Yuan precedent: transporting grain via the Yellow River upstream and then down along the Wei River (also known as the Imperial River) to Beijing.

However, due to the Yellow River's frequent breaches and flooding, this transport route was constantly shifting. By then, neither the waterway governed by Jia Lu nor the route opened at Tachangkou by Xu Da in the early Hongwu reign remained viable.

As previously noted, the Yellow River burst its banks at Yuanwu in the 24th year of the Hongwu reign (1391), diverting southeast from Kaifeng to flow into the Huai River at Shouzhou. This marked a major course change for the river. As a result, the former Jia Lu River channel silted up, and the Tachangkou transport route fell into disuse. Yet these changes were not the end of the river's volatility.

In the 30th year of Hongwu (1397), the Yellow River again breached its banks at Kaifeng. That winter, the course of the Cai River shifted near Chenzhou.[77] Although the Cai River had originally passed through Chenzhou, the reason for this "shift" warrants explanation. According to the *Ming History, "Treatise on Canals and Waterways,"* the river had previously flowed northeast from Kaifeng following a breach, but its lower reaches had since silted up, forcing the waters to breach again—this time heading south. Clearly, the Yellow River, in this episode, had usurped the Cai River's course, flowing southward beneath the walls of Chenzhou.

The Cai River had originally flowed southeast from Tongxu County, just south of Kaifeng. The Yellow River's new course in 1397 thus lay slightly west of its 1391 route and passed closer to Xihua County. Consequently, in the first year of Yongle (1403), the Sha River in Xihua overflowed and breached the dikes, allowing its waters to connect with the Yellow River.[78]

In the early years of the Yongle reign, northbound grain transport made use of this stretch of the Yellow River. The river's current was swift and perilous, with a particularly dangerous rapid located downstream of Chenzhou. To bypass this hazard, the route was redirected through the Sha River[79] to Yingqikou (modern-day southwest of Huaiyang County), where it merged back into the Yellow River. From there, vessels proceeded upstream to Baliyoushu (present-day southwest of Xinxiang

[77] Treatise on canals and waterways I: Yellow River, part 1. In *History of Ming*, vol. 83.

[78] *Veritable records of Emperor Taizong of Ming*, vol. 23, 9th month, 1st year of the Yongle reign.

[79] According to *The Comprehensive Gazetteer of the Qing Dynasty* (vol. 191, "Chenzhou" section) compiled during the Jiaqing reign: "The Sha River originates in the south of Huaining County (now Huaiyang County), receiving waters from the ancient Yellow River. It then flows eastward, taking in waters from the Cai Canal, continuing northeast of Xiangcheng County, further east past Xin'anjia in Shenqiu County, and again eastward receiving waters from the dried-up river (Kuhe), then flowing south of Jieshouji, before entering Taihe County in Yingzhou Prefecture, Anhui. It is in fact a distributary of the Cai and Ying Rivers." The gazetteer further notes that this Sha River is the same one excavated by Hu Zongyu of the Song dynasty based on the ancient Zhangbagou watercourse. According to modern maps (*Atlas of the People's Republic of China*, 1957 edition), a Sha River still exists between Xiangcheng County in Henan and Fuyang County in Anhui.

County), where the cargo was transferred to carts and transported overland to the Wei River.[80]

Although slightly shorter than the route used in the early Yuan dynasty, this overland segment remained arduous. Prior to the adoption of the Yellow–Wei River route, some officials had proposed excavating a short canal between the upper reaches of the Wei River and the Yellow River, reducing the overland distance to roughly a hundred paces and thereby easing the burdens of land transport. However, officials at the time only considered this idea briefly and never actually excavated such a canal.[81] Fortunately, within a few years, the Huitong River was successfully dredged, restoring the direct waterway and making the proposed Yellow-to-Wei canal unnecessary, so the matter was no longer raised.

The Dredging of the Huitong River

The Ming dynasty's project to dredge the Huitong River began in the ninth year of the Yongle reign (1411). Prior to this, numerous proposals had been made to restore the canal, yet none had materialized. The sole reason was that at the time, the imperial capital was still located in Jianning (present-day Nanjing), and the northern regions had not yet experienced significant concern over grain supply. After the capital was relocated to Beijing, however, despite the dual use of maritime and overland transportation, the sea route remained treacherous and the land route arduous, neither capable of resolving the underlying logistical difficulties. Under such circumstances, people naturally recalled the transportation routes of the Yuan dynasty.

In the ninth year of Yongle, Pan Shuzheng, the vice-prefect of Jining Prefecture, formally proposed the restoration of the Huitong River's old course. Emperor Yongle then commissioned Minister of Works Song Li and others to take charge of the project. A total of 385 *li* (approximately 192 km) was dredged from Jining to Linqing.

Song Li identified the Wen River as the main water source of the Huitong River, and therefore emphasized the necessity of managing the Wen River effectively to ensure a continuous and unimpeded flow. Among the regions traversed by the canal, the terrain of Nanwang was the most elevated, located approximately 35 *li* southwest of present-day Wenshang County. If the Wen River could be diverted to Nanwang and then split to flow both northward and southward, the entire course of the canal could be kept navigable.

---

80 *Veritable records of Emperor Taizong of Ming*, vol. 21, 9th month, 6th year of the Yongle reign.

81 According to *Veritable records of Emperor Taizong of Ming*, vol. 18, 3rd month, 1st year of the Yongle reign, "In the third month of the first year of the Yongle reign (1403), a military officer from Shenyang, Tang Shun, proposed: 'The Wei River flows into the sea at Zhigu. To its south lies the Yellow River, and the overland distance is merely over fifty li. If a canal could be opened from the Wei River, reducing the distance to only a hundred paces from the Yellow River, and granaries were built there to store the grain transported from the south, it would be convenient for both public and private purposes to transfer the supplies to the Wei River for shipment.' The emperor ordered the court ministers to deliberate the matter, but the plan was never carried out."

During the Yuan dynasty, the Wen River was also utilized in the construction of the Huitong River. At that time, a dam known as Gaocheng Dam was constructed northeast of Ningyang County to divert the Wen River into the Guang River. The Guang River flowed toward Jining, where it joined the Si River—a confluence situated far south of Nanwang. Therefore, if the Wen River was to be redirected to Nanwang, it could no longer be allowed to pass through the old channel of the Guang River.

Song Li adopted the advice of a local elder from Wenshang named Bai Ying, and once again constructed a dam at Gaocheng northeast of Ningyang. The newly constructed Gaocheng Dam differed entirely from that of the Yuan dynasty. Whereas the Yuan-era dam had been designed to divert the Wen River into the Guang River, the Ming-era dam was intended to block the Wen River from entering the Guang River, thereby forcing it to continue flowing westward.

It can be inferred that the Yuan dynasty's Gaocheng Dam had been built directly across the Wen River's course, whereas the Ming dynasty's dam was constructed at the divergence point where the Wen River split toward the Guang. By forcing the Wen River to flow westward from Gaocheng Dam, its original course was effectively restored.

The original course of the river turned north at Dongping Prefecture (present-day Dongping County) and flowed into the Daqing River. Following the advice of the elder Bai Ying from Wenshang, Song Li ordered the construction of the Daicun Dam on the Wen River, located sixty li east of Dongping Prefecture. This dam effectively blocked the river's original path, redirecting its flow southwest toward Nanwang, where the water was stored in Shushan and Mata Lakes to support canal transportation. As Nanwang occupied the highest elevation along the route, the Wen River bifurcated at this point: sixty percent of its volume flowed northward to Linqing and entered the Wei River, while the remaining forty percent flowed southward through Jining and into the Si River. In the same year, the old course of the Yellow River was dredged, restoring it to its path during the late Yuan and early Ming periods. Consequently, the segment from Tachangkou to Xuzhou once again became navigable via the Yellow River (See Fig. 7.7).[82]

The Distinctive Features of Lock-Based Canal Transport (Zhacao)

In the Ming dynasty, the Huitong River—a vital segment of the Grand Canal—was also known as *zhacao*, meaning "lock-based transport." This term arose from the uneven terrain along the canal, which necessitated the construction of numerous locks to ensure navigable water levels and the smooth passage of boats. The section stretching northward from Nanwang, the point of water bifurcation, to Linqing

[82] Annals of Chengzu II; Treatise on canals and waterways III: Grand Canal I; Biography of Song Li; Biography of Jin Chun. In *History of Ming*, vols. 6, 85, 153, 157.

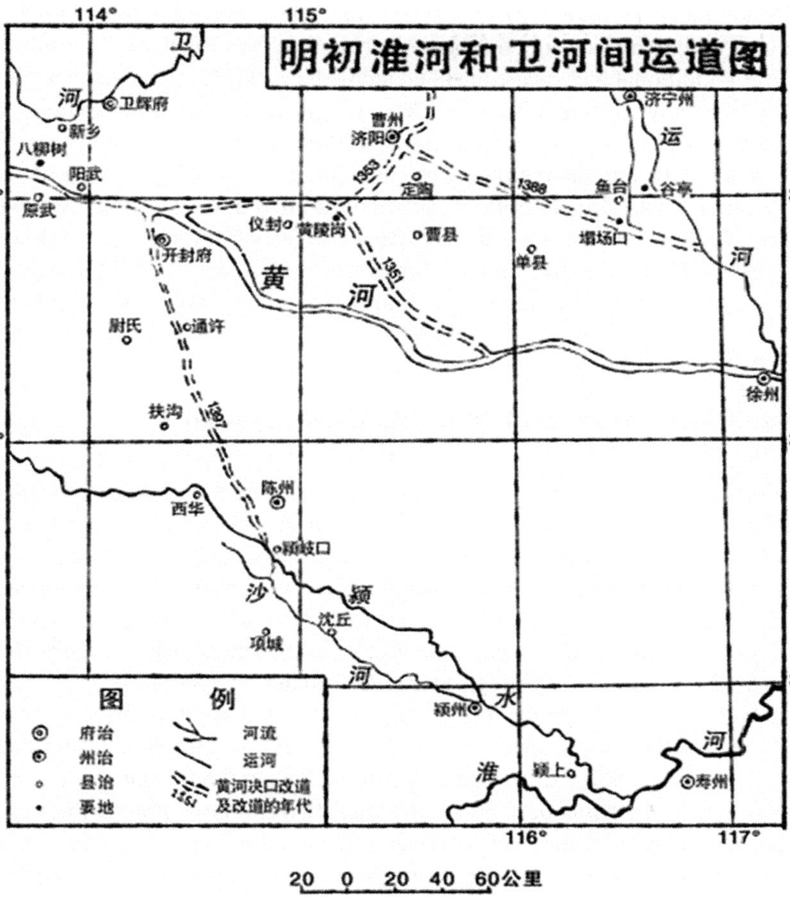

图例 - Legend	河流 Rivers
府治 Prefecture Seats	运河 Canals
州治 Commandery Seats	黄河决口改道及改道的年代 Historical Records of Yellow River Flood Breaches and Subsequent Course Alterations
县治 County Seats	
要地 Strategic Locations	

Fig. 7.7 Transportation routes between the Huai and Wei rivers in early Ming

spanned approximately 300 *li* and descended a total of 90 chi in elevation; the southward stretch from Nanwang to Zhenkou (north of present-day Xuzhou in Jiangsu Province) extended 390 li, with a total elevation drop of 116 *chi*. These significant changes in elevation made it essential to build locks at multiple points to regulate water levels and facilitate transport.

During the Ming period, 21 locks were constructed along the Nanwang–Linqing segment, and 54 more were built between Nanwang and Zhenkou. In a subsequent phase, 11 additional locks were installed along the 260-*li* Jia River section.[83] Owing to the unusually high number of locks throughout this area, this section of the Grand Canal came to be referred to collectively as *zhacao*. In fact, the practice of installing locks in this region of the canal predates the Ming dynasty. A considerable number had already been constructed during the Yuan period, as previously discussed. However, the scale and frequency of lock-building expanded substantially under Ming administration, to the extent that *zhacao* evolved into a distinct and specialized term, denoting this lock-intensive model of canal transport.

Chen Xuan's Management of the Canal System

With the successful dredging of the Huitong Canal, both maritime grain transport and the use of the Yellow and Wei Rivers for grain shipment were suspended. At this time, Chen Xuan, Earl of Pingjiang, was placed in overall charge of all matters related to grain transport. Wherever the transport routes were obstructed, he took steps to rectify the problems. Although he did not undertake large-scale excavation of new canals, his restoration and improvement of existing routes proved highly effective and established numerous lasting precedents for later generations.

In the thirteenth year of the Yongle reign (1415), finding that various dams at the confluence of the Grand Canal and the Huai River north of Huai'an hindered the passage of grain barges, Chen Xuan opened a channel from Guanjia Lake west of Huai'an city to Yachenkou, allowing the Huai River to flow directly through. This freed grain boats from the need to cross over the dams. The channel he reopened was originally a transport canal excavated by Qiao Weiyue during the Song dynasty, which had since become clogged with silt over time.

Chen Xuan continued to implement improvements. Near Xuzhou, the treacherous currents at Lüliang posed a significant navigational hazard, so he excavated a forty-*li* canal to the west, enabling water storage and smooth passage. At Gaoyou Lake, where strong winds and waves endangered boats, he constructed a long embankment along the lakeshore and opened another forty-*li* canal beneath it, thereby separating the canal from the lake and sparing the grain boats the risk of traversing its turbulent waters. He also excavated the Baita River in Taizhou, creating an alternative route for the canal to reach the Yangtze River, thus eliminating the previous reliance on the single outlet at Yangzhou.

Chen Xuan's work in regulating the canal system extended far beyond these few cases. However, since the other improvements pertain to technical aspects of

---

[83] Treatise on canals and waterways III: Grand Canal, part 1. In *History of Ming*, vol. 85.

hydraulic engineering, they need not be enumerated here.[84] With Song Li's restoration of the Huitong Canal and Chen Xuan's systematic regulation of the waterway, grain fleets from the southeast could thereafter proceed northward smoothly and without hindrance.

The Dredging of the Tonghui River

The early Ming court made great efforts to restore the canal system, yet one circumstance remains puzzling. During the Yuan, grain fleets from the southeast had sailed straight to the capital. Under the Ming, however, the waterway ended at Tongzhou and, from Tongzhou westward, the grain had to be carted into Beijing. The old Tonghui Canal was simply left to silt up and was not brought back into service. Though it was cleared several times between the Chenghua and Zhengde reigns, the work was slight and the results meagre.

The crux of the problem lay in the water supply. During the Yuan Dynasty, the Western Hills' Baifu Springs and other sources had been channeled to Jishuitan. Both the feeder channels and the lake lay outside the Imperial City, so grain barges could enter the capital. When Beijing was enlarged early in the Ming, Jishuitan was cut off to the north of the Forbidden City, leaving the barges no place to moor. Moreover, the water released from Jishuitan had to flow beneath the Forbidden City, then south through the Jinshui River. For this reason, the sluices along this route could only be worked by palace personnel. Since outsiders had no control over their opening and closing, the old course could not be restored.

Furthermore, the Yuan diversion of the Baifu Springs had run westward against a natural gradient. To avoid a reverse flow, the Ming engineers would have had to excavate a channel across the imperial tombs, which officials feared would disturb the geomantic veins of the site. The alternative source, Yimu Spring, had to pass through the ravine of Baiyangkou, where two torrents intersected, ensuring that it would be difficult to divert. Because of these challenges, reviving the Tonghui Canal proved to be an extremely hard project. It was not until the sixth year of the Jiajing reign (1527) that the censor Wu Zhong carried out a thorough renovation, once again allowing boats to reach the capital.[85] Wu Zhong's efforts, however, were confined to repairing locks and dams. The fundamental question of water supply remained unsolved. Moreover, after the enlargement of Beijing, the city's water consumption had risen sharply, so the canal's volume could scarcely be increased[86] and the Tonghui never regained its former importance (See Fig. 7.8).

---

[84] Annals of Chengzu III. In *History of Ming*, vol. 7; Treatise on rivers and canals III—Grand Canal, part 1. In *History of Ming*, vol. 85; Biography of Chen Xuan. In *History of Ming*, vol. 153.

[85] Treatise on rivers and canals VI—Grand Canal, part 2. In *History of Ming*, vol. 86.

[86] Hou Renzhi. The water-supply problem in the course of Beijing's urban development. In *Theory and Practice of Historical Geography*. Shanghai: Shanghai People's Publishing House, 1979.

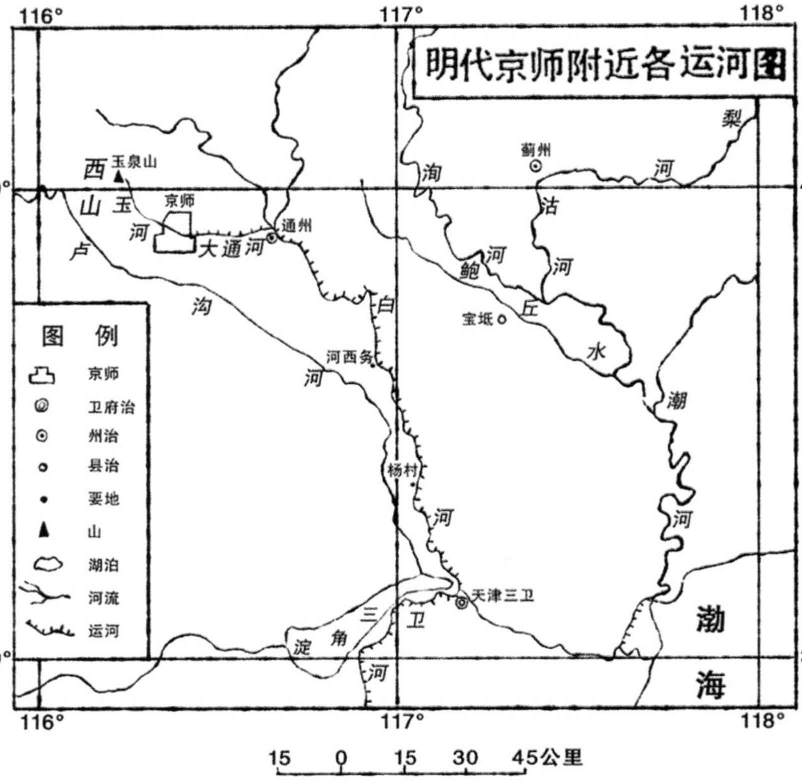

图例 - Legend
京师 Capital
卫府治 Military-Administrative Seats
州治 Prefecture Seats
县治 County Seats
要地 Strategic Locations
山 Mountains
湖泊 Lakes
河流 Rivers
运河 Canals

Fig. 7.8 Canals near the Capital in the Ming dynasty

Names of the Canal Sections in the Ming Dynasty

The Ming dynasty's north–south Grand Canal, in its points of origin and termination and in its general course, was much the same as that of the Yuan, yet the names given to its various sections differed. The nomenclature of the Ming canal was quite complex: the same stretch of water was called by different names in different places, according to the local river or to the function it served.

Although the waterway bore many different names along its course, it remained a single, continuous route. Tracing its path from north to south: the stretch that gathers runoff from the Western Hills of Changping, flows eastward to Tongzhou, and then enters the Bai River is known as the Datong River—this corresponds to the Tonghui Canal of the Yuan dynasty, renamed for the Datong Bridge it passes beneath. From Tongzhou southward to Zhigu, where it merges with the Wei River and flows out to sea, it is called the Bai River. Northward from Linqing to its confluence with the Bai River and onward to the sea, it is again referred to as the Wei River. Southward from Linqing, through Jining to Xuzhou, the canal draws water from the Wen, Guang, and Si Rivers. From Xuzhou south to the Huai River, it follows the course of the Yellow River. Between the Huai and the Yangtze, the route traverses the lakes of the Huaiyang region. South of the Yangtze, it becomes the transport artery serving Songjiang, Suzhou, and Zhejiang. The section from Huaiyang to the area south of Jingkou is commonly referred to as the Zhuan Canal. The segment from Zhangjiawan—fifteen li south of Tongzhou in the capital region—to Tianjin is also called the Tongji River. From Tianjin into Shandong, it is known as the North River; the stretch from the Feng and Pei Rivers to the Yellow River is often termed the Middle River; and the portion between the Yangtze and the Huai is typically called the South River.

In addition, there are the further designations of White Transport, Wei Transport, Sluice Transport, River Transport, Lake Transport, Yangtze Transport and Zhe Transport. White Transport refers to the Bai River; Wei Transport to the Wei River below Linqing; River Transport to the Yellow River south of Xuzhou; Lake Transport to the Huaiyang lakes between the Yangtze and the Huai; Yangtze Transport denotes the canal's entry into the river and its division into upper and lower routes (the term is not restricted to the canal itself); Zhe Transport signifies the waterways in Zhejiang. Sluice Transport, on the other hand, applies to the canal within Shandong, a section so thickly provided with locks that it takes its name from them. When the whole waterway is spoken of collectively it is most often called the Grain Canal; the name "Grand Canal" was comparatively little used—just as the Great Wall was commonly termed the Frontier Wall in Ming times. Naturally, popular usage sometimes preserved the old Yuan appellations, in which case no single rule can be laid down.[87]

Ming-Era Esteem for the Grand Canal

The Ming dynasty's reliance on and emphasis upon the Grand Canal far exceeded that of the Yuan. Under the Yuan, grain transport was primarily conducted via maritime routes, with the canal playing only a secondary role. Even at the height of Yuan maritime logistics, annual shipments did not exceed four million *dan*. In contrast, the situation in the Ming was entirely reversed. When the Yongle Emperor first

[87] Treatise on rivers and canals III—Grand Canal, I. In *History of Ming*, vol. 85.

relocated the capital to Beijing, maritime shipping was still employed alongside the canal, but the two handled roughly equal volumes. Following the successful dredging of the Huitong Canal, however, both maritime shipping and the Yellow–Wei River route were abandoned. Thereafter, the entirety of north–south grain transport came to rely exclusively on the Grand Canal.

In the early Ming period, no fixed grain transport quota had been established—the amount varied according to need, and no precise figures are available. Only in the eighth year of the Chenghua reign (1472) was a yearly quota of four million *dan* officially set, a figure that became standard thereafter.[88] This amount exceeded the peak maritime shipments of the Yuan and clearly attests to the Ming dynasty's more rigorous supervision and maintenance of the canal system.

The state placed the utmost responsibility on officials tasked with overseeing grain transport, given that the empire's stability depended upon the smooth operation of this system. During the Yongle reign, Chen Xuan was appointed as General Commander of Grain Transport, charged with administration of the entire canal. Early in the Yingzong reign, the canal was divided at Jining into northern and southern sections, each managed by a vice minister. In the seventh year of the Chenghua reign (1471), the canal was further subdivided into three sections: south of Pei County, north of Dezhou, and the Shandong section in between, each entrusted to a provincial surveillance commissioner. A Director-General of Grain Transport—commonly referred to as the "Director-General of River Works" (*Zonghe*)—was appointed to oversee the entire system. The prestige of this position reflects the Ming court's deep dependence on the canal and its pivotal role in maintaining state power.[89]

Disasters on the Huitong Canal

Running north–south, the Grand Canal happens to intersect the Yellow River—a configuration quite unlike that of the Tang and Song periods. The Yellow River is prone to burst its banks; the canal is prone to silt up. Either problem alone would be hard to manage, but when a river that easily breaks loose begins to affect a waterway that easily silts, the difficulties multiply. The Yuan dynasty lasted barely eighty years, yet even by its middle period the seeds of crisis had already sprouted: every Yellow River breach promptly choked the canal. During the Hongwu reign the Huitong Canal was completely blocked. Emperor Yongle's dredging of the canal restored the route and seemed an immediate blessing, yet at the same time it planted a hidden scourge that forever bound the Huitong Canal to the Yellow River and ultimately altered the destiny of both.

How did this come about? Because the Yuan canal's flow was too small to move large grain cargoes, the court turned to the sea route. The Ming, by contrast, entrusted

[88] Treatise on food and finance III: Grain transport. In *History of Ming*, vol. 79.

[89] Treatise on official posts. In *History of Ming*, vol. 73; Treatise on rivers and canals III: Grand Canal, part 1. In *History of Ming*, vol. 85.

the whole grain transport to the canal. While Ming management was meticulous and raised the canal's carrying capacity, the dynasty also resorted to a clever expedient that became the very root of later calamity. The "insufficient flow" actually referred to the Huitong Canal, whose only sources were the Wen River and a few minor streams, all dividing their waters between north and south. Such a scant supply could never float large craft. To augment the Huitong, the Ming engineers therefore decided to borrow water from the Yellow River itself. Between Kaifeng and Dong'e there had once been the old Yih River, and this disused channel was now employed to lead Yellow River water into the canal so that grain barges could proceed without hindrance. This was sanctioned in the fifth year of the Xuande reign (1430). In that year Grand Canal commander Chen Xuan submitted a memorial stating: "From Linqing to Anshan the channel is shallow in spring and summer, and boats can barely pass. Southwest of Zhangqiu (sixty *li* southwest of present-day Dong'e County, Shandong) the old Yi River once connected with Bian. The court formerly sent officials to maintain it; when the water was low, a weir at Jinlongkou (twenty *li* southwest of present-day Fengqiu County, Henan) raised the Yellow River and sent the flow down to Linqing to assist grain transport. In recent years, however, the post has been left vacant, the works have fallen into disrepair, and transport has become difficult. I beg that the former practice be restored."

The request was approved, and Yellow River water was again used to supplement the canal.[90] In the tenth year of Xuande (1435) the supreme commander of grain transport, together with the regional grand coordinators and vice-ministers, reviewed matters affecting both troops and civilians; their report for the following year (1436) noted: "At Shawen (thirty *li* northeast of present-day Shouzhang County, Shandong, and twelve *li* north of Zhangqiu) the Zhangqiu Canal formerly drew a branch of the Yellow River through Jinlongkou. Over the years sand has accumulated, the river has been blocked, and the canal has almost dried up; dredging is urgently required."[91] Thus the Yellow River water of the old Yi River continued to flow into the canal. Yet fortune and misfortune are intertwined: the convenience gained at the moment became, in time, the very cause of later disaster.

The Yellow River's breach that destroyed the Huitong Canal began in the thirteenth year of the Zhengtong reign (1448). That year the river burst its banks near Xingyang, turned northeast across Cao and Pu, followed the old Yih River, and plunged into Zhangqiu, where it tore open Shawen. The transport route was catastrophically ruined. At that moment saving the canal was judged even more urgent than plugging the breach itself: if the dike remained open, the worst would be the inundation of some people, livestock, fields, and houses; if the canal were blocked, the capital itself would be shaken to its foundations. Court and country alike focused their attention on the stretch around Zhangqiu and Shawen, leaving the breach at Xingyang to one side. The canal near Zhangqiu was dredged only to silt up again,

[90] *Veritable records of Emperor Xuanzong of Ming* (Ming xuanzong shilu), vol. 11, 10th month, 5th year of the Xuande reign.

[91] *Veritable records of Emperor Xuanzong of Ming*, vol. 9, 9th month, 10th year of the Xuande reign.

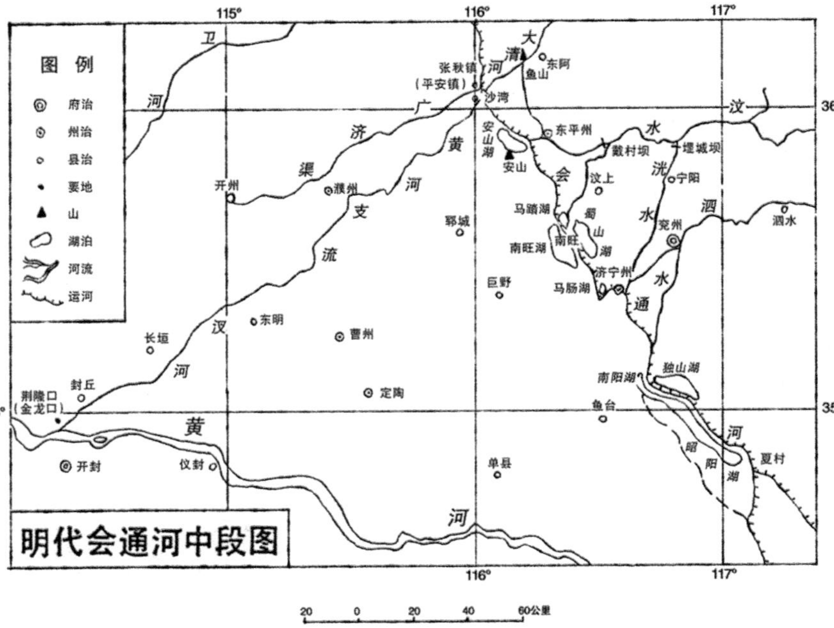

图例 - Legend
府治 Departmental Seats
州治 Prefecture Seats
县治 County Seats
要地 Strategic Locations
山 Mountains
湖泊 Lakes
河流 Rivers
运河 Canals

Fig. 7.9 Huitong Canal's middle section in the Ming dynasty

dammed only to break anew; not until the seventh year of the Hongzhi reign (1494) was the task finally completed—almost fifty years in all. Once the route was restored, in commemoration of the achievement Zhangqiu was renamed Anping, a gesture that speaks for the rejoicing of the age (See Fig. 7.9).[92]

[92] According to the *History of Ming*, vol. 85, in 1448, the dike at Shawan burst; it was soon sealed again. In 1452, it breached once more, and the following year the gap was closed; shortly afterwards it broke again, and was not finally completed until 1455. In 1489, Zhangqiu ruptured anew, and the breach was not closed until 1494.

The Perilous Route South of Xuzhou

While the Huitong Canal suffered grievously from the Yellow River's assaults, the stretch below Xuzhou—forced to borrow the river's own channel—was harassed even more often. The slightest upstream breach was felt here at once; should the river alter its course, the transport line was instantly severed. In 1411, when the Huitong Canal was reopened, the Yellow River had likewise been coaxed back to its early-Ming bed. Five years later it again burst its banks near Kaifeng, thrust southeast along the Guo River and reached the Huai at Huaiyuan, leaving only a feeble thread still running east past Xuzhou. The broad southern branch was now called the Great Yellow River, the meagre eastern thread the Small Yellow River. Each year the latter weakened, until in 1431 the censor Bai Gui petitioned to dredge the channel below Jinlongkou and lead the river's water to Xuzhou to keep the grain route open—clear evidence of how inadequate the flow had become.

In 1448, the river burst again at Xingyang and surged east toward Zhangqiu. One branch spread southeast from Yuanyang through Xiangfu (modern Kaifeng County), Fugou, Tongxu, Weichuan, Weishi, Linying, Yancheng, Chenzhou, Shangshui, Xihua, Xiangcheng, and Taikang, eventually entering the Huai River. Naturally, the old riverbed near Xuzhou silted up again. Though this breach was soon closed, similar disasters continued to occur.

During the Hongzhi reign, to prevent the Huitong Canal from further Yellow River interference, the long-discussed closure of the old Yi River course was finally implemented. Thereafter, the full force of the Yellow River swept eastward past Xuzhou—rapid and violent—so that the canal south of the city could scarcely enjoy a single day of safe navigation.[93]

The Diversion of the Canal Near Xuzhou

To spare the canal south of Xuzhou from the Yellow River's repeated ravages, its channel was rerouted several times to separate it from the river. The first proposals for a new cut were made in the Jiajing reign by Censor-in-Chief Hu Shining and the Director-General of River Works, Sheng Yingqi. In 1526–1527 the Yellow River burst its then banks at Cao and Shan counties (in present-day Shandong). The whole area north of Pei was flooded, the overflow surged east across the grain canal (i.e. the Grand Canal) into Zhaoyang Lake, and silt piled up until the route was almost impassable.[94] Hu Shining then recommended cutting a new channel east of Zhaoyang Lake through the districts of Teng, Pei, Yutai and Zou. This new route would be barely 100 li in total, running south to Liucheng (southeast of present-day Pei County, Jiangsu) and north to the Sha River (east of present-day Zou County, Shandong, flowing into

[93] Treatise on rivers and canals I: Yellow River, part I. In *History of Ming*, vol. 83.

[94] Disasters of Yellow River breaches. In *A comprehensive account of the Ming dynasty's major events*, vol. 34.

Zhaoyang Lake). Its western bank was to be strongly embanked to keep the lake waters (and any flood waters) from spilling into the canal.[95]

Sheng Yingqi offered a parallel proposal that envisaged an entirely new, 140 *li* river east of Zhaoyang Lake, starting at Wangjiakou on the lake's eastern shore, skirting Xia Village (in present-day Weishan County, Shandong) and rejoining the old channel at Liucheng in Pei County. Had it been completed, waterborne transit would have run along the eastern side of Zhaoyang Lake, sheltered by the lake itself from the Yellow River's incursions. However, while the work was under way, complaints reached the throne and the project was abruptly abandoned.[96]

Thirty years later, in the forty-fourth year of the Jiajing reign (1565), the Yellow River again burst its banks at Pei County, silting up two hundred li of the transport channel in either direction.[97] The following year the Minister Zhu Heng was ordered to take charge of both river and canal, reopening the abandoned trace and resuming excavation work. To hasten the completion of the project, Zhu himself conducted ground inspections. His enquiries revealed that the new channel begun under Sheng Yingqi—running south of Nanyang eastward to Xia Village and then southeast to Liucheng—was still discernible.[98] He therefore proposed that the section between Nanyang and Liucheng be reopened.[99] After some debate, the new waterway was finished and the grain route enjoyed a period of relative stability.[100] Once the new cut was in service, the old course west of Nanyang Lake and Zhaoyang Lake fell out of use and gradually silted up.

The second rerouting of the canal near Xuzhou involved an eastward shift to the Jia River. The Jia River has two sources: one originates in the southern valleys of Mount Fei in Fei County, Shandong; the other arises from Mount Jun in Yi County, also in Shandong. These are referred to as the eastern and western branches of the Jia River. The two streams converge near Jiakou Town—located thirty *li* northwest of Pi County in Jiangsu—where they meet the Yi River and flow onward to Suqian County before entering the Yellow River.

The plan to divert water from the Si River into the Jia River, merging with the Yi River to facilitate canal transport, was originally proposed by Weng Dali. In the third year of the Longqing reign of Emperor Muzong (1569 CE), a breach occurred in the Yellow River at Pei County, causing severe sedimentation that blocked the canal route. As a result, grain transport vessels were stranded near Pizhou (modern-day Pi County in northern Jiangsu). At the time, Weng Dali was serving as Director-General of River Works (*Zonghe*), and he advocated for opening a channel through the Jia River to restore canal navigation.

---

95 Treatise on rivers and canals I: Yellow River, part I. In *History of Ming*, vol. 83.

96 Treatise on rivers and canals III: Grand Canal, part I. In *History of Ming*, vol. 85; Biography of Sheng Yingqi. In *History of Ming*, vol. 223.

97 Treatise on rivers and canals I: Yellow River, part I. In *History of Ming*, vol. 83.

98 Biography of Zhu Heng. In *Collected writings of the Ming dynasty* (Ming wen), vol. 223.

99 Treatise on rivers and canals III: Grand Canal, part I. In *History of Ming*, vol. 85.

100 Biography of Pan Jixun. In *History of Ming*, vol. 223.

Weng's proposed route followed the base of Mount Zifang, passed Mount Liang and Mount Jing, and entered Dibanggou, proceeding straight toward Majiaqiao. This route spanned approximately eighty *li* and could serve as the course for a new canal.[101] Mount Jing lies about forty *li* north of present-day Xuzhou and is connected to Mount Liang.[102] Although the exact location of Mount Zifang is unclear, it could not have been far from Mount Jing. Weng's proposal was primarily motivated by a desire to avoid the dangers posed by Qingou and the Zhuo River.[103] However, just as he put forward his plan, the waters of the Yellow River receded, and grain boats resumed passage, leading to the plan's abandonment. The following year, another breach occurred at Pizhou, prompting Weng to renew his proposal to open the Jia River and redirect the Yellow River's course. Yet again, the plan was not implemented.[104] In the third year of the Wanli reign (1575 CE), the then Director-General of River Works, Fu Xizhi, also petitioned to excavate the Jia River to avoid the hazards of the Yellow River, but this too came to nothing.[105] It was not until the twenty-second year of the Wanli reign (1594 CE) that Director-General Shu Yinglong finally undertook practical action. At Hanzhuang (located sixty li southwest of Yicheng, in present-day Zaozhuang, Shandong), he excavated a canal to discharge lake waters—this was the so-called "Hanzhuang New River."[106]

The Hanzhuang New River, over forty li in length, lies south of Xingyi Ridge and was designed to divert water from the Peng River into the Jia River.[107] This excavation marked the first successful linkage of the upper Jia River with the Grand Canal, though the entire watercourse was not yet navigable. In the twenty-sixth year of the Wanli reign (1598 CE), Liu Dongxing, serving as Director-General of River Works, also endeavored to complete the connection but passed away from illness before his plans could be realized.[108]

Several years later, in the thirty-second year of the Wanli reign (1604 CE), Director-General Li Hualong succeeded in fully opening the new canal. Water was drawn from Lijiakou in Xiazhen, Pei County, merged with the Peng River, passed through the Hanzhuang Lake Mouth, and then combined with the waters of the Jia and Yi Rivers, finally exiting at Zhikou near Pizhou. The total length of this

---

[101] *Veritable records of Emperor Muzong of Ming*, vol. 37, *guiyou* day, 9th month, 3rd year of the Longqing reign.

[102] Xuzhou prefecture. In *Comprehensive gazetteer of Qing dynasty (Jiaqing edition)*, vol. 100.

[103] Treatise on rivers and canals III: Grand Canal, part I. In *History of Ming*, vol. 85.

[104] Ibid.

[105] *Veritable records of Emperor Shenzong of Ming*, vol. 35, *wuxu* day, 2nd month, 3rd year of the Wanli reign; and vol. 36, *dingsi* day, 3rd month.

[106] *Veritable records of Emperor Shenzong of Ming*, vol. 277, *wuxu* day, 9th month, 22nd year of the Wanli reign.

[107] *Golden mirror of water management*, vol. 127, citing *Complete study of the Shandong River works*.

[108] Treatise on rivers and canals III: Grand Canal, part I. In *History of Ming*, vol. 85; Treatise on rivers and canals III: Jia River. In *History of Ming*, vol. 87; Biography of Liu Dongxing. In *History of Ming*, vol. 223.

new canal exceeded 260 *li*.[109] The completion of this canal significantly distanced the main transport route from the Yellow River near Xuzhou, thereby avoiding the frequent disruptions caused by Yellow River breaches and siltation. At this point, roughly one-third of grain transport boats still passed beneath the walls of Xuzhou, but the majority had been rerouted to the Jia River. Gradually, the canal route through Xuzhou fell into disuse.

From Weng Dali's initial proposal to Li Hualong's completion of the project, more than thirty years had passed. This undertaking can be regarded as one of the major hydraulic engineering achievements of the Ming Dynasty.

Efforts to improve the Jia River waterway continued thereafter. In the fifth year of the Tianqi reign under Emperor Xizong (1625 CE), Assistant Commissioner for Grain Transport Zhu Guosheng initiated the excavation of a new canal beginning at Majiakou, on the eastern bank of Zhikou in Pizhou, and ending at the mouth of Luoma Lake in Suqian. This "Tongji New River," measuring fifty-seven *li* in length, connected with the Jia River upstream and the Yellow River downstream, bypassing the hazardous areas of Liukou, Mo'erzhuang, and Zhikou.

The following year, Director-General Li Congxin extended this work by excavating an additional ten *li* of canal near Luoma Lake, linking it via Chengou to the Yellow River.[110] Chengou lies two *li* west of Suqian County[111] and serves as the connection between Luoma Lake and the Yellow River. At that time, grain transport boats would navigate the Tongji New River, enter Luoma Lake, and then proceed to the Yellow River through Chengou. Following these two rounds of canal construction and rerouting, this section of the waterway remained largely free of major hazards for the remainder of the Ming dynasty (See Fig. 7.10).

Lakes Along the Sluice Canal

The purpose of excavating a canal is to compensate for the shortcomings of natural waterways. During construction, every effort is made to take advantage of existing natural channels, including nearby lakes, as this can greatly reduce the amount of excavation required. In the Spring and Autumn period, the states of Wu and Chu adopted precisely this approach when digging their canals. Even King Fuchai's Han Canal, built during Wu's time, was deliberately routed to the northeast to skirt Sheyang Lake.

Lakes adjacent to canals could serve a dual purpose: they functioned both as navigable waterways and as reservoirs for regulating water levels in the canals. Lian Lake, situated alongside the Jiangnan Canal, is a prime example of the latter. When

[109] *Veritable records of Emperor Shenzong of Ming* (Ming Shenzong shilu), vol. 292, yichou day, 1st month, 32nd year of the Wanli reign; Treatise on rivers and canals V: Jia River. In *History of Ming*, vol. 87; Biography of Li Hualong. In *History of Ming*, vol. 228.

[110] Treatise on rivers and canals III: Grand Canal, part I. In *History of Ming*, vol. 85.

[111] Pizhou. In *Essentials of historical geography*, vol. 22.

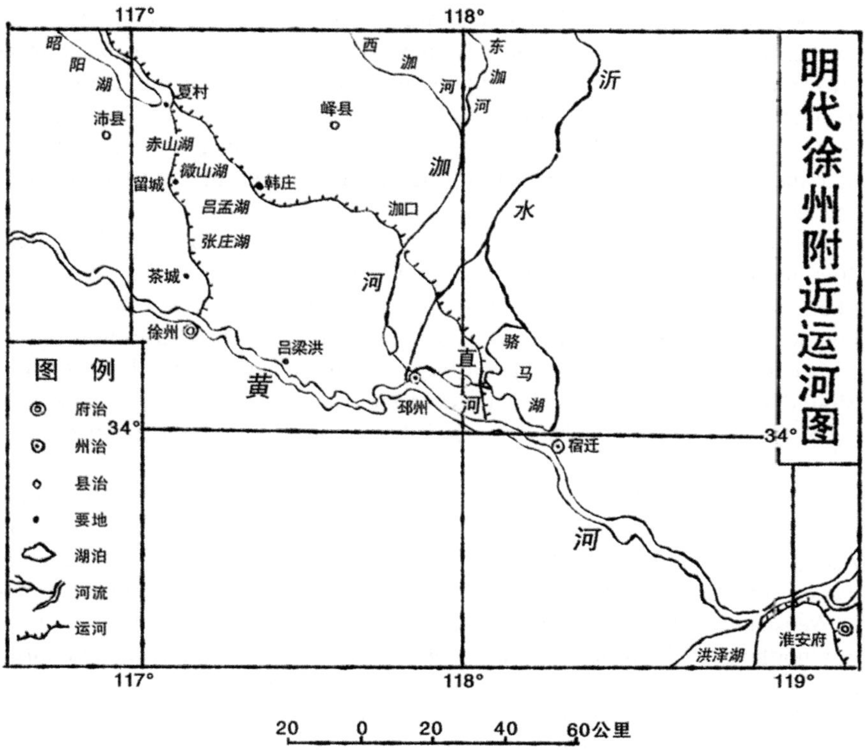

图例	- Legend	县治	County Seats
府治	Departmental Seats	要地	Strategic Locations
州治	Prefecture Seats	湖泊	Lakes
河流	Rivers	运河	Canals

Fig. 7.10 The Ming-era canal system near Xuzhou

the Jizhou River was excavated during the Yuan dynasty, it relied on the Wen and Si Rivers to supplement its transport function. Because the middle and lower reaches of the Si River were already serving as a major transport channel, diverting part of its waters into the Jizhou River inevitably reduced the volume flowing southward. To regulate this flow, numerous sluices and dams were built at intervals along the banks, which later earned this stretch of the canal the name *Sluice Canal*.

During the Yuan dynasty, most tribute grain was still transported by sea, so the issue of water supply in this section was not yet acute. By the Ming dynasty, however, when the canal system alone bore the entire burden of grain transport, this problem naturally became a major concern.

The Si River originally flowed southwest from Qufu and Yanzhou, then turned southeast at Luqiao. When the Yuan dynasty excavated the Jizhou Canal, it branched off from Luqiao to the north, passing through Jining and Dongping before joining the Daqing River. To the west of the Jizhou River lay the former Juye Marsh, known at the time as Liangshan Lake. East of Juye Marsh was another body of water called Maodudian. The Huangong Canal, opened by Huan Wen of the Eastern Jin, ran between these two lakes.[112]

Owing to repeated Yellow River floods, most of Liangshan Lake had already dried up by the Yuan dynasty, and Maodudian had likely disappeared even earlier. The Jizhou River was actually cut through the former bed of Maodudian. Below Luqiao, the Si River gathered the waters of numerous small tributaries flowing from the southeast of the old Juye Marsh basin and from the western slopes of Meng Mountain. This basin may once have contained several lakes, though probably not many. Given such topography, it was possible in this area to restore former lakes or to form new ones.

During the Yongle reign, when Song Li dredged the Huitong River, several lakes were already in existence: Zhaoyang, Nanyang, Mata, Shushan, and Anshan. Zhaoyang Lake lay northeast of present-day Pei County in Jiangsu; Nanyang Lake was east of present-day Yutai County in Shandong; Mata Lake was north of Nanwang in Wenshang County; Shushan Lake was southeast of Nanwang; and Anshan Lake was northeast of Liangshan County. Today, Mata Lake has dried up, and Anshan Lake has expanded into Dongping Lake. At the time, Song Li built sluices beside these lakes, which were called "water treasuries" (shuigui): when canal waters rose, excess water was stored in the lakes; when the canal ran low, water from the lakes was released back into it.[113]

In fact, Nanwang, Machang, and Shushan Lakes also existed at that time. Nanwang Lake was originally the collecting basin of Liangshan Lake. During the Yongle dredging, an embankment was constructed across it, dividing Nanwang into two sections. The western portion remained Nanwang Lake, while the eastern portion was further divided into two lakes: Shushan Lake to the south and Mata Lake to the north. The canal passed between Nanwang Lake and Shushan Lake.[114]

Among these lakes, Nanwang, Anshan, Machang, and Zhaoyang were most closely connected to canal transport and were collectively referred to as the "Four Water Treasuries." Shushan Lake, which also had a sluice gate, was likewise considered a water treasury.

Northeast of Nanyang Lake lies Dushan Lake. Together with the two lakes to their south—Zhaoyang and Weishan—Nanyang and Dushan are collectively referred to as the *Southern Four Lakes*. Nanyang and Zhaoyang Lakes already existed by the early Ming dynasty. Dushan Lake was originally just another name for Nanyang Lake, but later it became the proper name for the portion on the canal's western (left) bank, while Weishan Lake was formed somewhat later.

---

112 Commentary on the Ji River. In *Water classic*.

113 Treatise on rivers and canals III: Grand Canal I. In *History of Ming*, vol. 85.

114 Yanzhou prefecture. In *Comprehensive gazetteer of Qing dynasty (Jiaqing edition)*, vol. 166.

In the seventh year of the Jiajing reign (1528), when Sheng Yingqi proposed excavating a new canal on the western side of Zhaoyang Lake, Hu Shining observed that on the eastern shore of the lake, within the territories of Teng, Pei, Yutai, and Zou Counties, there was a place called Dushan. He noted that the proposed new waterway would pass through this area.[115] The canal advocated by Sheng Yingqi and Hu Shining would skirt the left banks of Nanyang and Zhaoyang Lakes and then rejoin the old grain-transport route at Liucheng. This new route closely followed the shoreline of Nanyang Lake, with Dushan Lake lying even farther to the left. Its configuration was reminiscent of the passage between Nanwang and Shushan Lakes, as the canal would run directly between the two lakes. The formation of Dushan Lake was clearly the result of the new canal's excavation, which allowed the waters of the Zou and Si Rivers to converge.

Weishan Lake came into being even later. By the time the new canal proposed by Sheng Yingqi and Hu Shining was finally opened by Zhu Heng in the forty-fourth year of the Jiajing reign (1565), Weishan Lake already existed. Besides Weishan, there were also Chishan, Zhangzhuang, Hanzhuang, and Lümeng Lakes. Hanzhuang lay farthest to the east, Zhangzhuang to its west, with Weishan and Chishan farther still to the west; collectively, these lakes were known as Lümeng Lake.[116] All were formed by Yellow River floodwaters that had failed to subside. Initially, though they stretched in a chain more than eighty *li* long, each was relatively small in size.

During the Longqing (1567–1572) and Wanli (1573–1619) reigns, the Yellow River breached its banks repeatedly, causing these lakes to merge into a single vast expanse. Gradually, Weishan became the largest of the Southern Four Lakes, its waters expanding ever farther south. Part of the southern section of the new canal excavated in the Jiajing period was eventually submerged beneath the lake, and even Liucheng itself became a lakeside town. To prevent disruption to the grain-transport route, the Jia River canal was subsequently excavated, as discussed earlier. The completion of this canal not only spared shipping from the dangerous waves of Weishan Lake but also allowed vessels to bypass the perilous Lüliang rapids near Xuzhou.[117]

From this it becomes clear that the lakes along this section of the canal possessed a character distinct from those found elsewhere. In most regions, lakes associated with canals were naturally formed. Here, however, the Wen and Si Rivers were too small to bear the burden of grain transport, so water had to be impounded to supplement navigation. The formation of Weishan Lake, which resulted from Yellow River floods, could certainly not have been foreseen. Thus, the situation must be attributed to human miscalculation rather than solely to natural changes.

Because the canal lacked sufficient water, it was compelled to rely on these lakes for regulation. Yet the winds and waves on the lakes made it impossible for grain barges to sail across them consistently. Unable to either fully bypass or fully traverse

[115] Waterways and ditches VI: transport canals. In *Essentials of historical geography*, vol. 129.

[116] Yanzhou prefecture. In *Comprehensive gazetteer of Qing dynasty (Jiaqing edition)*, vol. 166.

[117] Treatise on rivers and canals III: Grand Canal, part I; Treatise on rivers and canals V: Jia River. In *History of Ming*, vol. 85, 87.

the lakes, the canal was forced to run along their margins. A similar arrangement existed alongside the lakes of Baoying, Gaoyou, and Shaobo between the Yangtze and Huai Rivers, but in those places the routing was designed merely to avoid the lakes' storms and had little to do with the issue of water supply.[118]

Reorganizing the Nan River

During the Ming dynasty, the Yellow River not only harassed the Huitong River and the canal near Xuzhou, but also disrupted the Nan River—the Ming name for the section of canal between the Yangtze and the Huai Rivers. Two locations were most seriously affected: the canal mouth near Huai'an and the Gaojia Weir on the eastern shore of Hongze Lake.

Before the Yellow River shifted south, the canal's water level had always been higher than that of the Huai River. To prevent excessive water loss from the canal, dams and weirs were built at the entrance to regulate the flow. When the Yellow River shifted southward, its new outlet coincided with the canal mouth. The river's swift, high waters frequently surged in, silting up the entrance.

The Gaojia Weir, traditionally attributed to Chen Deng at the end of the Eastern Han, had long safeguarded Gaoyou, Baoying, and other districts east of Hongze Lake. Once the Yellow River captured the Huai River, the Huai could no longer discharge freely and backed up into the lake. Fortunately, the surrounding land was low-lying, so the lake could continue to expand. As it did so, the Gaojia Weir had to be repeatedly raised and strengthened to prevent failure; a breach would have submerged the entire region east of the lake—and destroyed the canal along with it. The canal between the Yangtze and the Huai had once been the safest and most trouble-free waterway. Even occasional silting could be easily remedied with dredging. By the Ming, however, it had become a point of constant concern.

Work on the Huai'an canal mouth began early: during the Yongle reign, Chen Xuan excavated a new channel. In 1582, Transport Commissioner Ling Yunyi, judging the entrance still dangerous, opened the Yongji River, forty-five *li* long. Starting south of Qingjiangpu at Yaowan, it passed Longjiang Sluice and Yangjiajian and debouched at Wujiadun into the Huai River, offering a slightly easier route. As for the Gaojia Weir, the Ming court spared no effort in heightening and reinforcing the great embankment that held back Hongze Lake. To reduce pressure on the dike, sluices were routinely opened to release water eastward into the chain of lakes along the canal. These lakes inevitably expanded, creating problems of embankment maintenance, wind-shelter, and flood control. While flood-prevention received unrelenting attention, the issue of wind exposure proved particularly challenging.

When the canal was first excavated, it had made full use of existing lakes to save on construction effort. But as these lakes received ever more overflow from Gaojia

[118] Treatise on rivers and canals III: Grand Canal I; Biography of Liu Dongxing. In *History of Ming*, vol. 85, 223.

Weir, their surfaces enlarged, and larger surfaces naturally meant stronger winds. To ensure the safety of grain junks, the only solution was to abandon the original lake channels and cut new side canals. The first of these was the Yue River, cut beside Gaoyou Lake by Chen Xuan. In 1489, the Vice Minister of Revenue Bai Ang excavated the Kangji River, forty *li* long, east of Pishe Lake at Gaoyou. This method of cutting bypass channels was called *fuhe* ("auxiliary river").

In 1585, the Transport Censor Li Shida opened the Yue River at Baoying to avoid the perils of Siguang Lake; this new cut, 1,700 *zhang* long, was named the Hongji River. In 1600, Director-General Liu Dongxing opened another Yue River, eighteen *li* long, at Shaobo and a further bypass at Jieshou, 1800 *zhang* long. With these works completed, the South Canal between the Yangtze and the Huai was finally freed from dependence on the lakes and became a direct north–south waterway.[119]

The original mouth of the South Canal was at Yangzhou. After Chen Xuan cut the Baita River at Taizhou, the entrance shifted eastward to link with junks crossing the Yangtze via the Mengdu River near Changzhou. Still farther east, near Taixing, the Beixin River was aligned with the Desheng New River from Changzhou. Junks could enter the Beixin River, pass the Taizhou Dam, and reach the canal at Yangzi Bay—a route even more convenient than the Baita River. Later, the Guabu mouth near Yangzhou was dredged as well. To the end of the dynasty, these three eastern and western routes were used alternately to great advantage (See Fig. 7.11).

Restoring the Jiaolai Canal—And Its Failure

Coastal transport in the Ming dynasty was used only in the earliest years of the dynasty. Although sea transport spared the state the immense labor of canal excavation, it carried the constant risk of storms and shipwrecks. The Grand Canal was frequently disrupted by the Yellow River, and interruptions in grain shipments were common. Whenever the canal was rendered impassable, proposals to revive sea transport would resurface; yet, as soon as the canal was repaired and reopened, these ideas would quickly fade.

Ming sea transport differed from that of the Yuan dynasty. Yuan grain junks, upon leaving the Yangtze River, would cross the open sea directly to Chengshan Head and then on to Zhigu. The Ming, however, followed the Yuan's earliest approach: ships hugged the coastline as they sailed northward. For this reason, proposals to re-excavate the Jiaolai Canal—a link between the inland canal network and the coast—were frequently raised. In essence, reviving sea transport and re-cutting the Jiaolai Canal were two aspects of the same scheme. Since sea transport was never permanently reinstated, the project to reopen the Jiaolai Canal never reached fruition.

The first to advocate reopening the Jiaolai Canal was Wang Tan of Changyi County, who submitted his proposal in the sixth year of the Zhengtong reign (1441). His

[119] Treatise on rivers and canals III: Grand Canal I; Treatise on rivers and canals IV: Grand Canal II; Biography of Li Shida. In *History of Ming*, vol. 85, 86, 220.

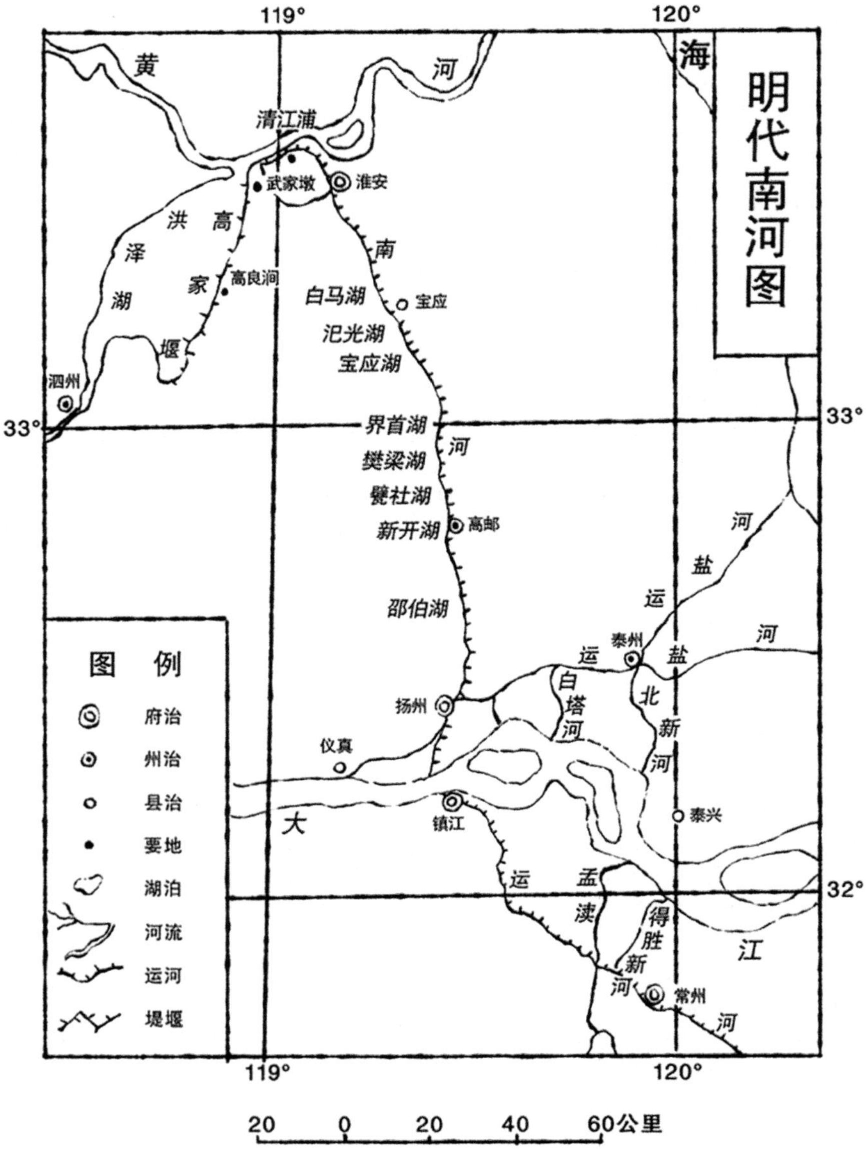

图例 - Legend

府治 Departmental Seats

州治 Prefecture Seats

县治 County Seats

要地 Strategic Locations

湖泊 Lakes

河流 Rivers

运河 Canals

堤堰 Weirs

Fig. 7.11 The Ming-era Nan River

suggestion met with no response. Nearly a century later, in the seventeenth year of the Jiajing reign (1538), the Shandong Grand Coordinator, Hu Zongzuan, finally initiated dredging work.

The following year, Vice Commissioner Wang Xian undertook the most difficult section of the work—the Mahao Cut, west of Xue Island near Jiao County—but before it was completed he was transferred, and the entire project was once again shelved. After Wang's departure, proposals to reopen the canal continued to surface. In the thirty-first year of the Jiajing reign (1552), Li Yongjing put forward the idea, followed by Li Guihe in the fifth year of the Longqing reign (1571), and Liu Yingjie and Xu Shi in the third year of the Wanli reign (1575). Yet none of these initiatives was carried out. It was not until the sixteenth year of the Chongzhen reign (1643) that Minister Ni Yuanlu revived the scheme, this time proposing that grain be transported by small boats from the mouth of the Jiao River to the watershed, then hauled overland for forty *li* to the Lai River, from which it could be floated down to the sea. But that same year the Ming dynasty collapsed, and Ni's plan never had the chance to be realized.[120]

The Ji Canal

In addition to the plans to restore the Jiaolai Canal, the Ming dynasty also excavated the Ji Canal. Jizhou (modern Ji County, Hebei), located on the northeastern frontier, was a vital military stronghold during the Ming. Much of the grain and fodder required by its garrison had long been transported by sea. After the successful dredging of the Huitong River led to the abandonment of maritime transport, supplies for Jizhou were shipped out from Zhigu, up the Chao River, and directly to the city walls of Jizhou. Although this sea passage was not especially long, it was nonetheless subject to the dangers of stormy weather.

To address this, in the second year of the Tianshun reign (1458), a canal was cut from the Gu River, running northeast along the coast to link with the Chao River. The canal was only about forty *li* in length and not considered a major engineering feat; however, because Jizhou was a strategic border outpost and its provisions could not be interrupted even briefly, the canal was frequently dredged to prevent silting.[121] In later years, when the canal fell into disuse, the Chao River flowing from Jizhou

[120] Treatise on rivers and canals V: Jiaolai Canal. In *History of Ming*, vol. 87.

[121] *Veritable Records of Emperor Yingzong of Ming*, vol. 298, 12th month, 2nd year of the Tianshun reign (1458): "Previously, Min Gong, a centurion of the Dahe Guard in the Metropolitan Area, submitted a memorial: 'Each year, Nanjing and the various guards of the Metropolitan Area ship 300,000 *shi* of grain to the granaries of Jizhou and other frontier garrisons. This requires 350 vessels and 6300 soldiers, who must cross more than seventy *li* of open sea. The violent winds and waves often delay the fleet for weeks. Even when a favourable wind arises, it may change suddenly mid-passage, and men, ships, and grain are often lost. Such transport is truly unsuitable. I have observed that the newly opened Gu River lies directly north of Jizhou, opposite the Shui-tao mouth of the Gu River, a mere forty-odd *li* apart. The waterway is deep enough for boats, but a ten-*li* stretch is blocked. If this section were excavated and grain shipments redirected through this route, the perils

began to be referred to as the Ji Canal. In reality, though, the Chao River was always a natural waterway and not an artificially excavated canal.

The Grand Canal Under the Qing: Inheriting the Ming Legacy

From the Ming to the Qing, dynasties changed, but the scale and structure of state-building remained essentially unchanged. The Qing capital was the same as the Ming—Beijing—so the existing infrastructure was simply inherited. With the seat of government unmoved, the transportation network naturally followed the old pattern.

Over more than two centuries of Qing rule, the great north–south artery was never fundamentally altered; only maintenance and reinforcement works were carried out. On rare occasions a realignment was made, but such adjustments were confined to stretches of no more than a hundred or so *li* and were matters of hydraulic engineering, not changes to the overall system.

The Influence of the Yellow River and the Huai River on the Canal

Qing maintenance efforts essentially continued the unfinished work of the late Ming. In the final decades of the Ming, the largest canal works had been concentrated at the junction of the Yellow River, the Huai River, and the canal itself. The primary issue was the unpredictable behavior of the Yellow River; to control the Yellow River was to control the Huai River, and the ultimate goal was to ensure smooth and uninterrupted canal transport.

In the early Qing, embankments along this critical stretch were continuously reinforced and the channels dredged for precisely this purpose. The first modification of the canal's watercourse occurred in the fifteenth year of the Shunzhi reign (1658), when River Conservator Zhu Zhixi excavated a new channel 250 *zhang* south of Shipai Mouth near Suqian. This new channel linked the Jia River to the Yellow River and, though only a little more than one *li* long, greatly facilitated grain transport after the old Dongkou entrance had silted up.

In the seventeenth year of the Kangxi reign (1678), Jin Fu, then Director-General of River Works, cut a crescent-shaped channel 840 *zhang* long west of Qingshuitan south of Gaoyou in order to close a breach there; this was named the Yong'an New River. Two years later, he excavated the 40-*li* Zao River through Luoma Lake, connecting the Jia River upstream with the Yellow River downstream, replacing Zhu Zhixi's Shipaikou cut of the Shunzhi era. A year later, the mouth of the Zao River itself silted

of the open sea could be avoided. Though it would require labor at one time, it would bring benefits for a thousand years.' … The proposal was approved."

up, prompting Jin to open a 3,000-*zhang* branch to the east, creating the now-famous Zhangzhuang Mouth, where grain junks could rejoin the Jia River.

In the twenty-fifth year of Kangxi (1686), Jin undertook the largest Qing canal project—the Zhong River. Starting from the Zhangzhuang Mouth, this new channel crossed Lake Luoma, passed Suqian and Taoyuan (modern Siyang County, Jiangsu), and reached Zhongjiazhuang, west of Qinghe County (modern Qingjiang, Jiangsu). This cut eliminated 180 *li* of hazardous Yellow River navigation, dramatically changing the canal's alignment. The Ming ideal of completely freeing the canal from the Yellow River had never been fully achieved: the so-called "Yellow River grain route" had remained a dangerous passage. With the completion of the Zhong River, the canal used the Yellow River for only a few *li*, and the term "Yellow River grain route" virtually disappeared.

The Zhong River itself underwent three subsequent modifications. In 1699, Yu Chenglong, then Director-General of River Works, judged the southern bank too close to the Yellow River and difficult to maintain, so he excavated a New Zhong River, 60 *li* long, from Shengjiadaokou in Taoyuan to Qinghe. The next year, his successor Zhang Penghe found Shengjiadaokou's course too tortuous and the new channel too shallow for boats, so he adopted a compromise: the upper reach of the old Zhong River and the lower reach of the New Zhong River were re-excavated and linked into a single channel. Finally, in the forty-second year of Kangxi (1703), the Zhangzhuang Mouth was moved to Yangzhuang. From that point onward, the canal and the Yellow River were fully separated (See Fig. 7.12).[122]

The canal north of the Huai River was at last freed from dependence on the Yellow River, eliminating many complications. Yet this solved only part of the problem. At the confluence of the Yellow River, the Huai River, and the Grand Canal, conditions remained extremely precarious. The swift current of the Yellow River entered the Huai only a short distance below the canal mouth at Huai'an, making it easy for Yellow River water to surge into the canal and choke it with silt.

The standard solution adopted by river conservancy experts of the time, apart from building dams at the canal mouth to block the Yellow River, was to use the clear waters of the Huai to scour away the accumulating silt. This required the Huai to stand higher than the Yellow River and to flow with sufficient force. To achieve this, the Gaojia Weir east of Hongze Lake was raised to elevate the lake's water level, and the outlet narrowed so that the outflow would be more forceful. Even these measures, however, sometimes proved ineffective: when the level of Hongze Lake dropped, Yellow River water would backflow into the lake.

In 1785, disaster struck when drought in the upper Huai basin caused Hongze Lake to shrink to a shallow expanse, just as the Yellow River swelled dramatically. The Yellow River poured back into the lake, and by the time the lake was full, the Huai mouth was completely silted up. With the mouth of the Huai blocked, the canal near Gaoyou and Baoying was severely affected, leaving officials with no choice but to borrow Yellow River water to sustain transport. Borrowing the Yellow River

122 Treatise on rivers and canals II: Grand Canal. In *Draft history of Qing*, vol. 108.

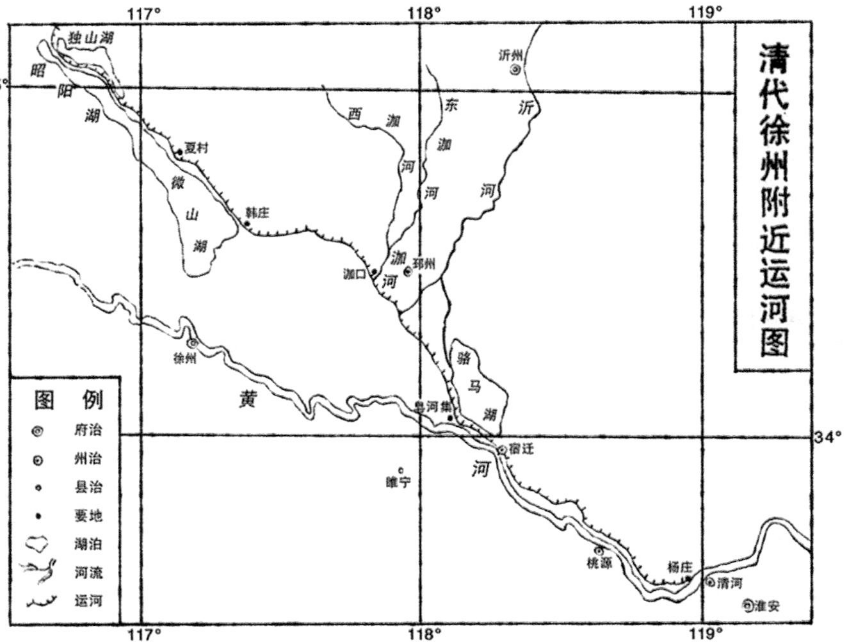

图例 - Legend | 要地 Strategic Locations
府治 Departmental Seats | 湖泊 Lakes
州治 Prefecture Seats | 河流 Rivers
县治 County Seats | 运河 Canals

Fig. 7.12 The Canal near Xuzhou during the Qing dynasty

was an inherently risky measure, yet it was adopted as a temporary expedient—"like drinking poison to quench thirst."

Even after the Huai River returned to its former volume, its mouth remained choked with silt. Only a thin and feeble stream continued eastward, utterly unable to resist the overwhelming force of the Yellow River. The excess water trapped in Hongze Lake, unable to flow east, turned southward into the Yangtze River. Many attempts were made to reverse this dependence on the Yellow River, but the Huai's volume was simply too small to scour away the silt and sustain the canal. For a time, some even attempted to divert the Huai's eastward flow into the Yellow River to avoid relying on borrowed Yellow River water, but by then the crisis had become so acute that such measures could not reverse the decline.

By the early Daoguang reign, the canal near Gaoyou and Baoying had silted up dramatically: its bed was raised by more than one *zhang* (over 3 m); stretches that had once been 30–40 *zhang* wide had narrowed to barely ten *zhang* or as little as five or six *zhang*; depths that had once measured one *zhang* and five or six *chi* now held barely three or four *chi* (about one meter) of water—and in some places, less than five inches. Under such conditions, grain transport became exceedingly difficult.

When no effective remedy could be found, attention turned once again to maritime transport, with the proposal that sea carriage should once more replace riverine transport. This was nothing less than a death sentence for the canal. Although some officials fought desperately to preserve the Grand Canal's role in grain transport, by this time river transport was already in its death throes and beyond salvation.[123]

Problems on the Lower Reaches of the Wei River

In the Qing dynasty, the canal north of Linqing frequently encountered problems, though these were far less severe than those at the confluence of the Yellow, Huai, and Grand Canal. The section of canal north of Linqing was in fact the lower reaches of the Wei River, whose water source derived from the Baimen Springs at Sumen Mountain in Hui County, Henan. As early as the Ming dynasty, concerns had already arisen that the Wei River's flow was too small to sustain navigation. In the thirteenth year of the Zhengtong reign of Emperor Yingzong (1448), efforts were made to divert water from the Zhang River into the Wei to supplement its inadequate volume.[124]

During the Longqing and Wanli reigns (1567–1620), the Zhang shifted its course northward into the Fuyang River, severing the branch that had previously flowed into the Wei. In the thirty-sixth year of the Kangxi reign (1697), however, the Zhang unexpectedly sent a branch southward once again, re-entering the Wei River at Guantao. By the forty-seventh year (1708), the entire flow of the Zhang River was entering the Wei at Guantao. Whereas water scarcity had once been a pressing issue, now there was too much water.

To prevent the canal below Linqing from bursting its banks, auxiliary channels were cut at several points, including Shaomaying in Dezhou (modern De County, Shandong) and Sinü Temple in En County. By diverting the water in this way, its force was diminished. Ultimately, this section of the canal posed relatively few difficulties: whenever water levels became excessive, opening several diversion channels was sufficient to ensure safety.[125]

---

[123] Ibid.

[124] Treatise on rivers and canals V: Zhang River. In *History of Ming*, vol. 87.

[125] Treatise on rivers and canals II: Grand Canal. In *Draft history of Qing*, vol. 108.

The Northward Shift of the Yellow River and the Obstruction of the Canal

The Grand Canal, spanning the empire from south to north, was inextricably bound to the fate of the Yellow River, and generations of hydraulic experts expended enormous effort to manage their relationship. Ultimately, when the Yellow River altered its course, the destiny of the canal was sealed. This change occurred during the reign of Emperor Wenzong. In the fifth year of the Xianfeng reign (1855), the Yellow River burst its banks at Tongwaxiang (northwest of present-day Lankao County, Henan), flowed past Zhangqiu, and seized the old course of the Daqing River to reach the sea. This was a major hydrological event for the Yellow River and a calamity for the Grand Canal. Previously, the canal's most perilous points were concentrated near the confluence of the Yellow and Huai rivers, while the stretches of the canal within Shandong Province had generally remained tranquil and safe. After the breach, the hazardous section shifted from the vicinity of Huai'an to that of Zhangqiu.[126]

The Huitong River section of the canal in Shandong had originally diverted the waters of the Wen River northward from Wenshang, eventually joining the Wei River at Linqing. Following the Yellow River's breach and its eastward flow past Zhangqiu into the sea, the Wen River was naturally subsumed by the Yellow River. As a result, the canal from Zhangqiu north to Linqing was left without any available water source. At the time, some proposed imitating the water-diversion scheme at Nanwang, suggesting that the Wei River be redirected southward near Linqing to alleviate the canal's shortage of water in this section. Yet the Wei's flow was insufficient to serve both regions. In the end, officials once again resorted to diverting water from the Yellow River—the familiar but dangerous expedient of "*borrowing the Yellow River to sustain transport*" (*jie Huang ji yun*). This stopgap measure, akin to "drinking poison to quench thirst," was repeatedly implemented. As had been the case near Huai'an, the canal's waterway became increasingly obstructed, and the same fate awaited the canal north of Zhangqiu.

In the eleventh year of the Tongzhi reign (1872), Emperor Muzong consulted Li Hongzhang, Governor-General of Zhili, for strategies to rehabilitate this section of the canal. Li bluntly responded that there was no other viable solution for the canal: "When extremity is reached, change is required, and with change comes adaptation. At present, steamships are abundant along the coast; it would be better to abandon the canal altogether and switch to maritime transport." Muzong accepted this proposal, transferring the majority of tribute grain shipments to sea routes while retaining only a small portion of river transport in order to preserve a vestige of canal operations.

The idea of reviving maritime transport had been floated as early as the Daoguang period. In the early years of the Xianfeng reign, with the rise of the Taiping Heavenly Kingdom and the disruption of inland routes, all southeastern tribute grain had already been shipped by sea without notable difficulty. Moreover, the use of steamships distinguished Qing maritime transport from the precarious sea voyages

---

[126] Treatise on rivers and canals I: Yellow River. In *Draft history of Qing*, vol. 107.

of the Yuan. Thus, Li Hongzhang was able to capitalize on the circumstances, and maritime transport rapidly supplanted river transport.[127]

Though borrowing water from the Yellow River was always a problematic solution, it had at least served as an emergency expedient. Yet even this became untenable when, on occasion, there was no water in the Yellow River to borrow. In the thirteenth year of the Guangxu reign (1887), the Yellow River burst its banks at Zhengzhou, and its main current shifted southward, entering the Huai River via the Jialu and Ying Rivers, cutting off the main channel and placing the canal under severe strain. With the Yellow River's principal course severed, the canal could no longer rely on its waters. Fortunately, the breach was repaired the following year, allowing canal navigation to resume—albeit only barely.

By the twenty-seventh year of Guangxu (1901), Li Hongzhang memorialized the throne, requesting that all north–south tribute grain shipments be commuted to monetary payments. As a result, both maritime and river transport were terminated. Thereafter, the government ceased to rely on the canal for grain transport and largely ignored whether it remained navigable. At this point, the historic function of the Grand Canal had effectively come to an end.[128]

The Huiji River

In addition to maintaining and reinforcing the Grand Canal that spanned the empire from north to south, the Qing dynasty also devoted considerable effort to other transportation waterways. Most of these efforts, however, focused on dredging existing channels, such as the Jialu River in Henan and the Ling Canal in Guangxi. New channels of major significance were relatively few, with the most notable project being the excavation of the Huiji River in the sixth year of the Qianlong reign (1741).

The Huiji River diverted water from the Jialu River at Zhongmou in Henan and carried it along the course of the Sha River to merge with the Qianya River, ultimately reaching the Guo River in Jiangnan. The total length of the Huiji River exceeded 65,000 *zhang* (approximately 216 km).[129] The "Sha River" mentioned here was most likely the former course of the Jinshui River from the Song dynasty, as no traces of the Jinshui River's old course remained by the Qing period.[130] Aside from this old course, there was no other waterway that could have been identified as the Sha River.

[127] Ibid.

[128] Treatise on rivers and canals II: Grand Canal. In *Draft history of Qing*, vol. 133.

[129] Treatise on rivers and canals IV: Hydraulic works of directly administered provinces. In *Draft history of Qing*, vol. 135.

[130] Kaifeng prefecture. In *Comprehensive gazetteer of Qing dynasty (Jiaqing edition)*, vol. 186.

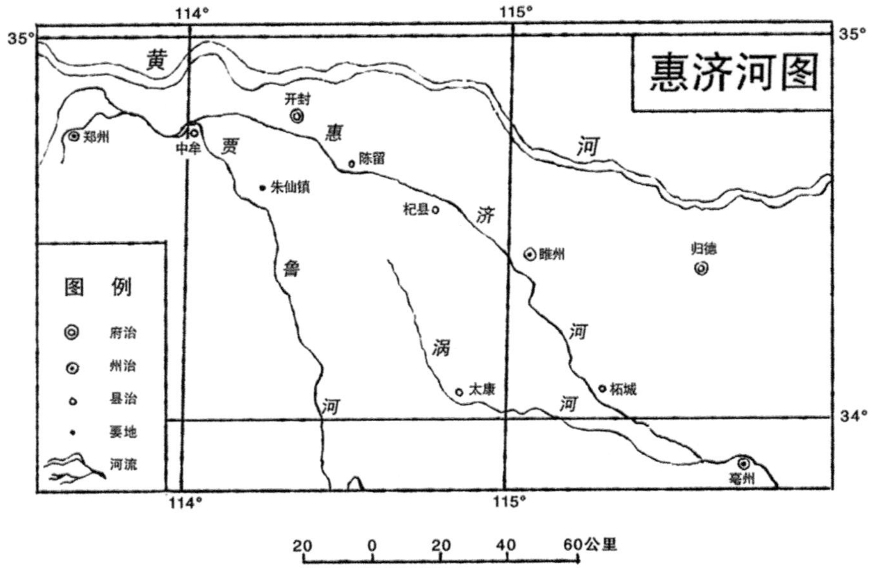

图例 - Legend
府治 Departmental Seats
州治 Prefecture Seats
县治 County Seats
要地 Strategic Locations
河流 Rivers

Fig. 7.13 The Huiji River

The Qianya River, located outside the city of Kaifeng,[131] was also known simply as the Qian River. Originating from the city moat outside the south gate of Xiangfu County (modern Kaifeng County), it extended only a short distance before entering the Huiji River at Gaojialou.[132] The Huiji River then passed through the contemporary counties of Xiangfu, Chenliu, and Qi County, as well as Suizhou (modern Sui County, Henan) and Zhecheng, before finally joining the Guo River at Bozhou (modern Bo County, Anhui).[133] From modern Kaifeng downstream, the river has not yet been completely obliterated. Nevertheless, its overall function was limited and did not produce any particularly significant impact (See Fig. 7.13).

[131] Treatise on rivers and canals IV: Hydraulic works of directly administered provinces. In *Draft history of Qing*, vol. 135.
[132] Kaifeng prefecture. In *Comprehensive gazetteer of Qing dynasty (Jiaqing edition)*, vol. 186.
[133] Ibid.

Minor Canals Between the Yangtze and Huai Rivers During the Ming and Qing Dynasties

During the Ming and Qing dynasties, the smaller canals located between the Yangtze and Huai Rivers were primarily concentrated east of the Jiangbei (North of the Yangtze) Canal, in the region traditionally known as the *Lixiahe* area. This area, characterized by its low-lying terrain and dense network of waterways, provided favorable conditions for canal construction. As early as the Southern and Northern Song dynasties, extensive canal-building efforts had already been undertaken here, and the Ming and Qing dynasties continued this legacy, further expanding the network.

As noted previously, the canal routes established during the Song period were organized into two main north–south branches. The southern branch began at Yangzhou, passed through Taizhou, Hai'an, Rugao, and Tongzhou, and extended eastward to Jinsha and Yuqing, possibly continuing as far as Lüsi Saltworks. The northern branch started at Gaoyou, passed through Xinghua, and reached Yancheng, before flowing further eastward into the sea. In addition to these two major branches, there were several connecting waterways: for example, the Yunyan River (Salt Transport River), which extended northwestward from Taizhou through Fancha to link with the Jiangbei Canal; and the Xixi River, which ran northeastward from Taizhou through Qintong to reach Xixi Town. The minor canals constructed during the Ming and Qing dynasties were developed largely on the basis of this earlier network (See Fig. 7.14).

The principal minor canals constructed in this region during the Ming and Qing dynasties were those passing through the prefectures and counties of Funing, Yancheng, Dongtai, Rugao, and Tongzhou. Although these waterways formed a continuous north–south line, each section bore a distinct name: some were called *Chuanchang River* (String-of- Saltworks River), while others were known as *Yunyan River* (Salt Transport River). In essence, they were all small-scale transport canals.

From Xixi Town northeast of Taizhou, the Chuanchang River flowed northeastward, passing the saltworks of Dongtai and Heduo, then turned northward through Dingxi, Caoyan, Baiju, and Liuzhuang, before reaching Wuyou Saltworks and finally Yancheng County, where it emptied into the sea. North of Yancheng, another section of the Chuanchang River ran through Xinxing Saltworks and Shanggang Town, continued through Funing County, and entered the sea via Sheyang Lake. A southern branch issued from Xixi Town, flowing southward through Liangduo, An'feng, and Fuan saltworks, where it became known as the Kuo River, continuing south to Hai'an Town.[134] South of Hai'an, the waterway was called the Yunyan River. From Rugao to Tongzhou, the Yunyan River gave off three branches in succession from north to south: the Lifa River, the Huangpu Stream, and the Shigang River, all flowing east into the sea. Upon reaching Tongzhou, it divided again into three outlets: one southwest of the city flowing into the Yangtze River, another southeast of the city leading north to connect with the Jinsha River and thence east through the former Haimen County into the sea, and a third northeast of the city, flowing through Xiting

[134] Yangzhou prefecture; Huan'an prefecture. In *Comprehensive gazetteer of Qing dynasty (Jiaqing edition)*, vol. 96, 93.

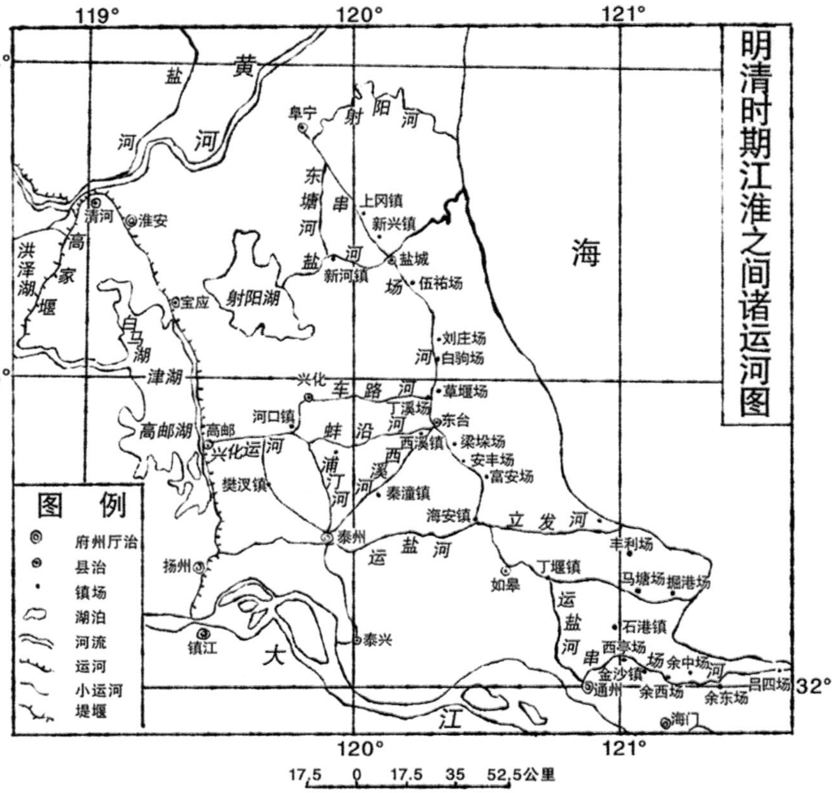

图例 – Legend		县治	County Seats
府州厅治	Prefecture, Department & Sub-prefecture Seats	镇场	Market Towns & Saltworks
		湖泊	Lakes
河流	Rivers	小运河	Minor Canals
运河	Grand Canal	堤堰	Embankment & Weirs

Fig. 7.14 Canals between the Yangtze and Huai Rivers during the Ming and Qing periods

Saltworks (thus called the Xiting River), east to Lüsi Saltworks, where it became the Chuanchang River before entering the sea.[135]

These segments of the Chuanchang and Yunyan Rivers, together with other minor canals, formed several regional hubs. Tongzhou was one such center, as were Taizhou and Xinghua. Taizhou, located midway on the canal route between Yangzhou and Hai'an, had long served as a nodal point; to its north ran canals connecting to Xixi and Fancha towns, routes already completed during the Song dynasty. In this period,

[135] Tongzhou prefecture. In *Comprehensive gazetteer of Qing dynasty (Jiaqing edition)*, vol. 106.

another canal—the Puting River—was added between Xixi and Fancha, linking northward to the Hailing Stream of Xinghua County. South of the city, the Jichuan River flowed through Taixing County before emptying into the Yangtze.[136] This Jichuan River generally corresponded to the canal cut by Chen Sunzhi in the Southern Song.

In Xinghua County, in addition to the Xinghua Canal and Puting River, there were also the Bangyan River and Chelu River. The Bangyan River branched off from the Xinghua Canal at Hekou Town in the southwest of Xinghua County and flowed east to meet the Chuanchang River at Xixi Town, which passed through Dongtai. The Chelu River flowed east from Xinghua County to join the Chuanchang River at Dingxi Saltworks.[137] Even Yancheng County possessed its own west–east waterway, the Yan River. South of the Yan River was another canal, the Guan River, which flowed northward into the Yan. At Xinhe Town, the Yan River split off a branch that joined the Dongtang River and flowed northeast, eventually merging with the Chuanchang River south of Funing County.[138] This Yan River greatly improved transport between Yancheng and the western hinterland.

The precise dates of construction for most of these small canals are no longer known. What can be confirmed is that the Chuanchang River north and south of Yancheng County was excavated during the Ming dynasty,[139] and that the Yunyan River in Tongzhou was opened during the Longqing reign of the Ming Emperor Muzong (1567–1572).[140] These two projects merit further explanation. The Chuanchang River's sections north and south of Yancheng bore different names: the northern section was called the Old Canal (*Jiu yunhe*), while the southern was known as the New Canal (*Xin yunhe*).[141] This distinction suggests that, although both segments were completed in the Ming dynasty, they were not constructed simultaneously.

As previously noted, the Yunyan River in Tongzhou had numerous branches. One of these flowed northeast from the city, passing through the Xiting Saltworks and then eastward through a succession of salt markets to reach the Lüsi Saltworks, where it became known as the Chuanchang River. Based on the distribution of saltworks in the region—Jinsha Saltworks lying east of Xiting, followed further east by Yuqing Saltworks—this branch was likely the canal excavated by Li Tingzhi during the Southern Song period. As discussed earlier, if Li indeed undertook this project, it is improbable that he would have left the stretch between Rugao and Tongzhou untouched. This suggests that at least some of the work traditionally attributed to the Longqing reign of the Ming dynasty was in fact dredging of an existing channel rather than entirely new construction. Some scholars argue that the canal opened by Li Tingzhi was the Chuanchang River itself. Located northeast of Tongzhou, this river connected the saltworks of Lüsi, Yudong, Yuxi, Jinsha, and Shigang, before exiting

[136] Yangzhou prefecture. In *Comprehensive gazetteer of Qing dynasty (Jiaqing edition)*, vol. 96.

[137] Ibid.

[138] Huan'an prefecture. In *Comprehensive gazetteer of Qing dynasty (Jiaqing edition)*, vol. 93.

[139] Ibid.

[140] Tongzhou prefecture. In *Comprehensive gazetteer of Qing dynasty (Jiaqing edition)*, vol. 106.

[141] Huan'an prefecture. In *Essentials of historical geography*, vol. 22.

through the Dingyan Sluice and merging with the Yunyan River. Another branch, entering via the Lifa River, linked the markets of Fengli, Juegang, and Matang, also emptying into the Yunyan River at the Dingyan Weir. The Dingyan Weir stood southeast of Rugao County, along the banks of the Yunyan River. If the Yunyan River indeed dates to the Longqing period, and if Li Tingzhi excavated the Chuanchang River, it is difficult to imagine that the latter would not have connected to Rugao and, ultimately, to Yangzhou.

Here it is worth briefly noting the Guan River of Haizhou. The administrative seat of Haizhou was located west of present-day Lianyungang in Jiangsu, already north of the Huai River. The Guan River lay south of Haizhou, and its headwaters could be traced back to Andong County (present-day Lianshui County). Today it is also called the Yunyan River (Salt Transport River). This Guan River is said to have been the grain-transport canal opened during the Chuigong reign of the Tang dynasty.[142]

Most of these minor canals served primarily for salt transport but also played important roles in flood control and irrigation. The so-called flood-control function allowed them to discharge excess water from the Jiangbei Canal. Should an accident occur at the Gaojia Weir east of Hongze Lake, these small canals could serve as drainage outlets. For this reason, local inhabitants regularly dredged these canals, and despite the many centuries that have passed, most of them remain navigable to this day.

The Development of Economic Metropolises Along the Grand Canal During the Ming and Qing Dynasties

The economic metropolises along the Grand Canal experienced continued prosperity and development during the Ming and Qing dynasties, building upon foundations laid in the Yuan dynasty. It is true that by the end of the Yuan period, the Grand Canal had suffered intermittent blockages, and grain transport had to be rerouted. Prolonged warfare led to the devastation of local economies: population decline, abandoned land, and, in some regions such as Shandong and Henan, areas described as "largely uninhabited."[143]

Following a period of recuperation and social recovery, productive forces gradually revived, and the Grand Canal underwent more comprehensive restoration, allowing it to resume its vital functions. During the Xuande reign of the Ming Xuanzong (1426–1435), in order to address the grain supply needs of the capital, it was decreed that: 1.5 million *dan* (石) of grain be transported by private citizens from Jiangxi, Huguang, and Zhejiang to the Huai'an granary; 2.74 million *dan* from Suzhou, Songjiang, Ningguo, Chizhou, Luzhou, Anqing, and Guangde to the Xuzhou granary; and 2.2 million *dan* from Yingtian, Changzhou, Zhenjiang, Huai'an, Yangzhou, Fengyang, Taiping, Chuzhou, Hezhou, and Xuzhou to

[142] Hizhou prefecture. In *Comprehensive gazetteer of Qing dynasty (Jiaqing edition)*, vol. 105.

[143] Gu Yanwu. Reclamation of wasteland. In *Notes on knowledge accumulated day by day*, vol. 10.

the Linqing granary.[144] The establishment of these grain depots along the canal not only served logistical purposes but also promoted the growth and prosperity of nearby economic hubs.

It was also during the fourth year of the Xuande reign that, in an effort to increase the tax revenues from market stalls and shopfronts across urban centers—including Beijing's Shuntian Prefecture, Nanjing's Yingtian Prefecture, and various prefectures such as Suzhou—a proposal was submitted to raise the existing tax rates tenfold[145] in thirty-three designated locations.[146] These included: Shuntian, Suzhou, Songjiang, Zhenjiang, Huai'an, Changzhou, Yangzhou, Yizhen, Zhejiang, Hangzhou, Jiaxing, Huzhou, Fuzhou, Jianning, Wuchang, Jingzhou, Nanchang, Ji'an, Linjiang, Qingjiang, Guangzhou, Kaifeng, Jinan, Jining, Dezhou, Linqing, Guilin, Taiyuan, Pingyang, Puzhou, Chengdu, Chongqing, and Luzhou. The imposition of such significantly increased levies indicates that these thirty-three prefectures and counties were the most vital economic centers of the time. Among them, eleven were situated directly along the Grand Canal: Shuntian, Dezhou, Linqing, Jining, Huai'an, Yangzhou, Zhenjiang, Changzhou, Suzhou, Jiaxing, and Hangzhou. In other words, one-third of the nation's leading economic metropolises were located along the canal—no small matter.

The canal's two termini—Beijing and Hangzhou—were both major urban centers. As the imperial capital, Beijing (Shuntian Prefecture) differed fundamentally from ordinary cities, with both commerce and industry enjoying significant growth. Its temple fairs (*miao shi*), which served as the focal point of its markets, were particularly large in scale. These fairs were held on fixed dates and the rows of market stalls could stretch for more than three *li*. The range of goods available was extensive; luxury items such as jewelry, ivory, jade, rare curios, and exquisite silks and brocades were transported from Yunnan, Guangdong, Fujian, Hunan, Jiangsu, and Zhejiang.[147] Without the transportation capabilities of the Grand Canal, the capital's markets would not have been so richly supplied. At the other end of the canal, Hangzhou was equally vibrant, known for having "360 trades, each with its own jargon,"[148] vividly illustrating its bustling commercial activity. It is also worth noting that Hangzhou's silk weaving industry was especially developed, contributing significantly to the prosperity of this economic metropolis.

Among the commercial cities along the Grand Canal, Yangzhou has long stood out as the most prosperous. One of Emperor Yang of the Sui dynasty's motivations for constructing the canal was precisely to connect it to Yangzhou. During the Tang and Song dynasties, Yangzhou's flourishing economy was unparalleled across the empire. Although the city suffered occasional devastation during the Ming and Qing periods due to warfare, it was nevertheless able to recover and regain its prosperity

[144] Grain transport. In *Collected statutes of the Ming dynasty*, vol. 56.

[145] *Veritable records of Emperor Xuanzong of Ming*, vol. 50, entry for the 1st month, Yichou day, 4th year of Xuande.

[146] Ibid.

[147] Liu Tong. *Scenes of the imperial capital: temple market at the Chenghuang temple.*

[148] Tian Rucheng. *Gazetteer of West Lake*, vol. 20.

in time. The thriving salt industry was likely a significant factor contributing to this economic vitality. In the city's markets, silk and spices were among the principal commodities traded. There were dedicated establishments for each: silk was handled by silk firms, and satin by satin guilds. Goods were first delivered to these commercial organizations before reaching the retail shops. Yangzhou's spices were renowned throughout the country, unmatched in quality.[149] The numerous gardens found both north and south of the Yangtze River often serve as indicators of regional wealth; in Yangzhou, these gardens significantly enhanced the city's splendor.[150]

Suzhou was another famous economic center along the Grand Canal. Its prosperity led to a rapid growth in urban population. For instance, Huangjiagang outside Suzhou's Changmen Gate was still an open area during the Ming dynasty, but by the Qing period, it had developed into a densely populated district with nearly a thousand households, their homes tightly packed together.[151] It is said that by the Qianlong reign, Suzhou already housed some 100,000 households.[152] The Qianlong era marked the height of Qing prosperity and overall economic development. Located along the canal and benefiting from its well-known transportation advantages, Suzhou appeared especially affluent. During this period, the artist Xu Yang, known for his painting skills, produced *Picture of a Flourishing Age*, a scroll depicting the bustling urban life of Suzhou's streets and waterways. According to estimates, over 230 shops bearing signboards are represented in the painting, categorized into fifty different trades of various sizes. The range of goods sold included not only local specialties from nine provinces—Sichuan, Guangdong, Yunnan, Guizhou, Fujian, Jiangxi, Zhejiang, Jiangsu, and Shandong—but also imported foreign goods.[153] Though Suzhou's markets drew on products from distant provinces and even overseas, the transportation of these goods necessarily relied on the canal. It may thus be said that it was precisely Suzhou's proximity to the Grand Canal that enabled it to become a nexus for such a rich assortment of goods, thereby supporting its growth as a thriving economic metropolis.

Moreover, Suzhou's wealth and prosperity, in addition to its abundant rice production, owed much to its flourishing silk industry. Local silk textiles were transported via the canal to markets across the empire. This is clearly illustrated by the development of Shengze Town in Wujiang County, located south of Suzhou. In the Hongzhi era of the Ming dynasty, Shengze was still a modest village, but by the early Kangxi reign of the Qing, it had grown to over ten thousand households,[154] all of whom depended on the production and sale of fine silk.[155] At that time, many towns and

---

149 Li Dou. Account of little Qinhuai. In *Records of Yangzhou painted boats*, vol. 9.

150 Qian Yong. Gardens. In *Random talks from the garden of treading flowers*, vol. 20.

151 Xu Xilin. *New Discourses of the glorious dynasty*, vol. 16.

152 Administration of officials: Magistrates II. In *Supplement to the imperial anthology of statecraft writings*, vol. 23; Shen Yu. *On governing Suzhou*.

153 Li, H. (1964) The prosperity of commerce and industry in early Qing Suzhou as seen from Xu Yang's Picture of a Flourishing Age. *Cultural Relics*, (1).

154 *Wujiang county gazetteer (Kangxi edition)*, vol. 1.

155 Livelihoods. In *Wujiang county gazetteer (Qianlong edition)*, vol. 38.

counties near Suzhou had economies centered on the textile industry, but none were as developed as Shengze—an advantage that can be attributed to its location near the canal.

What held true for the southern segment of the Grand Canal also applied to its northern stretches. Cities such as Linqing and Jining enjoyed great prosperity, both in their time and in historical memory. Jining was not only a crucial transportation hub where land and water routes converged,[156] but also a major center for the distribution of goods from the Jianghuai region.[157] Situated on the banks of the Tongji Canal—which was known as the Zha River) during the Ming dynasty—Jining lay between Linqing in the north and Chachengkou in the south. The terrain between Linqing and Chachengkou was uneven, requiring a system of locks to regulate the canal's water levels. Nanwang, to the north of Jining, was especially elevated, necessitating even more densely spaced locks in that section. Jining was also where the Si River joined the canal, making navigation especially difficult for boats. Thus, goods transported from the Jianghuai region often had to stop and be redistributed here, promoting the city's economic development.

As for Linqing, its location at the confluence of the Wei River and the Grand Canal was even more advantageous than that of Jining. As noted in historical records, "The city is nestled by hills, with hundreds of shops densely packed; where two rivers meet, a thousand masts rise like clouds. Merchants from five directions are examined at the customs, and canal traffic from seven provinces passes through its locks." It was also said to be "connected by water to the southeast, linking the Yangtze River and lakes; it guards a strategic position between the north and south."[158] These natural and geographical advantages explain how Linqing came to be a prominent economic center.

There were also cities that once held economic prominence but declined after the canal was rerouted and no longer passed through them. Xuzhou provides a clear example of such a transformation. Originally, the Grand Canal flowed directly beneath the city, but after the opening of the Jia River waterway, the number of grain transport vessels passing through Xuzhou sharply declined, and its prosperity waned accordingly. Local gazetteers thus described it only in general geographical terms—"the river winds northwest, flowing southward into the Huai River; it is connected to both the Yangtze and the Yellow Rivers"—but no longer emphasized the canal.

In the fourth year of the Xuande reign (1429), the Ming government issued an edict to collect stall taxes from all shops and markets along the canal, but Hexiwu was not included—likely because it was a town below the county level and thus

---

[156] Jining prefecture. In *Comprehensive gazetteer of Qing dynasty (Jiaqing edition)*, vol. 183, citing *Old gazetteer.*

[157] Treatise on geography: Customs. In *General gazetteer of Shandong*, vol. 40.

[158] Linqing prefecture. In *Comprehensive gazetteer of Qing dynasty (Jiaqing edition)*, vol. 184, citing *General gazetteer* and *Prefecture gazetteer.*

outside the scope of the edict. As mentioned earlier, Hexiwu's prosperity had been closely tied to maritime transport. Since the Ming dynasty relied more heavily on canal transport and paid less attention to sea routes, Hexiwu's economic fortunes may have declined. Nonetheless, it remained a canal-side market town, even if no longer comparable to centers like Linqing and Jining.

Chapter 8
The Decline and Restoration of the Grand Canal

Major Factors Behind the Grand Canal's Obstruction

The Grand Canal, which traverses China from north to south, had undergone continuous maintenance and improvement since its successful construction during the Yuan dynasty. By the Ming and Qing periods, it had developed into a highly efficient transportation route and served as the main artery of national communication. However, from the late Qing dynasty through the Republican era, the canal gradually fell into disrepair. The once-flourishing scene of boats in constant succession and sails rising and falling became increasingly rare. Moreover, several stretches of the canal became intermittently silted up. After the founding of the People's Republic of China, substantial efforts were made to restore navigation, and the canal is poised to surpass the accomplishments of past dynasties. Nonetheless, it remains necessary to briefly revisit the deterioration of the canal over the past century and the key factors that contributed to its obstruction.

The reasons for the canal's severe degradation and silting over the past hundred years are manifold. As the saying goes, "treading on frost precedes walking on thin ice"—this decline was not the result of a single day or event. While the relevant circumstances and the course of their evolution have been generally discussed earlier, the most critical moment came in the 27th year of the Guangxu reign (1901), when Li Hongzhang memorialized the throne requesting the abolition of the grain tribute system and its replacement with monetary equivalents. Once land taxes were monetized, the need for transporting grain tribute by water disappeared, thereby stripping the Grand Canal of its original function. Admittedly, the canal was not used solely for transporting tribute grain, but such transport accounted for the majority of its usage.

The government's repeated repairs and sometimes full-scale efforts to maintain the canal were driven entirely by the demands of the grain tribute system. The benefits afforded to commercial traffic were incidental at best, mere side-effects of the system. With the abolition of grain tribute transport, official attention to the canal ceased,

N. Shi, *A Historical Geography Research of Canals in China*,
https://doi.org/10.1007/978-981-95-3517-0_8

leaving it to decay on its own. When siltation occurred, commercial vessels naturally retreated as well. Thus, in examining the canal's misfortune over the past century, the abolition of grain transport by water must be seen as a decisive turning point—an issue that cannot be ignored.

The canal's initial replacement was maritime transport. The invention and introduction of steamships from the West significantly diminished the canal's importance. Ordinary vessels plying the canal were typically only two to three hundred *liao* in capacity. Larger ships exceeded the canal's depth and load-bearing capacity. Even these smaller vessels required elaborate lock navigation and towing assistance to traverse the canal—processes that were both time-consuming and labor-intensive. How could such methods compare with the efficiency of seagoing steamships? And this was during a time when the canal still functioned normally. If dangers or complications arose along the waterway, the difficulties only multiplied. Thus, maritime transport's supplanting of the canal was an inevitable development, especially given that their routes were more or less parallel.

In addition to maritime transport, the most direct competitor of the canal was the railway. The Beijing–Tianjin Railway (part of the Beijing–Shanxi Railway) was built in the 22nd year of the Guangxu reign (1896), running parallel to the old Tonghui and Bai Rivers. Soon thereafter, the Jinpu Railway was constructed, and by the first year of the Republic (1912), it connected north and south, paralleling the section of the canal from Tianjin to Yangzhou. South of the Yangtze River, the Shanghai–Nanjing and Shanghai–Hangzhou–Ningbo railways also followed a route broadly parallel to the Jiangnan (Southern Yangtze) section of the canal. These two railways were constructed in the late Qing period. Although the full Shanghai–Hangzhou–Ningbo line had not yet been completed at that time, the segment between Shanghai and Hangzhou had already been finished. By then, the canal had already deteriorated significantly, while the railway offered rapid and accessible transportation. Running parallel to the canal, it naturally diverted the canal's traffic.

Maritime transport, by contrast, could only replace canal shipping in areas close to seaports. Railways, however, paralleled the canal more closely, allowing them to compete with it all along the route. For example, the Beijing–Tianjin Railway bypassed Tongzhou, connecting it to Beijing only via a branch line. As a result, Tongzhou's prosperity waned significantly. Other canal-side towns such as Hexiwu and Zhangjiawan became even more obscure. The Tianjin–Pukou Railway reached as far as Linqing, and Linqing too fell into decline compared with its former glory.

An even clearer example is provided by the rise and fall of Daokou (in present-day Hua County, Henan). Though not located directly on the Grand Canal, Daokou lies on the upper reaches of the Wei River and owed its prosperity to the canal. Before the construction of railways, goods shipped from Tianjin to various parts of Henan went upstream via the Wei River, and goods from Henan destined for Tianjin traveled downstream the same way. Daokou served as a major transshipment center. Its prosperity once rivaled that of Zhuxianzhen or Zhoujiakou across the Yellow River. However, after the Beijing–Hankou Railway (now the northern section of the Beijing–Guangzhou Railway) began operations, Daokou suffered a severe blow. Its former prosperity quickly gave way to desolation. Even the construction of the

Daokou–Qinghua branch line (Qinghua Town, now Bo'ai County, Henan; the section from Xinxiang to Daokou was dismantled during the War of Resistance against Japanese did nothing to reverse its decline.

The Deterioration of the Grand Canal During the Republican Period

Amidst mounting competition and natural sedimentation, the fate of the Grand Canal was inevitable. During the Republican period, its condition deteriorated severely. A north-to-south survey reveals the extent of its ruin. From Beijing to Tong County (modern-day Tongzhou), the waterway was once known as the Tonghui River or Datong River. Although the river channel still existed, boats had long since disappeared. The stretch from Tong County to Tianjin, formerly the Bai River or Tongji River, is now referred to as the North Canal (*Bei Yunhe*). Due to prolonged neglect, the riverbed had silted up and risen, with many shallow areas along the route. In times of low water, navigation became impossible; during high water, breaches were frequent. Several such incidents occurred during the Republican era. At the time, authorities merely blocked the breaches without undertaking any comprehensive river restoration.

In 1939, an unprecedented breach in the North Canal caused flooding that submerged 814 villages and even inundated the center of Tong County.[1] From Tianjin to Linqing ran the historic course of the Wei River, now commonly called the South Canal (*Nan Yunhe*). Although part of the Grand Canal, this segment essentially formed its own hydrological system. Originating from Hui County in Henan Province, the Wei River's flow was variable, but it rarely dried up. Even when the canal south of Linqing was impassable, the Wei River could still transport goods to various parts of Henan. Within Hebei Province, this southern section of the canal—from Tianjin southward—remained the most navigable. Until the outbreak of the War of Resistance against Japanese, it could still accommodate freight boats carrying less than 1000 *dan*. Occasionally, shallow-draft steamships could also navigate the stretch between Tianjin and Dezhou.

South of Linqing to Tai'erzhuang lay the so-called Shandong Canal, entirely within Shandong Province. The Shandong Canal was bisected by the Yellow River into two parts: north of the Yellow River and south of Lingqing, the canal was called the North Canal (not to be confused with the North Canal between Tong County and Tianjin); south of the river and north of Tai'erzhuang, it was known as the South Canal (distinct from the South Canal between Tianjin and Linqing). Among all sections of the Grand Canal, the Shandong segment was in the worst condition.

Its main water source was the Wen River, which split at Nanwang to supply both northern and southern reaches. However, after the Yellow River changed course and diverted the Qing River into the sea, the northern flow of the Wen River followed the

[1] Liu Letang. (1959, July 26). The past and present of the North Canal. *Beijing Daily*.

Yellow River eastward, leaving the stretch from Linqing to the Yellow River without a stable water source. Late in the Qing dynasty, an attempt was made to divert water from the Wei River to maintain navigation, but the Wei's low discharge proved inadequate. Attempts to use the Yellow River as a substitute only exacerbated the problem, as its high silt content accelerated the canal's sedimentation. Eventually, this section became completely unnavigable. Although efforts were made in Shandong to reopen the canal through dredging, the work was so extensive it was akin to excavating a new river from scratch—an endeavor fraught with difficulty.

South of the Yellow River, the South Canal fared slightly better due to more reliable water sources. However, as a gated canal, it required careful water regulation. After the termination of grain transport by the state, maintenance ceased, locks fell into disrepair, and leaks led to depletion. A short segment from the Yellow River bank to Anshan became clogged due to proximity to the river and its frequent disturbances. Another segment from Anshan southward to Yuankou Town (in western Wenshang County, north of Nanwang) remained passable thanks to inflow from the Wen River and storage in Dongping Lake, but only for small boats carrying under 100 *dan*—larger vessels could no longer navigate. Between Yuankou and Nanwang to the Liulin Lock, a stretch of 30–40 *li* (approximately 15–20 km) had dried up and become impassable. This was mainly due to Nanwang's high elevation, inadequate water sources, and deteriorating lock infrastructure. From Liulin Lock south to Xiazhen, a region abundant in lakes, navigation was possible for boats carrying 100–200 *dan* during high water periods, but even then the route was frequently shallow and sluggish. South of Xiazhen to Tai'erzhuang, the canal was again cut off due to drying. In sum, roughly half of the Shandong Canal had become silted and impassable, and what remained functional was barely so.[2]

From Tai'erzhuang to Guazhou, just south of Jiangdu County, was the Jiangbei (North of the Yangtze) Canal. As the Huai River bisected this segment, it too was divided into two parts: the Middle Canal (*Zhong Yunhe*), north of the Huai River and south of Tai'erzhuang; and the Inner Canal (*Li Yunhe*), south of the Huai and north of Guazhou—also referred to as the Huainan Canal (*Huainan Yunhe*) due to its location. Before the Yellow River changed course northward, this region was notorious for breaches. Once the river migrated, the source of disruption was removed and the flow gradually stabilized. Comparatively, the condition of the Middle Canal was far worse than that of the Inner Canal. The Middle Canal relied partly on the southern flow of the Wen River south of Nanwang, and partly on water from the four southern lakes, especially Weishan Lake, but in practice it was sustained largely by the Yi, Si, and Jia rivers. The Si River entered the canal at Jining and was second only to the Wen River in importance. The Jia River flowed south from modern-day Cangshan County into the canal at Rukou Town in Pizhou, Jiangsu. The Yi River flowed through Linyi and Tancheng in Shandong before also entering the canal near Pizhou.

[2] Li Yizhi. (1932). Transportation on North China waterways. *Water Resources Monthly*, (2), 5–6; Shandong Construction Department. Achievements in water conservancy construction in Shandong Province. (1933). *Water Resources Monthly*, (3), 5–6; Wang Huzhen. (1934). Issues in canal rehabilitation. *Water Resources Monthly*, (7), 1.

When the canal was still functional, the Yi, Si, and Jia rivers played a vital role in ensuring navigability. These tributaries together drained much of the southern Shandong highlands. Since they all flowed into the canal, they posed a serious flood risk during the summer and autumn, threatening to breach the embankments. In ordinary times, however, poor lock maintenance and low water retention made navigation difficult. Nonetheless, small boats of 100–200 *dan* could still pass, and during flood periods, shallow-draft steamships could navigate between Suqian and Huaiyin.

The Inner Canal was the best-preserved segment north of the Yangtze. Though previously plagued by Yellow River sedimentation, the river's diversion north relieved the canal from such damage. Its only remaining threat was the excessive influx of water from the Huai River. During the late Qing and Republican periods, efforts to manage the Huai were inadequate, and the canal continued to suffer flooding. Fortunately, the Inner Canal was not only a vital north–south transportation corridor, but also served as the main distribution route for salt from the Lianghuai salt fields. As such, it received regular dredging and maintenance. Except in years of severe drought, it remained navigable to freight boats under 1,000 *dan* and to shallow-draft steamships, making it relatively convenient for transport.[3] During the War of Resistance against Japanese, the Yellow River breached at Huayuankou in Henan and flowed southeast into the Huai, eventually emptying into Hongze Lake and then the Yangtze—naturally impacting the Inner Canal.

Within the Grand Canal system, the Jiangnan (South of the Yangtze) Canal was the most intact. Apart from minor silting due to tidal intrusion at Zhenjiang's Jingkou and Hangzhou's Qiantang estuary, only the Lian Lake northwest of Danyang required regular dredging. Lian Lake served as the water source for canals running through Danyang, Zhenjiang, and Wujin, and historically great care was taken to maintain its water levels. During droughts, lowering Lian Lake by a single *cun* (寸, ~3 cm) could raise canal levels by a whole *chi* (尺, ~33 cm)—a remarkable feat of hydrological regulation. To maintain this balance, the lake had to be regularly dredged and protected from encroachment by lakeside residents. Thus, barring a few manageable locations, the Jiangnan Canal remained freely navigable. In fact, with its calm surface and uninterrupted vessel traffic, it surpassed natural waterways in ease of use.

Eastward along the southern coast of the Qiantang River, another segment—known as the Zhedong Canal—connected Hangzhou to Shanxi. Passing through Xixing on the Qiantang's southern bank, it was also referred to as the Xixing Canal. Though it was first excavated during the Southern Song dynasty, it was not part of the main north–south Grand Canal system. Nonetheless, it closely resembled the Jiangnan Canal in form and function, and suffered none of the silting problems at its inlets. Though artificial in origin, it had, over time, come to resemble a natural watercourse.[4]

[3] Li Yizhi. Transportation on North China waterways; Wang Huzhen. Issues in canal rehabilitation; Jiangsu Water Resources Bureau. (1931). Overview of Jiangsu North Canal projects. *Water Resources Monthly*, (1), 1.

[4] Wu Tongju. (1936). Perspectives on the Jiangnan Canal: Past and present. *Water Resources Monthly*, (9), 3.

The Necessity of the Grand Canal

While the opening of sea routes and the construction of parallel railways have undoubtedly dealt a heavy blow to the canal, its former course nonetheless ought to be restored without delay. Some may ask: since maritime transport and parallel railways have already compensated for the canal's function, why is it still necessary to revive it? Moreover, since the canal's decline was precisely due to the development of sea transport and railways, would not any attempt at restoration merely lead back to the same state of disrepair?

This line of reasoning, however, represents only one aspect of the issue. It cannot justify allowing the canal to continue to deteriorate without intervention. In the past, great attention was paid to the canal primarily for its role in grain transport for the state. At the same time, agricultural products, as well as industrial and mineral goods along its banks, also relied on the canal for transportation.

While maritime transport is indeed the most economical, its reach is limited to areas near seaports. The number of harbors capable of handling shipping between Shanghai and Tianjin is small—chiefly Lianyungang, Qingdao, Weihaiwei, and Yantai. These few seaports are clearly insufficient to accommodate the vast volume of goods produced along the canal. Furthermore, some of these ports are located at a considerable distance from the canal itself, limiting their effectiveness. If goods from various places along the canal must be routed through these few accessible seaports, this would necessitate additional modes of transport along the way. Take Qingdao, for example: if goods from the Shandong section of the canal are to be exported via Qingdao, they must first be transferred by the Jiaoji Railway, thereby increasing costs and undermining the economic advantage of maritime transport.

Moreover, not all goods produced along the canal are destined for export; some are intended for coastal or inland distribution. Thus, sea transport alone cannot fully address the logistical needs. Railways are indeed faster and clearly superior to canal barges in terms of speed, making up for the deficiencies of maritime shipping. Yet when it comes to cost, canal transport is far more economical. The current poor condition of the canal renders it incomparable with railways. However, if the canal were restored and its width expanded to increase capacity, it could complement the railway system—balancing speed with cost-efficiency. Passengers in need of speed may opt for rail, while goods requiring economical transport would better follow the waterway. In this way, the two modes of transport serve complementary roles rather than competing ones.

Western European countries, despite having highly developed railway systems, continue to attach great importance to canals. In addition to maintaining and improving existing ones, new canals are still regularly constructed, and many run parallel to railways. This demonstrates the close and mutually reinforcing relationship between the two modes of transport. These international experiences offer valuable lessons that China may draw upon.

If the canal's water supply lacks clear and abundant flow, siltation occurs easily. To ensure unimpeded navigation, regular dredging is essential. The main reason for the Grand Canal's silting has been the cessation of grain transport. The advent of sea transport and the building of parallel railways led the government at the time to neglect the canal. Yet grain transport may be dispensable, while the transport route itself is not. It would be wrong to assume that without a revival of grain shipments, restoring the canal serves no purpose.

One must recognize that although grain transport was once the canal's primary function, it was by no means its only value. In earlier times, both the state and the public focused too narrowly on this single function, overlooking its broader utility. The regions along the canal are rich in resources—Shandong's coal mines being especially notable—and these products must be transported outward. Goods from both northern and southern China also rely on the canal for mutual exchange. It is therefore unwise to allow the canal to remain in disrepair. Its restoration should be actively pursued, not only to the standards of the Ming and Qing dynasties, but beyond them—reaching the level of canal systems in contemporary countries around the world.

The Restoration of the Grand Canal

Since the founding of the People's Republic of China, the nation has undergone remarkable renewal, with rapid and vigorous development in economic construction and transportation. Railways and highways have continued to expand, air routes have become widespread across the country, and international exchanges have grown increasingly frequent. Navigable rivers are bustling with sailing vessels, and the seas teem with large merchant ships—a scene of great vitality and prosperity. The Grand Canal, with its long historical lineage, remains an indispensable component of the transportation network and ought to play its proper role in the modern era.

As noted earlier, by the late Qing and Republican periods, the Grand Canal had fallen into neglect due to the cessation of grain transport. The failing governments of those times had neither the will nor the capacity to develop infrastructure or promote public welfare. It was only after liberation that its condition began to improve, and the dilapidated canal gradually recovered. After the founding of New China, restoring and modernizing the Beijing–Hangzhou Grand Canal and transforming it into a major waterway became a key national policy. Relevant departments have since worked collectively to realize this goal.

The Grand Canal runs north–south across China and intersects with the Yangtze and Yellow Rivers. It is particularly closely connected to the Huai and Hai River systems. Since the Yellow River's course shifted southward and captured the Huai River's outlet to the sea, the canal has faced many new challenges. The northward migration of the Yellow River caused repeated disruptions, leaving behind a host of long-standing problems. Among them, Hongze Lake has been especially affected, as the Yellow and Huai Rivers pour into it, expanding its waters without bound. When the water level in Hongze Lake is unstable, the lower Huai River cannot flow freely into the sea, causing overflow in adjacent lakes such as Baima, Gaoyou, and Shaobo.

The Grand Canal, running northward from Yangzhou, skirts the eastern shores of these lakes. If the lakes overflow, canal traffic is inevitably obstructed.

The Hai River system likewise exerts a significant influence on the canal. South of Linqing, the canal aligns with the course of the Wei River. Tianjin, where multiple Hai River tributaries converge, serves as a key hub along the Grand Canal. Any disruption within the Hai River basin directly affects canal navigation. After 1949, the government launched large-scale initiatives to manage both the Huai and Hai Rivers, laying a solid foundation for the Grand Canal's rehabilitation. A major milestone was the completion of the Sanhe Sluice in 1953. Spanning 697.75 meters with a discharge capacity of 12,000 cubic meters per second, it stabilized the water level of Hongze Lake and maintained steady levels in the adjacent Baima, Gaoyou, and Shaobo Lakes—ensuring a continuous flow along the Inner Canal. Further south of Linqing, the South Canal depends on the clear waters of the Wei River. Repeated dredging, reinforced embankments, channel widening, and the construction of control and navigation sluices have greatly enhanced its navigability.

To preserve the integrity of the North Canal near Beijing and manage floodwaters when necessary, a diversion channel called the Yunchao Relief Canal was built in Tong County. It diverts part of the floodwaters from the North Canal into the Chaobai River. Beginning at the floating bridge at Tong County's northern gate, it runs east through Wangjiachang, Gucheng, Zhaoli, and Huzhuang before reaching Dongbu Village, where it joins the Chaobai River—a total length of 12 kilometers. Prior to this, the construction of the Shisanling Reservoir had already controlled water inflow from the Wenyu River, an upper tributary of the North Canal.

Nevertheless, the expansion of agricultural irrigation has increased water consumption, and even canal water has been diverted for other uses. The Inner Canal between the Yangtze and Huai Rivers, sustained by the lakes mentioned above, remains relatively unaffected. Similarly, the South Canal north of Linqing still has reliable flow from the Wei River, despite some diversion. The most problematic section lies between Xuzhou and Linqing, which depends on the Si and Wen Rivers for water—especially the Wen River, which must split its flow north and south. Diversion of these rivers has severely hindered navigation. South of Jining lie the so-called "Four Southern Lakes"—Nanyang, Dushan, Zhaoyang, and Weishan—which historically functioned as water reservoirs for the canal, regulating its flow. With falling water levels, the lakebeds have become exposed and overgrown with reeds, reducing navigability. As a result, the over-200-kilometer stretch from Xuzhou to the Yellow River has become semi-navigable, while the section north of the Yellow River to Linqing is essentially impassable.

In 1982, the Inner Canal was dredged; in 1983, the Bulao River was similarly improved. The Bulao River flows southwest from the western shore of Weishan Lake, branching off near Maocun northwest of Xuzhou. Originally flowing south, it turned east below Maocun and joined the Tongji New Canal near northwest of Pi County. The Tongji New Canal refers to a 56-*li* (approximately 28-kilometer) section running from the Jia River to Luoma Lake. Once dredged, the Bulao River effectively became a new stretch of the canal, offering a shorter route than the older path around Xiazhen (north of Weishan County, Shandong) via the Jia River. The section from Linqing to Jining was also dredged and improved.

The most difficult stretch for drawing water lies between Xuzhou and the Yellow River, which relies on the Si and Wen Rivers; similarly, from the Yellow River north to Linqing, water shortages persist. Indeed, in the Beijing–Tianjin–Hebei region, where rainfall is scarce and industrial and agricultural development is intense, water shortages have become acute. To address these issues, the state has begun implementing a strategic project: the South-to-North Water Diversion (SNWD). Among its three routes, the Eastern Route is most closely associated with the Grand Canal. The Eastern Route begins at the Jiangdu Pumping Station near Yangzhou in Jiangsu Province, drawing water from the Yangtze River and sending it northward along the Grand Canal. In 1983, the government approved the first phase of the project, which has already achieved preliminary results.

Although parts of the Eastern Route follow new paths, most of it runs northward along the Grand Canal. From the Jiangdu station to Huai'an, it follows the canal closely. From there, it diverts west along the Huai River, and at Gaoliangjian it turns down the Wanfu River to reenter the Grand Canal. A new double-line ship lock is under construction in Huai'an, and the main navigation route is expected to follow the original canal alignment. North of Huai'an, the route continues along the canal to Pi County, then follows the Bulao River and joins the Jia River near Hanzhuang (southeast of Weishan Lake), eventually flowing into Weishan Lake. From there, it continues through the Four Southern Lakes to the vicinity of Jining. Near the Yellow River, it shifts into Dongping Lake rather than following the canal. From the Yellow River to Linqing, the route again deviates slightly west of the canal's path. North of Linqing, the canal follows the Wei River, and the SNWD route similarly traces the Wei River northward. While there are some deviations, these are relatively minor and do not present significant obstacles.

In January 1985, the dredging of the Huai–Si section of the Grand Canal was completed and navigation resumed. This section had long been a bottleneck on the Beijing–Hangzhou Canal. With its reopening, ports such as Xuzhou, Pi County, and Tai'erzhuang began exporting coal, sand, and other goods in large volumes. On December 22, 1985, Xinhua News Agency reported the successful completion and acceptance of the widening and dredging project of the Lincheng section in Gaoyou County—formerly another major bottleneck. Soon after, an article titled "*Restoration of the Northern Jiangsu Section of the Grand Canal Nearing Completion*" was published. Thus, during the Sixth Five-Year Plan period, one of the key national infrastructure projects—the restoration of the main navigation channel of the Northern Jiangsu section of the Grand Canal—was successfully completed. The 404-kilometer section from Xuzhou to Yangzhou has reached the standards of a Class II navigation channel. Thousand-ton barge convoys loaded with coal can now travel directly from the Xuzhou coal base to the Yangtze River. This section, now comprehensively rehabilitated, has seen its shipping capacity greatly enhanced, with its economic and logistical benefits increasingly evident.

Since 1982, 158 kilometers of the previously narrow and congested canal channel have been widened, and two new ship locks—Huai'an and Suqian—have been constructed. As a result, the 800-*li* section of the Grand Canal has reached the specifications of a Class II navigable waterway, with a bottom width of 60–70 meters and a water depth of 4 meters, capable of accommodating convoys of vessels up to 2000 tons. The annual throughput capacity of the canal's northern Jiangsu section has increased by approximately five million tons compared to pre-rehabilitation levels, while the volume of water diverted from the Yangtze River to the north has also risen significantly—an unprecedented achievement.

During this period, eight new water-lifting stations were constructed along the canal, supplementing existing facilities such as the Jiangdu pumping station. Together, they enable the stepwise elevation of Yangtze River water to be delivered to water-scarce regions. Upon completion of the dike-cutting project along the 50-kilometer stretch south of Huai'an, the water transfer capacity increased by 80 cubic meters per second. This has significantly improved irrigation for nearly 20 million *mu* of farmland along the canal, as well as urban and industrial water supplies and domestic use. To the east of the canal lies a low-lying area colloquially known as the "Pot Bottom Depression," where the canal's water level is higher than the city walls of Xinghua County. Historically, breaches of the canal's embankments have occurred repeatedly in this area. In the current phase of the canal's rehabilitation in northern Jiangsu, soil dredged from the channel was used to reinforce embankments, while vulnerable sections were protected with riprap slopes. These measures have improved drainage conditions across an area of 100 square kilometers along the canal.

These developments have successively demonstrated the comprehensive benefits of the project in terms of water management and redistribution. It is truly encouraging news. Restoration efforts for other sections of the canal are underway, and more positive reports are expected to follow. Not only has the Grand Canal been comprehensively restored and significantly improved, but thanks to modern construction methods, it is being transformed into a waterway bustling with ships and increasingly intensive freight activity.

The Excavation of Several Minor Canals

Since the founding of the People's Republic of China, in addition to the restoration and rehabilitation of the Grand Canal, several smaller-scale canals have also been constructed. One such example is the Qingnian Canal (Youth Canal) on the Leizhou Peninsula in Guangdong Province. In the northern part of the peninsula runs the Jiuzhou River, which originates in the eastern part of Luchuan County, Guangxi. Flowing southward through northern Lianjiang County, it then turns southwest and empties into Anpu Port. The Jiuzhou River forms the Hedi Reservoir along the northern ridge of Lianjiang County, from which the Qingnian Canal draws its water. Flowing southward, it passes east of Lianjiang and north of Haikang County, and discharges into Leizhou Bay to the east. A tributary branches off north of Kelu in Haikang County, flowing southwest and entering the sea south of Jiajia in western

Haikang. The areas traversed by the Qingnian Canal are largely flat plains, with only a few elevated areas in southern Suixi County, around which the canal makes a detour to the west. Even so, the terrain posed little difficulty for construction (See Fig. 8.1).

Also worthy of mention is northern Jiangsu, historically a region dense with minor canals. Since the founding of New China, even more canals have been built, with the Xintongyang Canal being the longest. An older canal once connected Yangzhou and Nantong—a route previously discussed. The old waterway passed through places like Taizhou, Hai'an, and Rugao, but was characterized by many meanders. The section of the Xintongyang Canal between Yangzhou and Hai'an, however, was constructed anew, diverging from the original channel. From Yangzhou, it leads northeast to Yiling, and from there runs directly east to Hai'an. The stretch between Yiling and Hai'an is nearly straight along an east-west axis. From Hai'an to Nantong, the canal continues along the original route, which so far has seen no major alteration. However, just west of Hai'an, a new canal—known as the Ru-Hai Canal—was constructed, flowing southeast through Motou and entering the Yangtze River west of Tiansheng Port in Nantong. This canal runs roughly parallel to the old waterway between Hai'an and Nantong, but is straighter and less sinuous. Travelers heading from Nantong to Hai'an need only traverse a short section of the river before entering the canal and proceeding directly to Hai'an (See Fig. 8.2).

From Nantong to Lüsi Saltworks, the original transport route consisted of the Yunyan River and the Chuanchang River. A new waterway, the Tonglü Canal, has now been built, running directly from Nantong to Lüsi Saltworks. Such a straight canal is unprecedented in this region. North of the Tonglü Canal lies the former Lifa River, which ran from Hai'an to Bencha Saltworks in the east. After renovation, it is now called the Bencha Canal.

Also commendable is the Tongyu Canal, which runs from Nantong through Hai'an to Funing. The section from Nantong to Hai'an follows the same course as the previously mentioned Xintongyang Canal. From Hai'an to Funing, the route follows the historical Fangong Dike—an anti-flood embankment constructed by the Song Dynasty official Fan Zhongyan. Over time, the coastline gradually extended eastward, and counties such as Sheyang and Dafeng were established east of the dike. To the west of the dike once lay the Chuanchang River, while near Funing, a new canal was built east of the dike, rather than west of it. Today, the entire northern stretch of the Tongyu Canal, starting from Hai'an, is located east of the Fangong Dike, creating more favorable conditions for transportation along the coastal region.

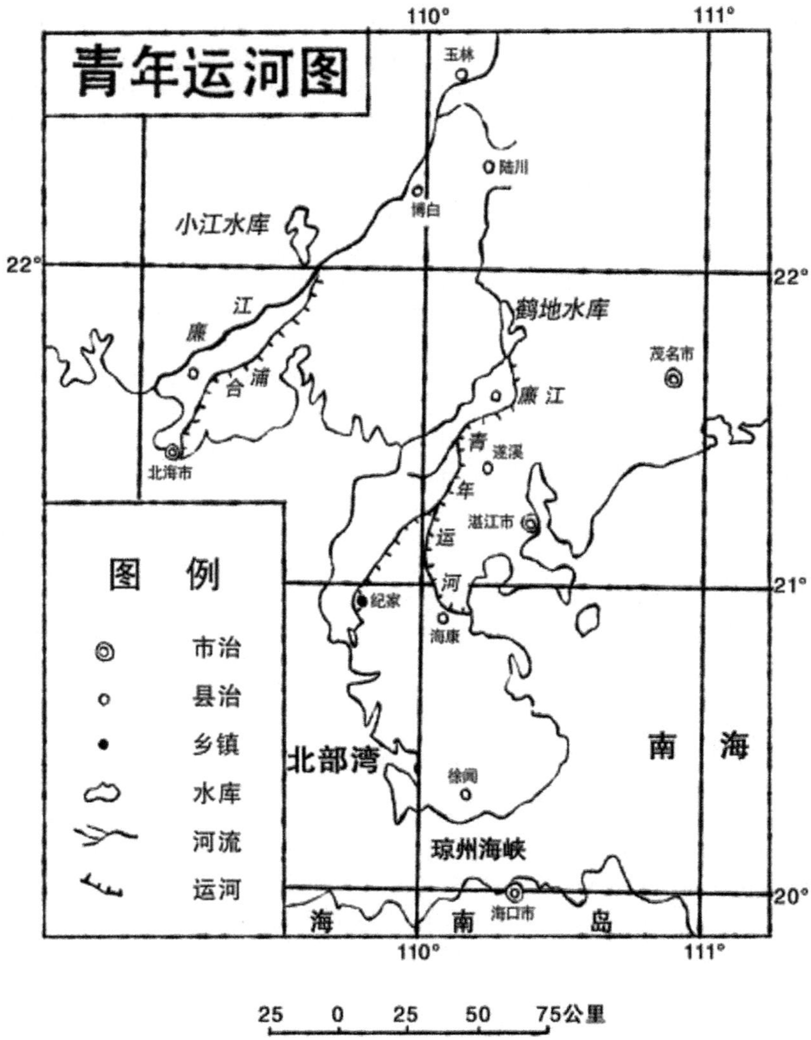

图例 Legend

市治 City Seats

县治 County Seats

乡镇 Townships

水库 Reservoirs

河流 Rivers

运河 Canals

Fig. 8.1 The Qingnian Canal

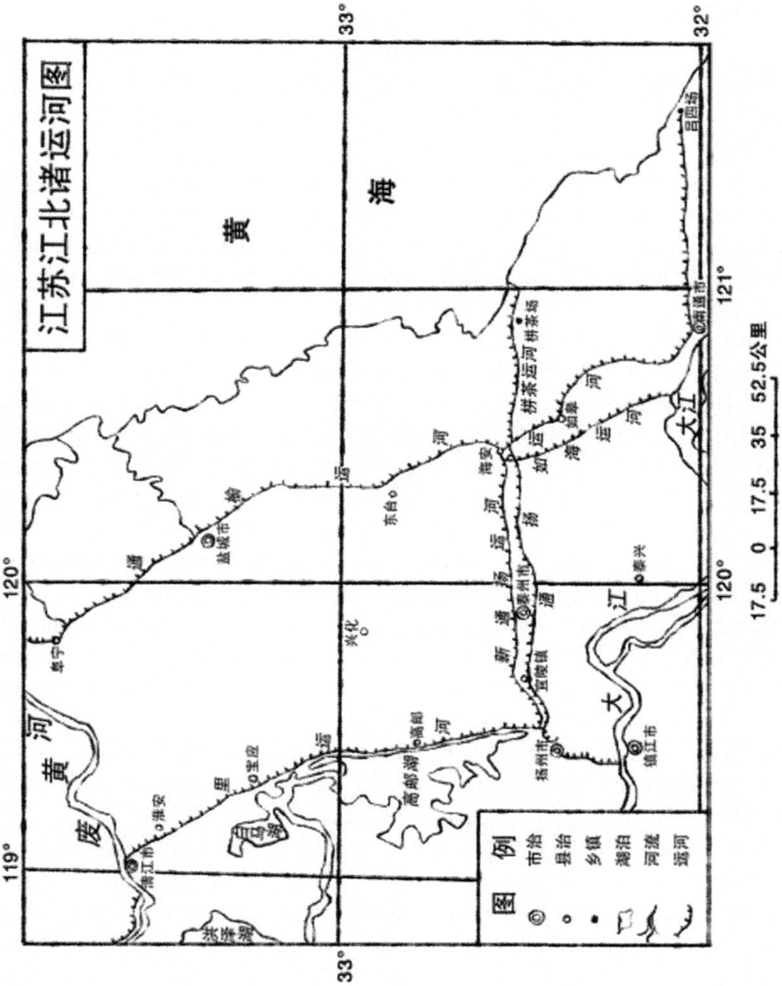

图例 Legend

市治 City Seats

县治 County Seats

乡镇 Townships

湖泊 Lakes

河流 Rivers

运河 Canals

Fig. 8.2 Canal transport in northern Jiangsu

Concluding Remarks

Reflecting on the long history of canal development in China, one observes an unbroken tradition of construction and innovation spanning more than two millennia. Nearly every river and region amenable to canal excavation has, at one time or another, been utilized for that purpose. Each completed canal yielded tangible benefits: some became lifelines of transportation during their respective eras and exerted enduring influence on subsequent generations. These canals stand as monuments to the ingenuity of our ancestors—achievements that continue to inspire admiration and reverence.

Some canals have long since dried up or become obstructed by silt, their traces difficult to detect. Others still flow with clear waters and support bustling vessel traffic. The southern section of the Grand Canal is one such example, particularly the segment stretching from Baoying and Gaoyou in Jiangsu Province down to Hangzhou in Zhejiang. In addition to the Grand Canal, several smaller waterways—such as the Eastern Zhejiang Canal, the salt-transport canals in the Lixiahe region of Jiangsu, and the Lingqu connecting the Xiang and Lijiang rivers—continue to serve valuable roles. Today, the Jiangsu section of the Grand Canal has already undergone successful rehabilitation, and other damaged segments are being restored. Once the entire canal is fully navigable again, reestablishing a continuous north–south inland waterway along China's eastern corridor, it will represent a remarkable national achievement—one that will significantly advance the broader objectives of modernization. The progressive construction of smaller canals further enhances these gains, creating a cumulative effect of growing benefit. That the enduring fruits of generations of labor and wisdom can now be revitalized and extended—exceeding the accomplishments of the past—is a legacy of enduring significance.[5]

The restoration of the Grand Canal is essentially a project to recover and expand upon historical achievements. Naturally, this leads to the question: what should be done with other ancient canals beyond the Grand Canal itself? History is marked by continuous change, and transformations in topography and hydrology are not uncommon. Certain canals—or segments of them—have been so drastically altered by natural forces that restoration is no longer feasible. However, others should not be hastily discarded and left to languish beneath wild vegetation. Canals were originally built to facilitate transport and commerce. While times have changed, their foundational utility has not been entirely eclipsed. Water transport, though generally slower than road or rail today, remains more cost-effective in many cases. Otherwise, the effort to restore the Grand Canal—despite the existence of major north–south rail lines such as the Tianjin–Pukou, Shanghai–Nanjing, and Shanghai–Hangzhou routes—would hardly be justified.

From a historical perspective, the broad plains to the north and south of the lower Yellow River have been the most densely developed areas for canal construction.

[5] Shi Naihai. (1983). Exploring the reasons for the most advanced cultural development in the Yellow River basin during ancient times from a geographical perspective. *Historical Geography*, (3).

Since the inception of the Hong Canal system, canal works have been carried out in phases, culminating in the large-scale projects of the Tang and Song dynasties, such as the Bian River and the Yongji Canal east of the Taihang Mountains. Over centuries, frequent breaches of the Yellow River and silt-laden floods have caused the elevation of these plains to rise. Nonetheless, plains remain optimal terrain for canal construction, just as they were in ancient times. If we are to prioritize the restoration of ancient canals, this region deserves foremost attention. It was the cradle of Chinese civilization and the ancestral homeland of our forebears. While its early cultural flourishing was due to multiple factors, the abundance of rivers was surely among the most critical. Reviving these canals would not only bolster the region's modernization efforts but also serve to enrich its cultural development.

That said, the task of restoring these ancient waterways is far from simple. The most formidable obstacle is the Yellow River itself. Without effective and lasting management of the Yellow River, not only will efforts to restore Central Plains canals remain precarious, but even the long-term security of the modern Grand Canal cannot be assured. At present, the Grand Canal intersects the Yellow River. With contemporary engineering methods, it is possible to construct bridges to span the river or tunnels to pass beneath it. However, in the event of a major flood, the Yellow River's torrents—laden with silt—could quickly destroy canal infrastructure. All the major historical canals in the Central Plains—the Hong Canal, the Bian River, the Yongji Canal—shared a common dependence on the Yellow River for water. Any future attempts at restoration will face the same constraint. No other river can fulfill this role on a comparable scale. If the Yellow River remains poorly managed, even if its intake channels are preserved, any breach would still inundate vast canal systems and render them inoperable.

Yet these challenges must not deter us. Both the Hong Canal and Bian River remained navigable for centuries after their construction. While human neglect contributed to their eventual decline, natural causes also played a decisive role. Historical records suggest that the Hong Canal required little maintenance following its initial excavation. Its decline began only in the mid–Western Han, when the Yellow River breached its banks at Huzi (in present-day southern Puyang County, Henan) and flooded southeastward into the Juye region of southwestern Shandong, eventually merging with the Huai and Si rivers.[6] This deluge overwhelmed the lower Hong Canal and disrupted the entire watercourse. The Bian River, built under Emperor Yang of the Sui, similarly remained navigable throughout the Tang dynasty with little reported need for dredging. By the Song dynasty, however, dredging had become a routine necessity, and the government institutionalized maintenance efforts. Still, by the mid–Northern Song, the riverbed had risen significantly—more than ten feet above the surrounding terrain in some areas. Following the dynasty's collapse, dredging ceased entirely, and what was once a major shipping route became a thoroughfare for carts and horses.

The root of these issues remains the Yellow River. It has long been recognized for its heavy sediment load. Yet sediment levels have not remained constant throughout

[6] Book of canals and waterways. In *Records of the grand historian*, vol. 29.

history. When vegetation coverage in the river's upper and middle reaches was healthy, erosion was relatively limited, and sediment content low. However, as forests and grasslands were destroyed—particularly during the expansionist policies of the Qin state and later under Qin Shi Huang and Han Wudi—the transformation of frontier grasslands into farmland greatly increased erosion and sedimentation, leading to more frequent breaches. Even the Hong Canal could not remain unaffected. From the Eastern Han through the Wei and Jin periods, however, grassland areas began to expand and forests recovered, reducing erosion and stabilizing the soil. From the early Eastern Han, the Yellow River remained calm for nearly 800 years—a testament to the effectiveness of ecological balance. The Bian River, built on this stable foundation, remained navigable for a long time. Once vegetation cover again declined, however, sedimentation problems quickly resurfaced, and the canal became unsustainable.

These historical facts confirm one thing: the Yellow River is not beyond control; its management lies within the scope of human capability. If effectively governed, not only can the canals of the Central Plains be revived, but they may also remain viable for generations to come.

At present, the entire nation is united in its resolve to achieve modernization. Canals are a vital component of national transportation and thus an essential link in the modernization effort. By honoring past achievements while embracing innovation, the continued development of our inland waterways is entirely within reach.

Postscript

When this book was originally printed in the 1940s, due to the poor printing conditions of the time, it was regrettably not possible to include the relevant maps. Last year, as it was prepared for publication by the Shaanxi People's Publishing House, the text was revised and expanded, and maps were finally added—thus fulfilling a long-cherished wish, however modestly.

In the process of preparing the maps, I owe special thanks to Professor Tan Qixiang. He kindly gifted me a complete set of the *Historical Atlas of China* (folio edition), which served as a highly convenient and invaluable base map for this work. For such generosity, I can only say: "I hold it close to my heart; how could I ever forget it?".

I am also deeply grateful to Tian Zhihui and Zhang Daoyi. It is thanks to their meticulous and skillful redrawing of the maps that these illustrations could be completed. I am profoundly moved by their gracious assistance and hereby record my heartfelt appreciation.

Nianhai Shi
1987

N. Shi, *A Historical Geography Research of Canals in China*,
https://doi.org/10.1007/978-981-95-3517-0

Appendix I
Glossary of Geographic and Hydraulic Terms

Chinese Term	English Translation
安丰场	An'feng Saltworks
安山湖	Anshan Lake
霸水	Ba River
白渠	Bai Canal
白沟	Bai Canal
白驹场	Baiju Saltworks
白马湖	Baima Lake
白塔河	Baita River
褒水	Bao River
鲍丘水	Baoqiu River
宝应湖	Baoying Lake
北新河	Beixin River
蚌沿河	Bengyan River
比水	Bi River
沘水	Bi River
汴渠	Bian Canal
汳水	Bian River
渤海	Bohai Sea
蔡水	Cai River
蔡河	Cai River
漕河	Cao River
草堰场	Caoyan Saltworks
浐水	Chan River
瀍水	Chan River
长荡湖	Changdang Lake

(continued)

N. Shi, *A Historical Geography Research of Canals in China*,
https://doi.org/10.1007/978-981-95-3517-0

(continued)

Chinese Term	English Translation
长丰渠	Changfeng Canal
潮河	Chao River
车路河	Chelu River
成国渠	Chengguo Canal
赤山湖	Chishan Lake
串场河	Chuanchang River
大白渠水	Dabai Canal
大陆泽	Dalu Marsh
丹水	Dan River
荡水	Dang River
丹阳湖	Danyang Lake
大清河	Daqing Rier
大野泽	Daye Marsh
得胜新河	Desheng New River
荻塘	Di Pond
砥柱	Dizhu Rapids
洞庭	Dongting Lake
堵水	Du River
独山湖	Dushan Lake
东海	East China Sea
东泇河	East Jia River
房山	Fang Mountain
樊梁湖	Fanliang Lake
汾水	Fen River
丰水	Feng River
封大水	Fengda River
丰利场	Fengli Saltworks
沣水渠	Fengshui Canal
五湖	Five Lakes
浮水	Fu River
富安场	Fu'an Saltworks
高家堰	Gaojia Weir
高邮湖	Gaoyou Lake
滆湖	Ge Lake
漕渠	Grain Canal
大运河	Grand Canal
谷水	Gu River

(continued)

(continued)

Chinese Term	English Translation
泒水	Gu River
沽河	Gu River
洸水	Guang River
广通渠	Guangtong Canal
广武山	Guangwu Mountain
固城湖	Gucheng Lake
龟山运河	Guishan Canal
涡河	Guo River
故漳河	Guzhang River
邗沟	Han Canal
河东	Hedong
黑水洋	Heishui Ocean
菏水	He River
鸿沟	Hong Canal
洪水	Hong River
淮水	Huai River
洹水	Huan River
涣水即濊水	Huan River (Hui River)
黄池河	Huangchi River
桓公沟	Huangong Canal
惠济河	Huiji River
惠民河	Huimin River
胡芦河	Hulu River
获水	Huo River
滹沱河	Hutuo River
御河	Imperial River (Yuhe)
济水	Ji River
洳河	Jia River
贾侯渠即广漕渠	Jiahou Canal (Guangcao Canal)
贾鲁河	Jialu River
涧水	Jian River
将渠	Jiang Canal
江南运河	Jiangnan Canal
胶水	Jiao River
胶莱运河	Jiaolai Canal
界首湖	Jieshou Lake
津湖	Jin Lake

(continued)

(continued)

Chinese Term	English Translation
泾水	Jing River
京水	Jing River
京水河	Jingshui River
荆溪	Jingxi River
金口河	Jinkou River
九嵕山	Jiuzong Mountain
济州河	Jizhou River
沟河	Ju River
巨马河	Juma River
巨野泽	Juye Marsh
滱水	Kou River
涞水	Lai River
莱州大洋	Laizhou Dayang
狼汤渠	Langtang Canal
儽水	Lei River
雷夏泽	Lexia Marsh
利漕	Li Canal
漓水	Li River
梨河	Li River
梁垛场	Liangduo Saltworks
立发河	Lifa River
灵轵渠	Lingzhi Canal
六辅渠	Liufu Canal
六门堰	Liumen Weir
刘庄场	Liuzhuang Saltworks
龙首渠	Longshou Canal
潞河	Lu River
鲁沟水	Lugou River
卢沟河	Lugou River
吕孟湖	Lumeng Lake
洛水	Luo River
骆马湖	Luoma Lake
吕四场	Lüsi Saltworks
马肠湖	Machang Lake
马颊河	Majia River
茂都淀	Maodu Marsh
马踏湖	Mata Lake

(continued)

(continued)

Chinese Term	English Translation
马塘场	Matang Saltworks
孟渎	Mengdu River
蒙笼渠	Menglong Canal
沔水	Mian River
绵曼水	Mianman River
洺水	Ming River
泰山	Mount Tai
梁山	Mount Liang
南河	Nan River
南旺湖	Nanwang Lake
南阳湖	Nanyang Lake
牛饮山	Niuyin Mountain
北济故渎	Old Course of the Beiji Canal
汴水故渎	Old Course of the Bian River
瓠子河故渎	Old Course of the Huzi River
绛水故渎	Old Course of the Jiang River
南济故渎	Old Course of the Nanji Canal
庞官陂	Pangguan Reservoir
平虏渠	Pinglu Canal
甓社湖	Pishe Lake
濮水	Pu River
浦汀河	Putting River
浦阳江	Puyang River
淇水	Qi River
汧水	Qian River
钱塘江	Qiantang River
溱水	Qin River
沁水	Qin River
清水	Qing River
清河	Qing River
清池渠	Qingchi Canal
青水洋	Qingshui Ocean
秦淮水	Qinhuai River
秦岭山	Qinling Mountain
泉州渠	Quanzhou Canal
濡水	Ru River
汝水	Ru River

(continued)

(continued)

Chinese Term	English Translation
桑干水	Sanggan River
三角淀	Sanjiao Lake
沙河	Sha River
商水	Shang River
苕溪	Shao Stream
邵伯湖	Shaobo Lake
圣水	Sheng River
昇原渠	Shengyuan Canal
射阳湖	Sheyang Lake
石臼湖	Shijiu Lake
石塘河	Shitang River
蜀山湖	Shushan Lake
泗水	Si River
笥河	Si River
汜水	Si River
汜光湖	Siguang Lake
长安故城	Site of Chang'an City
斯洨水	Sixiao River
嵩山	Song Mountain
松江	Song River
嵩高山	Songgao Mountain
松阳山	Songyang Mountain
涑水	Su River
睢水	Sui River
索水	Suo River
太湖	Tai Lake
太行山	Taihang Mountain
堂阳渠	Tangyang Canal
讨虏渠	Taolu Canal
桐水	Tong River
通惠河	Tonghui River
通济渠	Tongji Canal
通利渠	Tongli Canal
通阳河	Tongyang River
黄河支流	Tributary of the Yellow River
橐水	Tuo River
王屋山	Wangwu Mountain

(continued)

(continued)

Chinese Term	English Translation
沣渠	Wei Canal
沣水	Wei River
渭水	Wei River
洧水	Wei River
卫河	Wei River
汶水	Wen River
西泇河	West Jia River
舞水	Wu River
无棣沟	Wudi Canal
武亭川	Wuting River
五姓湖	Wuxing Lake
吴兴塘	Wuxing Pond
伍祐场	Wuyou Saltworks
五丈河	Wuzhang River
夏水	Xia River
湘水	Xiang River
潇水	Xiao River
小清河	Xiaoqing River
斜水	Xie River
西蠡河	Xili River
兴化运河	Xinghua Canal
新河	Xinhe Canal
新开湖	Xinkai Lake
西亭场	Xiting Saltworks
西溪河	Xixi River
旬水	Xun River
洵河	Xun River
盐池	Yanchi Lake
阳渠	Yang Canal
阳城山	Yangcheng Mountain
阳朔山	Yangshuo Mountain
江水	Yangtze River
长江	Yangtze River
大江	Yangtze River
阳羡溪	Yangxian Stream

(continued)

(continued)

Chinese Term	English Translation
胭脂河	Yanzhi River
姚暹渠	Yaoxian Canal
河水	Yellow River
黄河东流	Yellow River's Eastward Course
伊水	Yi River
潩水	Yi River
沂水	Yi River
颖水	Ying River
阴沟水即涡水	Yingou River (Guo River)
永昌河	Yongchang River
永济渠	Yongji Canal
潏水	Yu River
榆河	Yu River
玉河	Yu River
余东场	Yudong Saltworks
遇明河	Yuming River
运盐河	Yunyan River
玉泉山	Yuquan Mountain
余西场	Yuxi Saltworks
余姚江	Yuyao River
余中场	Yuzhong Saltworks
湛渠即白沟	Zhan Canal (Bai Canal)
漳水	Zhang River
张甲河	Zhangjia River
张庄湖	Zhangzhuang Lake
昭阳湖	Zhaoyang Lake
柘山	Zhe Mountain
郑国渠	Zhengguo Canal
震泽	Zhenze Lake
直河	Zhi River
终南山	Zhongnan Mountain
中条山	Zhongtiao Mountain
潴泽	Zhu Marsh
泚水	Zi River
滋水	Zi River
甾获渠	Zihuo Canal

Appendix II
Table of Chinese Dynasties

Period of the Three Sovereigns (unverified) 三皇时代	Fu Xi 伏羲	about 2852–2697 BC
	Nüwa 女娲	
	Shen Nong 神农	
Period of the Five Legendary Rulers (unverified) 五帝时代	Huang Di 黄帝 (Yellow Emperor)	about 2697–2184 BC
	Zhuan Xu 颛顼	
	Di Ku 帝喾 (Emperor Ku)	
	Yao 尧(Emperor Yao)	
	Shun 舜 (Emperor Shun)	
Xia Dynasty 夏		2070–1600 BC
Shang Dynasty 商		1600–1046 BC
Western Zhou Dynasty 西周		1046–771 BC
Eastern Zhou dynasty 东周 770–256 BC	Spring and Autumn Period 春秋	770–476 BC
	Warring States period 战国	475–221 BC
Qin Dynasty秦		221–206 BC
Han Dynasty 汉 206 BC–AD 220	Western Han 西汉	206 BC–AD 25
	Eastern Han 东汉	AD 25–220
Three Kingdoms Period 三国		AD 220–80
Jin Dynasty 晋 AD 265–420	Western Jin 西晋	AD 265–317
	Eastern Jin 东晋	AD 317–420
Southern and Northern Dynasties 南北朝 AD 386 (or 420)–589	Southern Dynasties 南朝	Song 宋 AD 420–79
		Qi 齐 AD 479–502
		Liang 梁 AD 502–57

(continued)

N. Shi, *A Historical Geography Research of Canals in China*,
https://doi.org/10.1007/978-981-95-3517-0

(continued)

<table>
<tr><td></td><td></td><td>Chen 陈
AD 557–89</td></tr>
<tr><td></td><td rowspan="5">Northern Dynasties 北朝</td><td>Northern Wei 北魏
AD 386–534</td></tr>
<tr><td></td><td>Eastern Wei
东魏
AD 534–50</td></tr>
<tr><td></td><td>Northern Qi
北齐
AD 550–77</td></tr>
<tr><td></td><td>Western Wei 西魏
AD 535–56</td></tr>
<tr><td></td><td>Northern Zhou 北周
AD 557–81</td></tr>
<tr><td colspan="2">Sui Dynasty 隋</td><td>AD 581–618</td></tr>
<tr><td colspan="2">Tang Dynasty 唐</td><td>AD 618–907</td></tr>
<tr><td rowspan="2">Five Dynasties and Ten Kingdoms 五代十国
AD 907–79</td><td>Five Dynasties 五代</td><td>AD 907–60</td></tr>
<tr><td>Ten Kingdoms 十国</td><td>AD 902–79</td></tr>
<tr><td rowspan="2">Song Dynasty 宋
AD 960–1279</td><td>Northern Song 北宋</td><td>AD 960–1127</td></tr>
<tr><td>Southern Song 南宋</td><td>AD 1127–1279</td></tr>
<tr><td colspan="2">Liao Dynasty 辽</td><td>AD 907–1125</td></tr>
<tr><td colspan="2">Jin Dynasty 金</td><td>AD 1115–1234</td></tr>
<tr><td colspan="2">Xixia Dynasty 西夏</td><td>AD 1038–1227</td></tr>
<tr><td colspan="2">Yuan Dynasty元</td><td>AD 1206–1368</td></tr>
<tr><td colspan="2">Ming Dynasty 明</td><td>AD 1368–1644</td></tr>
<tr><td colspan="2">Qing dynasty 清</td><td>AD 1616–1911</td></tr>
</table>

Translator's Note

The publication of this English edition was made possible with the support from the 2024 *China Classics International* Grant.

This book stands as a landmark work by Professor Nianhai Shi, marking his transition from traditional studies of diachronic geography to a deeper exploration of the dynamic interplay between human activity and geographical transformation across history. Through the lens of canal construction across successive Chinese dynasties, the work invites readers to reflect on the enduring philosophy of coexistence between humanity and nature.

Because this book was originally written decades ago, some geographic terms may differ from contemporary usage. Where such discrepancies do not hinder comprehension, the editors of the Chinese edition have chosen to preserve the original terminology. This English translation follows the same editorial principle to remain faithful to the source. All place names are rendered in *Hanyu Pinyin*, and the translation of terms for mountains, rivers, lakes, canals, saltworks, dikes, and sluices is provided in Appendix I for reference.

We extend our heartfelt gratitude to Shandong People's Publishing House for entrusting this project to our translation team, and to Shanghai Insource Culture Company for their overall coordination and liaison work throughout the process. We are especially indebted to HE Jun and Phillip Neel for their meticulous copyediting and proofreading.

We also gratefully acknowledge the contributions of MA Ruixue, PENG Fei, HU Peizhi, TANG Zhujuan, WU Xiaolian, and ZHUANG Kaixuan from China University of Petroleum–Beijing, whose initial translation drafts provided a crucial foundation for this edition.

Dianying Wang
School of Literature and Journalism & Communication
Yantai University
Yantai, China
July 2025

N. Shi, *A Historical Geography Research of Canals in China*,
https://doi.org/10.1007/978-981-95-3517-0

GPSR Compliance

The European Union's (EU) General Product Safety Regulation (GPSR) is a set of rules that requires consumer products to be safe and our obligations to ensure this.

If you have any concerns about our products, you can contact us on ProductSafety@springernature.com

In case Publisher is established outside the EU, the EU authorized representative is:

Springer Nature Customer Service Center GmbH
Europaplatz 3
69115 Heidelberg, Germany

Batch number: 09286991

Printed by Printforce, the Netherlands